Western Turf Wars

Western Turf Wars
The Politics of Public Lands Ranching

Mike Hudak

Biome Books
Binghamton, New York

Published by
Biome Books
38 Oliver Street
Binghamton, NY 13904-1516 USA
http://www.biomebooks.com

Copyright © 2007 by Mike Hudak. All rights reserved. No part of this book may be reproduced or transmitted in any form or by any means, electronic or mechanical, including photocopying, recording or by any information storage and retrieval system, without written permission from the author, except for the inclusion of brief quotations in a review.

Biome Books is committed to preserving ancient forests and natural resources. We elected to print *Western Turf Wars* on 50% post-consumer recycled acid-free paper, processed chlorine free. As a result, for this printing, we have saved 15 trees (40' tall and 6-8" diameter), 6,440 gallons of wastewater, 2,590 kilowatt hours of electricity, 710 pounds of solid waste, and 1,395 pounds of greenhouse gases.

Biome Books made this paper choice because our printer, Thomson-Shore, Inc., is a member of Green Press Initiative, a nonprofit program dedicated to supporting authors, publishers, and suppliers in their efforts to reduce their use of fiber obtained from endangered forests. For more information, visit http://www.greenpressinitiative.org

Publisher's Cataloging-in-Publication

    Hudak, Michael John, 1952-
        Western turf wars : the politics of public lands
    ranching / Mike Hudak.
        p. cm.
        Includes bibliographical references and index.
        LCCN 2006909580
        ISBN-13: 978-0-9790816-1-3
        ISBN-10: 0-9790816-1-0
        ISBN-13: 978-0-9790816-2-0
        ISBN-10: 0-9790816-2-9

      1. Grazing--Environmental aspects--West (U.S.)--Anecdotes. 2. Range policy--West (U.S.)--Anecdotes. 3. Range management--West (U.S.)--Anecdotes. 4. Public lands--West (U.S.)--Anecdotes. I. Title.

    HD241.H83 2007   333.74'130978
                              QBI07-600193

Cover photography: Mike Hudak
Manufactured in the United States of America
10 9 8 7 6 5 4 3 2 1

Contents

Acknowledgments	ix
Preface	xi
Acronyms	xv
Introduction	xvii

Part I: Governmental Personnel	1
Douglas K. Barber—*(USFS) district ranger, deputy forest supervisor, assistant regional engineer*	2
Clait E. Braun—*(CDOW) program manager, researcher*	8
Leon Fager—*(USFS) wildlife biologist, regional fisheries biologist, wildlife program manager*	30
Renee Galeano-Popp—*(USFS) range conservationist, regional botanist, program manager*	50
Steve Gallizioli—*(ADGF) chief of wildlife management*	66
Dave Gilman—*(USFS) soil scientist*	77
Martha Hahn—*(BLM) associate state director, state director*	83
David A. Koehler—*(USFS) range conservationist; (BLM) supervisory range conservationist, resource area manager, range ecologist*	97

Don Oman—*(USFS) district ranger, ecosystems staff officer* — 110

Robert W. Phillips—*(ODFW) assistant chief (basin investigations); (USFS) regional fisheries biologist, program planner* — 125

Jim Prunty—*(USFS) fire management officer* — 130

Doug Troutman—*(NPS) ranger; (BLM) wilderness ranger, wilderness specialist, disabled access coordinator* — 139

Larry Walker—*(BLM) range conservationist* — 153

Pat Ward—*(USFS) wildlife biologist (research)* — 158

Bill Worf—*(USFS) assistant ranger, regional staff, forest supervisor, regional director* — 165

Part II: Nongovernmental Conservationists — 183

Joy Belsky (1945–2001)—*grassland ecologist at ONRC and ONDA* — 184
    Robert Amundson—*Joy Belsky's husband* — 185
    Jonathan Gelbard—*(Conservation Value) co-founder/executive director* — 192
    Bill Marlett—*(ONDA) executive director* — 201

Patrick Diehl—*(Escalante Wilderness Project) co-founder* — 205

Julian Hatch—*(Boulder Regional Group) director* — 231

Steven G. Herman—*(The Evergreen State College) faculty* — 255

Steve Johnson—*(Defenders of Wildlife) southwestern field representative* — 271

Ralph Maughan—*(Wolf Recovery Foundation) president* — 282

Bobbi Royle—*(Wild Horse Spirit Sanctuary) co-founder*   297

Mike Sauber—*(Gila Watch) co-founder*   308

Todd Shuman—*(Sierra Club) volunteer*   321

Charmaine White Face—*(Defenders of the Black Hills) coordinator*   335

Notes   344

Glossary   371

References   377

Index   385

About Mike Hudak   396

Acknowledgments

This book grew out of an educational video project that I began in the summer of 2003 and continued during summer 2004. I am grateful to the people whose financial contributions partially underwrote the research costs for that effort: Anne Ambler, Robert Amundson, Kenneth R. Anderson, Marshall E. Atwell, Carol Hee and Ted D. Barnett, Joyce Blumenshine, Jeff Boghosian, James and Carolyn Boyden, Brian Brademeyer, Kate Cunningham, Steve Ember, Siva and Richard Heiman, Alice E. Hudak, Andy Kerr/American Lands Alliance, Linda Leas, Eileen and David Patch, Doris Reed, Julian Shepherd, Sunil Somalwar, Sperling Foundation, Thomas G. Thompson, Fred and Alycemae Townsend, Cynthia S. Westerman, Gina Wilson, and Robert Witzeman.

My profound thanks to the people without whom this project would have not been possible—those who not only donated their time to be interviewed for the videos and ultimately for this book, but who also reviewed the edited transcripts of their interviews.

Considerable research went into annotating the interviews. My thanks to the people (other than the interviewees) who assisted me in that endeavor: John Horning, Beth Painter, Mark Salvo, Kieran Suckling, John Weisheit, Victoria Woodard, and George Wuerthner.

And finally, my thanks to the people who provided editorial comments on sections of the text: Carol Hee Barnett, Christopher B. Bedford, James Boyden, Kate Cunningham, Eileen Patch, Pamela Rice, and Joan Zacharias.

Preface

When I began hiking on public lands of the American West during the early 1990s, I expected to encounter majestic landscapes relatively untouched by civilization. Yes, I already knew about environmental impacts from logging, but I had chosen to visit places that had escaped the logger's saw—deserts, grasslands, subalpine meadows, and designated wilderness areas. Imagine my dismay when I encountered eroded gullies, trampled meadows, and shallow, slime-covered streams—all consequences, I came to learn, of livestock grazing.

By 1997 I had become sufficiently disturbed by these experiences that I changed the focus of my summer hiking trips from that of recreation to self-education. Rather than continuing to encounter, by chance, the impacts of grazing in the pursuit of picturesque hiking adventures, I now began to investigate those public lands notoriously degraded by livestock grazing. Visits to the Gila Wilderness in New Mexico, the Granite Mountain Open Allotment in Wyoming, and the Grand Staircase-Escalante National Monument in Utah, among many other locations, provided raw material for my articles, lectures, and Internet essays.

During my western travels I encountered many environmental activists and former resource managers who suggested places for me to investigate. Sometimes they would provide transportation to a grazing allotment along with expert commentary about its history and environmental status.

Through my association with these people I began to envision a new type of educational project about environmental damage caused by raising livestock on western public lands. Up to that time, several books had been written about the topic: *Sacred Cows at the Public Trough*,[1] *Waste of the West*,[2] *The Western Range Revisited*,[3] and *Welfare Ranching*[4] being the best known. There were also hundreds of articles, some Internet websites, and a small number of videos.

But unlike the authors of these works, who had primarily relied upon scientific studies, I chose to follow in the tradition of oral historians such as Stan Steiner[5] and Studs Terkel.[6]

Specifically, I would interview people who had first-hand knowledge of the environmental conditions on public lands and had an understanding of the manner in which livestock had been managed there. In wishing to retain the immediacy of those conversations and to make them available to the largest possible audience, I decided to produce a series of Internet videos based on those interviews.

In July 2003 I set out for two months of travel during which I interviewed public lands experts throughout the West. Following my return home, I became increasingly troubled by the amount of valuable information I had acquired that could not be effectively presented in the form of videos. Eventually, I overcame my reluctance to pursue a more ambitious project that would make use of that information—a book of edited transcripts from the interviews.

I also realized that such a book would require a greater breadth of information than I had acquired in the interviews from 2003. To remedy this deficiency I returned to the West in July 2004 to conduct additional interviews. When I returned home that October, I had accumulated, from two summers of effort, forty videotaped interviews totaling approximately sixty hours.

The people I interviewed represent a wide range of expertise, both within and outside of management agencies. Careers of former governmental employees span the period from the early 1950s to the early years of the twenty-first century. Among this group are people who had served in the role of botanist, biologist, soil scientist, fire specialist, wildlife researcher, wilderness specialist, range conservationist, district ranger, forest supervisor, and district, state, and regional director. Many of these people had served in multiple roles. Individually they had fulfilled from thirteen to more than thirty years of service with the management agencies.

The nongovernmental conservationists had begun their activity more recently—not before the early 1970s. Most of them were continuing in their activism at the time of their interview. Many of these people had formed small organizations to enhance their effectiveness as environmental advocates. A few worked for, or with, environmental organizations founded by other people.

Western Turf Wars is really a collection of short independent books—each book being the edited transcript of an interview. To further highlight this structure I have subdivided each interview into titled chapters.

The transcripts are grouped into two sections: the first consisting of governmental employees; the second of citizen advocates. Within each section the transcripts are ordered alphabetically by surname of the interviewee. The only exception to this approach occurs with three interviews focused on the accomplishments of the late range ecologist Joy Belsky. In this case I have collected those three edited transcripts under Belsky's name and then ordered them alphabetically by surname of the interviewee.

Western Turf Wars is edited in accordance with principles presented in the fifteenth edition of *The Chicago Manual of Style* (University of Chicago Press, 2003).

The interview transcripts assembled here comprise the first extensive collection of first-person conservationist perspectives on the management of livestock by governmental agencies in the United States. More importantly, the transcripts lay bare the politics both inside and outside the agencies that have driven, and continue to drive, that management.

Acronyms

ACLU	American Civil Liberties Union
ADC	Animal Damage Control
ADGF	Arizona Department of Game and Fish
ALMRS	Automated Land and Mineral Record System
AML	appropriate management level
AOU	American Ornithologists' Union
APHIS	Animal and Plant Health Inspection Service
ATV	all terrain vehicle
AUM	animal unit month
BIA	Bureau of Indian Affairs
BLM	Bureau of Land Management
BP	before present
CBD	Center for Biological Diversity
CDOW	Colorado Division of Wildlife
DA	district attorney
DC	District of Columbia
DDD	dichloro-diphenyl-dichloroethane
DDT	dichloro-diphenyl-trichloroethane
DOI	Department of Interior
EA	environmental assessment
EIS	environmental impact statement
EPA	Environmental Protection Agency
ESA	Endangered Species Act of 1973
EU	European Union
FLPMA	Federal Land Policy and Management Act of 1976
FOIA	Freedom of Information Act
FONSI	Finding of No Significant Impact
GS	General Schedule
GSENM	Grand Staircase-Escalante National Monument
ICBEMP	Interior Columbia Basin Ecosystem Management Project
IWP	Idaho Watersheds Project

NAR	national antelope refuge
NEPA	National Environmental Policy Act
NF	national forest
NFMA	National Forest Management Act
NMSU	New Mexico State University
NPLGC	National Public Lands Grazing Campaign
NPR	National Public Radio
NPS	National Park Service
NRCS	Natural Resources Conservation Service
NRDC	Natural Resources Defense Council
NSF	National Science Foundation
NWF	National Wildlife Federation
NWR	national wildlife refuge
ODFW	Oregon Department of Fish and Wildlife
ONDA	Oregon Natural Desert Association
ONRC	Oregon Natural Resources Council
ORV	off-road vehicle
OSU	Oregon State University
OWC	Oregon Wildlife Commission
PEER	Public Employees for Environmental Responsibility
RA	research assistantship
RAC	resource advisory council
RMP	resource management plan
RNA	research natural area
SCS	Soil Conservation Service
SPCA	Society for the Prevention of Cruelty to Animals
SUWA	Southern Utah Wilderness Alliance
SVIM	Soil Vegetation Inventory Method
TA	teaching assistant
UN	United Nations
USAID	United States Agency for International Development
USFS	United States Forest Service
USFWS	United States Fish and Wildlife Service
USGS	United States Geological Survey
WAE	when actually employed
WAFWA	Western Association of Fish and Wildlife Agencies
WSA	wilderness study area
WWP	Western Watersheds Project

Introduction

The American cowboy today stands as a heroic icon of our national identity. But until the early 1880s these lowly ranch hands suffered from a markedly unfavorable public image. Published reports at the time typically described cowboys as unsavory in appearance; uncivil in behavior. Then, in 1884, when William F. Cody needed a new act for his Wild West Show, he selected for the role, "Buck" Taylor, a former employee of his Nebraska ranch. In Taylor the first cowboy superstar was born.

Within a few years the cowboy mystique was bolstered by publication of Theodore Roosevelt's *Ranch Life and the Hunting-Trail*, a collection of stories about his life and adventures on his North Dakota ranch.[1] And even though "for every Roosevelt who sang his praises there were a dozen journalists to damn his bad manners,"[2] the cowboy was well on his way to becoming a much admired symbol of American culture.

As the technology of entertainment and advertising has advanced, so too have the ways in which the cowboy has been marketed. Books, articles, motion pictures, radio, musical recordings, television, and even clothing and rodeos have all contributed to the cowboy's popularity.

Early on, cowboy films made stars of Tom Mix, John Wayne, and Roy Rogers ("King of the Cowboys"), along with many others. So popular was the character created by Tom Mix, in fact, that the radio show on which it was based continued with various actors portraying him for ten years after his death.

When television came on the scene in the 1950s, cowboys soon appeared there as well. William Boyd, already famous from films and radio as Hopalong Cassidy, brought his edited films to television in 1951. That same year Roy Rogers also made the transition from cinema to television and remained there for six years.

Marketing of the cowboy changed significantly in the 1950s. Previously, cowboy characters had been fashioned to appeal mostly to young boys. But in 1952, *Gunsmoke*, appearing on CBS radio, became the first western drama targeting adults. The series was soon brought to television where it thrived for twenty years, thus becoming to date the longest running prime time dramatic television series with a consistent setting and recurring characters.

From the 1950s to the 1970s more than one hundred western drama series appeared on American television. Many of these shows enjoyed runs of several years and then continued to play for many more in syndication.

Why has the American cowboy achieved such popularity? William W. Savage Jr. has noted that "the cowboy is a symbol for many things—courage, honor, chivalry, individualism—few of which have much foundation in fact."[3]

If Savage is correct on both counts, then the cowboy is loved not so much for what he is, or even was, but for what marketers have made of him—the public's admiration stemming largely from the mythical cowboy embodying those qualities they most admire or strive to emulate in their own lives.

Were the widespread acceptance of this "cowboy myth" not potentially fraught with adverse consequences for America's natural environment I would have investigated it no further. But Lynn Jacobs, in *Waste of the West*, suggested that the myth itself undercuts efforts to protect public lands. "Perhaps our biggest obstacle," Jacobs wrote, "to ending public lands ranching is social/cultural—our unconditional worship and support for The Cowboy."[4] A decade later, Jacobs' sentiments were echoed by George Wuerthner in *Welfare Ranching* when he wrote of the resistance to "regulate, modify, or restrict public lands livestock production."[5] "This resistance can be attributed," Wuerthner claimed, "in large part, to the significant symbolic value of beef in American culture, combined with the powerful iconic status of the cowboy in American culture and mythology."[6]

That livestock production has damaged western public lands, and continues to damage them in many locations, is amply demonstrated in the peer-reviewed scientific literature, as well as in such popular books as *Waste of the West* and *Welfare Ranching*. I choose to not duplicate that effort.

Nonetheless, the reader of the present work should possess at least a cursory understanding of the environmental damage and economic costs that public lands ranching entails, so as to better appreciate the value of examining the management, and more basically the politics, from which those consequences follow.

In the seventeen western states the US Forest Service and Bureau of Land Management (BLM) manage a combined 230 million acres of land[7] upon which approximately 24,548 operators[8] graze livestock. Much of this land is arid, rendering the meager riparian areas—springs, streams, ponds, and the vegetation surrounding them—vital to the survival of native wildlife.

According to one government report of the late 1980s, "poorly managed livestock grazing is the major cause of degraded riparian habitat on federal rangelands."[9] And another such report stated that "extensive field observations in the late 1980s suggest riparian areas throughout much of the West were in the worst condition in history."[10] No significant widespread improvement in riparian conditions has occurred since these reports were issued.

More general assessments of lands managed by federal agencies indicate similarly dire conditions. In the late 1980s, 64 percent of Forest Service rangelands were either of unknown condition or contained less than 51 percent of the expected plant species.[11] A more recent report about lands managed by the BLM found that 78.2 percent of those lands shared these characteristics.[12]

Livestock, through their degradation or even destruction of native vegetation, and by their impacts on the earth itself, have damaged wildlife habitat. Adversely affected by these conditions are an estimated 82 species of mammals, 58 species of birds, 114 species of fish, 35 species of amphibians and reptiles, 23 species of mollusks, and 12 species of insects.[13] Another survey, this one focused on the list of endangered, threatened, and candidate species designated under the Endangered Species Act, determined that livestock grazing is a "significant factor in the decline of 76 listed and candidate fish and wildlife species."[14]

In addition to its environmental impacts, livestock production on federal public lands exacts an economic burden on taxpayers. A GAO study found that in fiscal year 2004 the BLM's and Forest Service's grazing receipts fell short of their expenditures by almost $115 million.[15]

Economic and production benefits provided by public lands ranching are minuscule, particularly when viewed in light of the environmental degradations and financial costs just described. Across the eleven western states, income and jobs derived from federal forage are only 0.04 percent and 0.07 percent respectively,[16] while the livestock feed provided by federal lands is a scant 2 percent of the national total.[17]

Although many of these statistics have been available for several years, public awareness—even among members of environmental organizations—has been, and remains, quite low. A related matter, the virtual invisibility of ranching-industry goals among the public, has made it easier for ranchers to "maintain (or strengthen, if possible) their privileges to use public lands, to keep fees low, and to minimize federal regulation of their use of public lands."[18]

Further benefiting the ranching industry are the structural and procedural weaknesses built into the land management agencies. Thus, although numerous laws intended to protect public lands have been on the books for several decades, we find that they are insufficient to insure that the agencies manage livestock production in an environmentally sustainable manner.

Western Turf Wars addresses these deficiencies by directly confronting the cowboy myth on its own terms, not through the presentation of scientific studies, but through entertaining narrative and personal anecdote.[19]

Environmental professionals from land management agencies and citizen activists alike speak from personal experience about how ranching industry pressure exerted on the agencies has led to mismanagement of our public lands. We learn, for instance, that many ranchers are astute businessmen expert in working the political system to their economic advantage. But in the course of such pursuit, environmental laws are often subverted, environmental studies are ignored, government employees are persecuted. And ultimately the natural environment suffers.

The cowboy myth, rooted as it is in the nineteenth century, has avoided scrutiny in regard to government land management, endangered species, and taxpayer subsidies. *Western Turf Wars* compels us to examine the myth within the context of these present-day realities.

Part I

Governmental Personnel

Everybody wants good science. It's just that they want the good science to agree with what they want.
—Larry Walker, range conservationist, BLM (retired)

Even during an administration that is pretty pro-environment, and wanting to solve problems for the environment—for the people of the United States—there's only so much the government agencies can do, because they're still under the gun of all the politicians. And most of them bow to the pressure from the ranchers.
—Don Oman, district ranger, USFS (retired)

I'd sit in meeting after meeting where the Forest Service leadership at the time was developing strategies to deal with Arizona Game and Fish, and to keep them out.
—Leon Fager, wildlife program manager, USFS (retired)

The politics of the West is something that's awfully hard to understand, because the worst possible use for this western country is livestock grazing. Absolutely the highest use for planning in the West should be to preserve it as a watershed. Yet it's hardly ever considered.
—Jim Prunty, fire management officer, USFS (retired)

Douglas K. Barber

Upon receiving his BS degree in civil engineering from the University of California at Davis in 1974, Doug Barber joined the US Forest Service building roads on the Sierra National Forest in California. Subsequently, he transferred to Juneau, Alaska, where he served as district engineer, later being promoted to regional facilities engineer. Barber received his MS degree in engineering and science management from the University of Alaska–Juneau in 1979. He went on to hold positions as district ranger (1981–83) on the Wasatch-Cache National Forest (Utah), and as deputy forest supervisor (1989–94) of the Apache-Sitgreaves National Forest (Arizona). He served as assistant regional engineer in the Southwestern Region Office from 1994 until taking early retirement in 1995. Since then he has worked as a Naval base management contractor and completed five-and-a-half years of active duty in the Navy Reserve as a captain in Korea, Alaska, New Mexico, and Washington, DC. Since granting this interview, Barber worked as a civilian Navy employee for a year before rejoining the US Forest Service as forest engineer of the Sierra National Forest.

Doug Barber made his remarks on the 10th of June 2004 in Springfield, Virginia.

Chapters
1. Cattle removed from Sandrock Allotment, Apache-Sitgreaves NF (1984)
2. Barber's letter to Senator Domenici about the West Fork Allotment (11 February 1998)
3. Maintenance costs of livestock grazing on public lands
4. Barber's proposed changes to federal grazing permits
5. "Blue sky" value of federal grazing permits

CHAPTER 1

Cattle removed from Sandrock Allotment, Apache-Sitgreaves NF (1984)
The Sandrock Allotment was huge. It covered a good portion of the Blue Range Primitive Area.[1] It's at least 40,000 acres.

They closed that allotment in 1984. Bought it out, actually. The Forest Service did a land exchange for the base property. The old man that had it—his name was Freddie Fritz—had the Triple X Ranch and some other properties that were scattered around throughout this allotment. When Freddie got old, he couldn't manage it very well—it was kind of let go for years. This was fairly low elevation area that had been overgrazed to the point where the soil horizon was gone for a lot of it.[2]

The allotment had been permitted for 275 head year-round. When they rounded up the cattle,[3] they got 875 head off of it. No wonder it was hammered.

Well, it really bothered the cattlemen that the Sandrock was closed. And during the negotiations over the forest plan in 1989, it kept coming up, "Gee, when are you gonna reopen the Sandrock?"

I did a lot of backpacking in that country. And I really enjoyed it down there, particularly in the fall. It's just beautiful. And I saw that this country will need a long time for recovery. In 1994 it was just starting to recover. But I also saw all the "improvements"—the fences. There must be hundreds of miles of fences on this allotment that Freddie hadn't been taking care of for a long time, most of which are beyond repair. Now that it's been closed for twenty years, will the government pay to put all those fences back up so we can have 200 head of cows on 40,000 acres? It's absurd.

To date there haven't been any cattle put back on the Sandrock. I hope there never will, because it's economically crazy. And it's environmentally not sound.

CHAPTER 2

***Barber's letter to Senator Domenici about the West Fork Allotment (11 February 1998)*[4]**
The Apache trout was named as a threatened species due to not only grazing impacts, but because of intermixing with other fish that had been stocked.[5] Consequently, Arizona Game and Fish Department was going into a number of the high country streams on the West Fork Allotment[6] and cleaning out the planted fish. They'd build a barrier and then they'd kill them upstream. The grazing permit on the allotment was up for renewal in 1993, and so they were doing the environmental assessment for it.

As I said in the letter to Senator Domenici, it was roughly 370 head for five months a year. And it was high country area where there really isn't a whole lot of good grass except for the riparian zones near the streams, though a lot of the area is very heavily timbered. Consequently, cows tend to congregate in the streams and break down the banks.

You go back a hundred years to when we had meandering streams and perennial streams. Then, due primarily to overgrazing before the turn of the twentieth century, a lot of them have become downcut and widened. So we were trying to do some restoration there for this trout, and we had to get the cattle out of the streams. The solution that we came to after a lot of agonizing over it was to fence the streams.[7]

But I kept asking the question, "Why do we have to fence them? Are we fencing the stream to protect the stream? Or are we fencing the stream to protect the cows?"

The mindset in the Forest Service is "multiple use" at all costs. Everywhere on the forest, if there's timber growing, you must manage timber. If there's grass, you must manage range, and so on throughout the various resources. They believe strongly in multiple use. And they don't really look at the costs of doing that. You're just stuck with some historical use.

The Forest Service came in to get control of grazing in that area, and they've been working on it for about a hundred years. They keep tightening up on the use and bringing in more scientific management to make it better. But at some point you have to question the cost of doing this. What is that cost compared to the remaining value of the grazing that's there?

My contention was that we had passed that point a long time ago, and that we should really be looking at the cost of the fencing that was needed to protect the trout—because protecting the trout is a given—versus just removing the cattle. And we weren't, at that point, willing to take that step.

My writing the letter to Senator Domenici was prompted by his proposing a grazing reform bill, which essentially was going to make it more difficult for the Forest Service to change the rancher's permits. That was my understanding of the way it was going. It was more power to the ranchers; less to the administrators. So I said, "I think you're going in the wrong direction here."

Senator Domenici had led the fight for a balanced federal budget, and here he was supporting the continuation of an old industry that was costing us more than it was worth. And thereby driving up the budget deficit.

Domenici sent me a reply, which I didn't retain, unfortunately. But essentially all he said was something like, we obviously disagree on this issue. Thank you for your interest.

It was very short, which I thought was fine. At least he stated his position. But I felt it was just a dichotomy. Here you've got this guy who says he wants to balance the federal budget. And then he's supporting an obvious loser. But that's politics.

CHAPTER 3

Maintenance costs of livestock grazing on public lands
The cattle numbers on the West Fork Allotment were such that we could expect an income of $2,800 a year, and we were spending $100,000 to build fences. And we don't even get all the $2,800. Half of that goes into the rangeland improvement fund, which goes to building improvements like new fences.[8] Again, if the cows weren't there, you wouldn't need those improvements. They're not of value to anything else. The wildlife may use some of the water developments, but if the cattle weren't there the wildlife would be fine. And they would have a lot fewer impediments to their travel because of fences, and so would other users of the national forest.

Livestock grazing imposes costs on other Forest Service activities
In 1994 the Forest Service began doing environmental analysis up in that high country that included more Alpine Ranger District allotments, and some on the Springerville Ranger District. There was discussion about a quarter of a million dollars to build fencing. The Forest Service recognized that all of the Apache trout streams needed to be protected from grazing. But again, under the assumption that the grazing would continue.

It's not just the direct grazing stuff that's expensive. If there's a campground, and people start to get upset that cows are in there, you build a fence to keep the cows out. Who pays for the fence? The Forest Service divides up their money into the various pots. It's not range dollars that pay for that fence around the campground. It's recreation dollars because it's a recreation improvement. But again, if the cows weren't there, you wouldn't need the fence. If you could get to all the bottom-line costs, you'd find that they amount to a huge subsidy.

CHAPTER 4

Barber's proposed changes to federal grazing permits
We should stop issuing ten-year grazing permits. Instead, we should put them out to bid every five years. And at that point, they would sink or swim on their own. If there was enough value in the permit, and it was good enough grazing with all the environmental constraints that we need to do it right, somebody will bid on that. If not, the allotment will be abandoned.

Under the current system, a rancher gets to build a "blue sky" value into that permit. And selling that to somebody else is, I think, the main reason a lot of them are hanging on, because there really isn't that much money in the running of cattle. That needs to stop. And I said in my letter to Senator Domenici[9] that it amounts to a form of welfare.

CHAPTER 5

"Blue sky" value of federal grazing permits

When Aldo Leopold[10] was on the Apache National Forest in 1910,[11] they had rampant overgrazing. They knew they had to do something about it for watershed protection. So they started the permit system. The first step was just to get the cows under permit. Then they tried to get the permit numbers down. It's been a continuing iterative process for a hundred years now.

In order to get a permit, a rancher had to have a base property that would be his home base. And then he would get adjoining federal land under the permit for grazing. These permits were all for ten-year terms. If we had put them out for some sort of competition when they expired, a person would pay for that what it's worth for ten years without having an expectation that it's going to continue.

But the Forest Service was also interested in community stability. Always has been. Not only with grazing, but also with timber, which is why they supported the continuing timber harvest even when it wasn't economically viable. Rural communities depended on the mill. Well, the ranchers depended on that national forest grass. So we tended to help them to have that permit. When it came up for renewal, we always renewed it.

And what the ranchers are paying the federal government for their grazing privilege per cow is less than they would pay to graze on private land.[12] So there's something of a built-in subsidy there.[13]

When it finally comes time for a person to sell his ranch, he sells it with the expectation that he's also selling the right for the new owner to graze his cows on the national forest. Because the Forest Service always transfers the permit to the new owner, they sort of build the expectation that "Yes, I can sell that permit."

The Forest Service always insisted, "No, you're not selling the permit. We are transferring the permit." But then they always transfer it.

So the ranch sells for much more than the land alone would be worth because the permit is included. The only trouble is, it's not the taxpayers that sell it to this new owner. It's the old owner that sells it to the new owner. And that's what I refer to as a "blue sky" value.[14] He's selling blue sky. He's selling the good will of the American public to the new owner, and he gets a fairly good return for that. That's why he hangs on so fiercely to that permit, even if he's not really making any money with it. The permit is "sold" on the per cow basis, and so when the Forest Service sees environmental problems caused by the grazing, and they want to reduce the number of cattle, that's real money out of that rancher's pocket.[15]

Clait E. Braun

Clait E. Braun received his BS in technical agronomy from Kansas State University (1962), his MS in wildlife management from the University of Montana (1965), and his PhD in wildlife biology from Colorado State University (1969). During his student years, he was at times employed by the US Department of Agriculture in Kansas and Montana, and by the Montana Department of Game and Fish. In 1969 Braun joined the Colorado Division of Wildlife where, during his thirty-year career, he held various positions including program manager, wildlife research leader, and researcher. His studies of sage-grouse in Colorado led to his discovery of the Gunnison sage-grouse in 1977. He was also instrumental in naming the species.

Braun has published over 290 scientific peer-reviewed and technical publications, mostly on birds, including sage-grouse. He is a past president of the Wildlife Society, past president of the Wilson Ornithological Society, past president of the Colorado-Wyoming Academy of Science, and past editor of the **Journal of Wildlife Management**. *He has been an invited lecturer at over twenty-five colleges and universities in the United States and Canada. Since January 2000 he has headed the consulting firm Grouse Inc.*

Clait Braun made his remarks on the 25th of September 2004 in Tucson, Arizona.

Chapters
1. Braun's early life and formal education
2. Discovery of the Gunnison sage-grouse
3. The Gunnison Sage-Grouse Working Group
4. Naming of the Gunnison sage-grouse (1995–2000)
5. The petition to list the Gunnison sage-grouse as a threatened or endangered species (January 2000)
6. Decline of the Gunnison sage-grouse
7. Tactics to protect sage-grouse
8. Prospects for the Gunnison sage-grouse if listed as a threatened species
9. Does the Uncompahgre Plateau Project benefit Gunnison sage-grouse?
10. Does holistic management benefit sage-grouse?
11. WAFWA report: "Conservation Assessment of Greater Sage-grouse and Sagebrush Habitats" (June 2004)

12. Predictions about the listing of sage-grouse
13. West Nile virus kills sage-grouse
14. Negative impacts of livestock production on sage-grouse habitat

CHAPTER 1

Braun's early life and formal education

I grew up on a farm in eastern Kansas. Wanted to be farmer. I loved farming—the equipment. I liked livestock. But there was no way I'd be able to make it in farming. I could see that early on.

And so I initially went to junior college to study history, and then transferred to Kansas State University (Manhattan) to get a BS degree in soil science. My basic training is all in technical agronomy—soils, crops, how to grow things. How to make soils produce things. Very useful background.

During that time I also worked as a soil scientist for the Soil Conservation Service (SCS), which is now the Natural Resources Conservation Service. I described and mapped soils, and conducted block survey of soils in Shawnee County, Kansas, near Topeka, the state capitol.

I went on to get a master's at the University of Montana. While I was there I also worked on soil surveys in Montana, again with the SCS. Saw a lot of wildlife. Had a great time. Got a master's degree at Montana on waterfowl economics.

During that time I also taught at the University of Montana. I was the assistant in the Soils Lab, General Conservation, and Big Game Management—a number of different courses. I even taught Forest Entomology.

I always worked for Montana Game and Fish in those days. Now it's "Montana Fish, Wildlife and Parks." And there are other less desirable terms that are applied to those agencies that include parks.[1] I worked on a study for Montana Game and Fish on the effects of spraying insecticides to kill certain types of insects, and the effects of the spraying on blue grouse.[2] These pesticides were primarily carbamates and organo-phosphorus type of compounds. Having worked on blue grouse, I decided that I wanted to get a PhD, which I did.

At that point I was looking for a project. White-tailed ptarmigan[3] came to mind because there was very little known about them. For a PhD study you need to do something that's novel, innovative, creative, and that takes science further than where it was. That's why I decided on white-tailed ptarmigan.

I also looked at where I could have access to these birds year-round because the only previous work on white-tailed ptarmigan had been done in Montana where there was very limited access in winter. So I chose Colorado because there are lots of roads there. That's how I ended up going from the University of Montana to Colorado State for my PhD.

CHAPTER 2

Discovery of the Gunnison sage-grouse

When I completed college I was offered a teaching position at a university. I was also offered a research position with the Colorado Division of Wildlife in Fort Collins working with waterfowl, mourning doves, and band-tailed pigeons. We also had a side agreement that I could continue to work with ptarmigan, which was very attractive. I declined the teaching position and went to work for the Colorado Division of Wildlife thinking that I would only work for that agency for several years. My work with them continued for thirty years during which I always maintained a small study on white-tailed ptarmigan.

In 1973 the director of the Colorado Division of Wildlife was in a bind in terms of reports coming due for money received from the BLM for research on sage-grouse. The division director essentially assigned me to be the state's representative on sage-grouse research, which also turned out to be sage-grouse management. This led me into a number of different avenues, some of them very management oriented, which led directly to discovery of the Gunnison sage-grouse in 1977.

I didn't get to the Gunnison Basin until 1977, when I finally convinced the people there that we needed to collect systematic data on the sage-grouse that lived there. I didn't have a clue that they were different at that time. But in October 1977 I knew they were different because I had the wings in my hand that hunters returned to us in a wing collection program. The only question was, how much different would they be?

At that time, though, I was focused on what is now called the "greater" or the "northern" sage-grouse in northern Colorado because that's where the money had been obligated initially in the 1960s before I came on board. The research on sage-grouse that began in the mid-1960s focused on their responses to treatments designed to eliminate or greatly reduce sagebrush. I expanded that effort in the 1970s to look at effects of coal mining, of harvests on population size, as well as methods to estimate population size, and the ways that sage-grouse use sagebrush for food and nesting. I put the Gunnison sage-grouse issue on the back burner until I had the opportunity to bring a PhD student on board, which I did in 1982–1983 with Jerry Hupp.

I hoped in that initial year with Jerry that we would have a terrible winter at some point during his PhD project.[4] We were very fortunate the severe winter came the first year. It was wonderful. It wasn't wonderful for the birds. It wasn't wonderful for animals in general, but it was wonderful in terms of data collection.

Jerry's work very clearly indicated that the birds had a number of differing patterns—habitat use patterns and behavior patterns. Jerry was not a behaviorist, but he finished up his PhD with a series of papers, one of them on morphometrics of sage-grouse in Colorado, where we clearly showed that the birds at Gunnison were significantly smaller, with no overlap with the birds in northern Colorado or elsewhere, which is a step towards species recognition.

At the same time, I had a student, Jessica Young, inquire from California about coming to look at the Gunnison sage-grouse. She was an undergraduate in San Diego with Jack Bradbury, and she had worked with sage-grouse in the Mono Lake area of California. She came to Colorado and took a quick look at the Gunnison sage-grouse. She did some videotaping of what she saw, not only there, but at a couple of locations in northern and northwestern Colorado. She was quick to recognize that the birds were different behaviorally, just like Jerry had told her, which led to her PhD work with us.

Jessica did her PhD (at Purdue University) looking at habitat, but especially behavior.[5] She pretty well worked out that these birds, in addition to being morphometrically different, were also behaviorally different. When she played the calls of the Gunnison sage-grouse to other sage-grouse, they ran away from the taped calls. When she played the calls of the larger sage-grouse to the larger sage-grouse, they came right back and tried to essentially mate with the speaker.

She did the same thing at Gunnison. She played the calls of the larger-bodied sage-grouse to the Gunnison sage-grouse, and they ran away from the speaker. She then played the calls of the Gunnison sage-grouse to them, and they came right to the speaker, clearly showing that there were mechanisms in place to prevent breeding between the large- and the small-bodied sage-grouse.

Immediately after Jessica's work, I was deeply getting into the Gunnison sage-grouse issue because Jessica and I realized that there weren't very many of them. They were also very thinly distributed and very scattered in their distribution across southwest Colorado. So I set out in those days to learn where they were. This led to a publication on historic and present distribution and status of all sage-grouse in Colorado.[6]

Jessica kept hammering on me that this was going to be a new species. I was not convinced. I'm pretty conservative. But with her behavior data and morphometrics data, plus the difference in habitat use data from Jerry Hupp, it was clear that they had really diverged quite a bit from the larger birds.

Then we brought on a student in late 1994 by the name of Sara Oyler, who became Sara Oyler-McCance, for her PhD study at Colorado State University through the Colorado Cooperative Wildlife Research Unit. And like the students before her, Sara was supported by the Division of Wildlife. She worked on the genetics of these populations, plus habitat patch size and modeling.

Her work, coupled with some of the early genetics work done by Jessica, clearly demonstrated that these birds were completely different. By this time I knew that we were in a bind. "We" meaning the state of Colorado. There weren't that many of the birds left, and I knew that as soon as we named this species, it was likely that there would be a petition to list them.[7]

In the meantime I enlisted another graduate student, by the name of Michelle Commons, and funded her MS study at the University of Manitoba. She worked on small populations of what became the Gunnison sage-grouse—trying to understand these populations like in Dolores County where there were maybe 150 birds at that time; Dry Creek Basin, where there might have been 200 to 300 birds; Crawford, where there might have been 100 birds. Her work was very pioneering in that respect, especially regarding what needed to be done in terms of habitat manipulations.

We also decided that we needed to join hands with others to develop some kind of plans to prevent a listing. In late 1994 the Colorado Division of Wildlife, and the Bureau of Land Management concurred that we needed to prepare a conservation plan for sage-grouse in the Gunnison Basin. But there is some background to that. The chief biologist at the BLM in Washington, DC, came out to Gunnison because Jessica and I were after him to help the Gunnison BLM people become a wee bit more "green"—more interested in sage-grouse. Jessica took him fishing. By the time he got through fishing, several decisions had been made, including that the local BLM office would either work with us, or people would go elsewhere. That basically was it. No one will admit that. But I know what the deal was.

Thus, the BLM in Gunnison became very green, which was wonderful in many respects. We started developing a conservation plan for Gunnison sage-grouse in the Gunnison Basin. Earlier I had held several meetings with all the biologists in the state of Colorado responsible for large areas with sage-grouse. We discussed developing a statewide conservation plan for sage-grouse. We kicked that back and forth at several meetings. But I could see that we weren't going anywhere because all the biologists were in denial.

"There is no problem."

"We don't really need to do this."

"It's not time."

Consequently, we put that on the back shelf. It's in the process now of being done, but it's a wee bit late in the game.

CHAPTER 3

The Gunnison Sage-Grouse Working Group

What we did with the Gunnison sage-grouse process was that the BLM, out of the Gunnison Resource Area, and the Division of Wildlife, through my office in Fort Collins (with the backing of the director and the people in Denver), went forward with the development of a plan for Gunnison sage-grouse for the Gunnison Basin. And we did it. It was very long. Took a long time.

We went through a series of stages with all the people there. Initially I gave a presentation in January 1995 to whoever showed up. It was a standing-room-only-type crowd. There was a lot of interest in what was going to happen. People had been hearing about this grouse. Jessica, bless her heart, was behind a lot of that, which was really good.

Well, once we got into the meeting (and subsequent meetings) we went through a lot of denial. There's anger. There's fear.

"This isn't my problem."

"It's somebody else's problem."

"There is no problem."

"You made this problem up."

Lots of issues like that. It took us about a year to get through that process to "acceptance."

"Yes, we have a problem. This is where we are. This is where we'd like to be."

Then it took another year or so to actually work through the issues because everybody had their own issue. Everybody thought it was somebody else's fault. No one wanted to accept any piece of the blame even though we had reached a consensus that we had a problem, and the problem was best solved at the local level.

When the Work Group plan was finalized in June 1997, it was a great step forward, as it was the first conservation plan prepared for sage-grouse. But all acceptance and implementation was voluntary. There were a number of conservation actions that were to be done: monitoring, evaluation, and research.

Well, they did an evaluation of that project in the early 2000s.[8] And the evaluation essentially pointed out that they had failed. A lot of finger pointing was done in the early 2000s about the Gunnison Work Group and how they had failed to deliver. And they did. They did very well at certain things such as publicizing the sage-grouse issue. In terms of doing anything for sage-grouse on the ground, they did poorly. The population declined. Some people say that it was drought. It wasn't drought. These birds evolved with drought. It was a combination of factors. Not just one.

One of the big problems in the Gunnison Basin originally was the county commissioners. If you don't get county commissioners to buy in and support a plan, it's very likely you're not going to have success. The county commissioners in Gunnison finally sent a planner to the meetings. But the county commissioners, who were mostly livestock and development interests, never really "signed on" even though they signed the plan. Consequently, they allowed development of house sites, dump sites, whatever, right into the heart of Gunnison sage-grouse range. All these things are negative.

The Natural Resources Conservation Service (NRCS) has been very negative in the Gunnison Basin. They have tried to say how great they are, while at the same time they're helping private ranchers fragment the habitat by killing sagebrush, making the patches smaller—less useful. So the NRCS, which I used to work for, is one of the guilty parties.

But they're not the only guilty party. The BLM has done some things—they're trying hard to their credit—but they haven't really been able to effectively change allotment management plans, what they call "AMPs." The Tomichi Allotment is a good example where they said they made changes, but they really didn't. They were unable to take advantage of public support for Gunnison sage-grouse by effectively changing AMPs or eliminating some grazing. Inappropriate livestock grazing, especially stocking rates and timing of use, continued.

I think back to the Uranium Mill Tailings Remedial Action (UMTRA) Project, where the government essentially paid $100,000 to take cows off one allotment for ten years in the 1990s. Ten years were up. Guess what? They put the cows right back in there.[9]

If they had bought that allotment in perpetuity, it would have been really good for sage-grouse. But it wasn't purchased, and the BLM, and the state health department, and the people who were behind UMTRA botched it. I told them they botched it before they botched it, and I have the correspondence to demonstrate that.

Sage-grouse are a sensitive species for the BLM, which they are supposed to treat as they would candidate species that have been listed by the US Fish and Wildlife Service. If they did, you wouldn't know it because things haven't really progressed properly for Gunnison sage-grouse anywhere where BLM-managed lands occur.

I can commiserate with people who say, "Hey, we got this great plan." I agree we have a workable plan. But there's very little money behind it. And it's all voluntary on the part of everybody—the Division of Wildlife, the BLM, the Forest Service.

The Forest Service doesn't have clean hands at all. They've been up burning on Flat Top Mountain—burning out winter and nesting habitat, trying to create brood habitat for sage-grouse. Gunnison sage-grouse north of Tomichi Creek and the Gunnison River do not need brood habitat. They need winter and nesting habitat. What does the Forest Service do? They burn the winter and nesting habitat to create brood habitat. Makes good sense doesn't it? So everybody is a little bit screwy with this.

There was a lot of anger in the Gunnison Valley about Gunnison sage-grouse numbers going down.

"Oh, woe is us. They're gonna be listed. Life is going to change. It's going to be terrible."

Even a listing wouldn't make life terrible unless they defined critical habitat[10] and enforced it. I can't remember the last time that the Fish and Wildlife Service volunteered to list critical habitat or enforce provisions of the Endangered Species Act on a scale large enough to be effective. They might do it when they have thirty shrews left. Or five hundred bats. Or thirteen owls. Whatever. But even then they typically will turn and run when there is pressure from the developers.

The only way you can get them to do anything really is to sue them. It's terrible. Unprofessional. Species will disappear before things happen. That's real life.

CHAPTER 4

Naming of the Gunnison sage-grouse (1995–2000)

Jessica Young and I sat down and debated how to approach the issue of naming the species. We gave a joint presentation in 1995 at the Wilson Ornithological Society meeting in Williamsburg, Virginia, where we did name it. The taxonomists didn't like that because we weren't taxonomists. They were completely unhappy with the way we did it. But we got the name out there: the "Gunnison sage-grouse." And we named it *Centrocercus minimus*, which is still the name to this day.

I had a very long chat with the person involved with the AOU nomenclature committee, the committee on taxonomy, at a meeting at Colby College in Maine in June 1999, where we hammered out the details as to what the naming would be on the Gunnison sage-grouse. We also tried to hammer out the details on the naming of the other sage-grouse. In other words, there was just one sage-grouse prior to the Gunnison sage-grouse. So what were we going to call the other sage-grouse?

Jessica and I in this paper,[11] which is co-authored with Sara Oyler-McCance, Jerry Hupp, and Tom Quinn, had argued for the use of the name "northern sage-grouse" because we had been using that for several years and it was in the literature already. It referred to the area where sage-grouse were originally discovered and described in 1805 by the Lewis and Clark Expedition—north of the Great Falls of the Missouri River in northern Montana. That's the reason that we chose "northern."

The AOU, in their infinite wisdom, told us essentially that "taxonomists describe birds; biologists don't; we will name this grouse whatever we decide." The chairman of the committee and I had this very heated discussion at Colby College, where I carried the day on the Gunnison sage-grouse as *Centrocercus minimus* as the scientific name, and we disagreed on what the other sage-grouse would be named. He wanted to go with "greater sage-grouse." I didn't like "greater sage-grouse" because that would suggest that the Gunnison sage-grouse would be the "lesser sage-grouse." I didn't like that.

The AOU jumped the gun in the race to name and rename sage-grouse, and published their preference in the July 2000 issue of *The Auk* even though they knew our descriptive paper would be published later in 2000. This was clearly an effort to gain the upper hand in the naming issue.

We have a greater prairie chicken; we have a lesser prairie chicken. I don't really care for "greater" and "lesser." It's not really very useful. We didn't call the heath hen the "greater heath hen," or the Atwater's prairie-chicken the "Atwater's greater prairie-chicken," when both are subspecies of greater prairie-chicken.

We could have gotten away probably with "*Centrocercus gunnisonii.*" But we were saving "*gunnisonii*" for a trinomial in case that a different race of Gunnison sage-grouse had had existed historically in New Mexico, possibly in Texas or Arizona. Certainly, the Gunnison sage-grouse occurred at least in New Mexico, very likely in Arizona. Very likely into the Panhandle of Texas, at which time we thought it would be "*Centrocercus minimus gunnisonii.*" Hasn't worked that way.

The AOU came out in July of 2000, before we published our definitive paper, saying that we had two sage-grouse. We had the greater sage-grouse (*Centrocercus urophasianus*), and we had the Gunnison sage-grouse (*Centrocercus minimus*). And they accepted that. So that is how the Gunnison sage-grouse came to be.

CHAPTER 5

The petition to list the Gunnison sage-grouse as a threatened or endangered species (January 2000)

In January 2000 a petition was filed to list the Gunnison sage-grouse.[12] The Fish and Wildlife Service had been looking at this, scrutinizing it, so to speak. They wanted to beat the petitioners to the punch. Thus, they dated a finding in December 1999 to make Gunnison sage-grouse a candidate species before they actually approved it, so that they could say it was a candidate species when the petition was filed.[13]

Very interesting slight of hand at that point. And I knew the people involved with that. That's where the Gunnison sage-grouse situation stands today, as it is still a candidate species.[14]

CHAPTER 6

Decline of the Gunnison sage-grouse

Probably the biggest change that occurred for the southern populations was in the 1930s.[15] And that was primarily fire control. There was tremendous expansion of pinyon and juniper. There was tremendous overgrazing by livestock. There were tremendous increases of sagebrush abundance because of loss of grasses and forbs. All of those things were negative for sage-grouse. It caught up with them in the 1940s and many populations disappeared.

Different things happened to the grouse in the 1990s—primarily people-related things. The landscape got chopped into smaller and smaller pieces. We had Interstate 70 built all the way across Colorado. We had populations that existed in Eagle County, Pitkin County, Garfield County, and Ouray County, which then became more accessible because of increased mobility of people and more second homes. Those pieces got chopped until they were so small they wouldn't support sage-grouse. It's still habitat, but it's really people causing the habitat to become less and less useful. That's what will happen as we go on. Although there will be other confounding factors such as more intensive land use for oil and gas developments, more vegetation treatments to increase domestic livestock grazing capacity, and more power lines.

People argue that there's less livestock grazing now than from the 1940s to 1960s. But that's a simplistic response. Livestock grazing has changed the composition of the habitats that sage-grouse once used. We have more sagebrush. We don't need more sagebrush. We have less grass. We have less forbs. We have less height of grasses. So you can't just say, "It can't be livestock grazing as one of the causes because we have less livestock out there." This is partially true. But livestock have changed a lot of the habitat. They're continuing to change it with more fences, more water in pipes and tanks, more even use of landscapes. Even if you took the livestock off, you would still have a terrible time in maintaining some populations. It could be done, but it would take time and money.

Population size has decreased since 1999 for a lot of reasons. In part because of drought. In part because of research efforts in which radio-marking of sage-grouse reduced their survival and productivity.

Some people say, "Well, it was a drought."

My comment is *horsefeathers*. Sage-grouse evolved with drought. They've been through hundreds of thousands of years with periodic drought. Can't get too excited about it. It's in the gene code. They evolved special mechanisms to handle that situation.[16]

What they haven't evolved are special characteristics, or behavior patterns, or habitat use patterns to handle habitat fragmentation, habitat loss, habitat degradation, increased predation risk, telephone lines, reservoirs—lots of things that they didn't evolve with. They evolved with all the predators that are presently there with maybe an exception of one or two, such as red fox and raccoon, in localized areas.

It really is a complexity of factors why numbers have decreased. Certainly drought has been one of them.

The evidence this year indicates that some populations have held their own. Some populations have decreased. One population is down to maybe six or eight birds. It's functionally extirpated. That does happen.

I just gave a paper at a meeting of the Wildlife Society in Calgary, Alberta, where I actually put numbers to these things, and the year of extirpation for all the populations past and present.[17]

The Dolores County[18] population is disappearing this year right in front of our eyes. It's a goner. There may be a bird or two running around next year, but essentially it's done. All of these small populations are going to blink out.

Presently we're down to less than 3,000 birds. Maybe it's 2,400. Maybe it's 2,000. But less than 3,000 for sure. Six, eight populations, if you can call eight birds a population.

CHAPTER 7

Tactics to protect sage-grouse

With all these situations there's both private and public lands involved. Gunnison Basin is the last big block of Gunnison sage-grouse habitat that is mostly public—60 percent. All the other pieces are mostly private with the exception of Crawford and a portion of the Dry Creek/Miramonte area.

When I was there, I applied for $2.5 million with the help of another gentleman.[19] We got it to buy land—the main lek site and surrounding area, which was up for subdivision at Miramonte.

A lot of things that we did then, you couldn't do now because of politics. The agency has been told by the political masters that conservation easements are the best way to go. "Don't worry about ten-year leases. Don't worry about buying land. Go for perpetual easements."

Well, unfortunately, perpetual easements aren't going to help a short-term problem. There's both a short-term problem and a long-term problem. In the short term we've got to make sure the birds are still there five years from now. Then we can think about long term.

Thus, there needs to be a multi-faceted approach. A simplistic conservation easement approach isn't going to work. And it's not working. The Gunnison sage-grouse, as far as I'm concerned, is highly endangered. Of course, the Fish and Wildlife Service will not move anything until this lawsuit is decided.[20]

The only species that get listed by the Fish and Wildlife Service onto the Federal Threatened and Endangered Species List are those for which there's public support for lawsuits. And Fish and Wildlife Service does respond to that. Normally they respond when the population is so small it can't be recovered. And definitely one of their present plans is to dilly-dally—delay as long as possible. If they have to list these things, they don't want them to be recoverable, because that negatively affects economic interests. That's my view of it.

CHAPTER 8

Prospects for the Gunnison sage-grouse if listed as a threatened species

There's a better chance of significant changes at the local level in terms of county zoning, denial of developments that fragment existing habitats, removal of livestock grazing pressure in some areas, improvements of nesting cover and brood habitats, and protection of winter habitat, if the Gunnison sage-grouse is listed. The BLM's Gunnison Resource Area, to their credit, has gotten much better about timing of grazing. However, duration of grazing and intensity of grazing have not really changed, even though they may turn out now in June instead of May. Of course, the ranchers would like to turn out in April as soon as they can get the fences open through the snow. They'd like to have the cows on the upper range, so to speak, to get them off their hay meadows where the livestock are wintered.

Livestock grazing is never going to go away. My view. It will always be with us at least on private lands, and most likely on public lands too. But we need to change a number of things about it if we want to make sure that some rangeland wildlife species continue to thrive.

No one in their right mind right now will tell you that Gunnison sage-grouse are thriving. They are not. It's a downhill slide. Some people say a "glide." But it looks pretty steep to me, when I look at the big picture. Not nice.

CHAPTER 9

Does the Uncompahgre Plateau Project benefit Gunnison sage-grouse?

The Uncompahgre Plateau runs approximately southeast to northwest in about three different Colorado counties—Mesa and Montrose, possibly into San Miguel County, depending on how you define it.

The habitat there was very patchy initially. The best area at Sanborn Park was ploughed up and planted to wheat. Sage-grouse haven't existed there for many years.

Another area, west, southwest of Delta, Colorado, was sprayed with herbicide by the BLM in the mid-1960s. All the Gunnison sage-grouse there disappeared. This is before we knew they were Gunnison sage-grouse. But obviously they were there.

There's actually a report by Allen E. Anderson on mule deer,[21] which talks about sage-grouse being there. They were very thinly distributed all across the Uncompahgre. The Gunnison sage-grouse was very adept at using very small patches of suitable habitat over large expanses, which suggests that at one time they were much more mobile, and dispersed better than they are at the present time.

Most of the populations on the Uncompahgre are gone. We knew from a wing collection program, which I started in the 1970s, that we'd get a few wings from Gunnison sage-grouse in those locations (what we know now are Gunnison sage-grouse). So they were there, but we didn't know much about them. We didn't focus on them at all.

The Uncompahgre Plateau was one of those interesting higher elevation areas better suited for blue grouse and Columbian sharp-tailed grouse than it was for sage-grouse. But certainly there were Gunnison sage-grouse there. It was the link between the populations at Glade Park and Pinon Mesa, Miramonte, and at Gunnison Basin and Cimarron and Cerro Summit. The Miramonte population was linked then with the population at Dolores and Dove Creek. The Uncompahgre, at one time, was a historical linkage between some of these now isolated, and small and vanishing populations.

There's been a lot of effort placed on mule deer in Colorado, which is the bread and butter (or one of the pieces of bread that used to be buttered) in Colorado in terms of income—license revenue. Elk now are the biggest thing. But there's been a lot of concern about mule deer across the Uncompahgre. The Division has put a lot of money into the UP Project, as has the BLM. I think the Forest Service is putting some money into it. They're doing some habitat manipulations. None of it's on a scale large enough to really be positive for Gunnison sage-grouse. My professional view is that they have no chance. Well, there's always a chance, but it's a very slim chance because the core populations are so reduced that to get migrants from the core population would be unusual now. We just don't get enough birds to get the population started.

What is the effective population size? Is it 20 birds? Is it 50 birds? Is it 150 birds? At least 150 birds. I'm not even convinced *then* it's effective. More likely it's in the neighborhood of 300-plus birds to be an effective population size for sage-grouse.

I'm not a big fan of Uncompahgre-type projects. I think they're too local. They'll do okay for certain species like mule deer, which are a lot more adaptable than Gunnison sage-grouse. They'll do reasonable things for blue grouse. But in terms of Columbian sharp-tailed grouse, which historically were there, and for Gunnison sage-grouse, which historically were there, it's not large enough. It's not long term enough to make a difference.

Now, if you burned off the whole Uncompahgre Plateau on a landscape basis, that might be wonderful.[22] That isn't going to happen.

CHAPTER 10

Does holistic management benefit sage-grouse?
I've heard a lot about the Allan Savory methods, holistic management, so to speak. The system was developed in Africa where there is rhizomatous grass—plants that have rhizomatous roots. The more you graze them, the more they spread. Most of the grasses in North America aren't like that. There are some examples, exotic grasses especially, that do that. But they're not the native grasses that developed here. Consequently, the Savory method—short-duration, high-intensity—works best in irrigated pastures and farmland where you can move the animals every seven days or three days or whatever.

But short-duration, high-intensity grazing is negative for birds that are ground nesters, or birds that live in the vegetation at knee-height or lower.[23] It's negative and it's deadly for them. It decreases numbers markedly. You'd be hard-pressed to find any benefit for any gallinaceous bird, and most grassland birds.

There may be one or two exceptions where that's not true. Maybe horned larks. Maybe. Don't know. Haven't seen the data.

In general, holistic management contains very little useful substance in terms of managing western rangelands. But there are lots of people who swear by it. It's kind of like being religious. You have to really buy in, and not everybody has bought in. The data don't support their beliefs—"all the trampling will put in more seeds, all the manure will fertilize those seeds, and the heavy trampling will create a smooth system where we don't have erosion." No matter how you cut it, it's pretty far-fetched.

CHAPTER 11

***WAFWA report: "Conservation Assessment of Greater Sage-grouse and Sagebrush Habitats" (June 2004)*[24]**
The report was written to satisfy the Fish and Wildlife Service's need for a document to assess the status of sage-grouse across the West. The Fish and Wildlife Service agreed that the petition[25] had merit, and they had to do a twelve-month assessment. That drove this document even though it was in progress before then. The Western Association set up a group just to prepare this document. They detailed some very competent people to work on it. There's Jack Connelly from Idaho, and Mike Schroeder from Washington. San Stiver from Nevada, now retired and living in Prescott. Also, Steve Knick of the USGS out of Boise, Idaho. So there were very competent people involved with it. It's a massive document.

Unfortunately, it's only as good as the data that the states provided for certain chapters. Although, some chapters stand by themselves very nicely.

The population numbers are really one of the things I'm interested in, because a lot of this controversy goes back to 1998 when the Western Association invited me to give a paper in Jackson Hole, Wyoming, on the status of sage-grouse, which I did.[26]

They didn't ask me to put numbers to population estimates, which no one had done.

I did.

I compiled and presented a table. It's published in that paper showing the actual numbers of sage-grouse in all these western states based on the best data we had for the spring 1998, and a couple of critical assumptions. The estimate was in the neighborhood of 140,000-plus total sage-grouse.

When I gave that paper in Jackson Hole in July 1998 there was stunned silence from the directors and their seconds and their thirds (their staff people). They were shocked that there were so few sage-grouse.

Maybe we had two million earlier. Some people say we had seven to ten million earlier.

Ever since then, people have been trying to figure out, "Was Braun correct or not correct?" What some really wanted to do was to prove that he was wrong.[27]

The people who prepared the WAFWA assessment asked all the states to provide numbers. They then "massaged" those numbers and came up with an estimate of how many sage-grouse there really are.

All the states but two said that populations were down and declining. The two states reporting that populations weren't down—were stable or increasing—were California and Colorado. I don't know a lot about California, but I know people in California. They assure me that things aren't rosy for the sage-grouse there.

I know a lot about Colorado because I was there. I have copies of all the data which I can look at. When I teased apart Colorado data, it was clear that either they had entered the data incorrectly, or the data were misused. That's saying it fairly mildly.

Whether that was done deliberately, or whether it was done by some technician who had entered the data incorrectly, I don't have a clue. But I know for a fact that the numbers for Eagle County and for south Routt County, which are in this assessment, are incorrect. I know, for a fact, that the data for Middle Park, Colorado, (Grand County) are incorrect.

I could not tease apart the data for Moffat County and North Park because they are incorporated into Wyoming and Utah populations.

The North Park data are probably correct because that population is doing well. It was coming to a high in the cycle when I was departing in 1999. That high will continue for several years and then will go down again. But right now things are fairly rosy.

I also know that the data from Moffat County do appear to show that the population is up. I did an analysis for the 1998 paper[28] where I showed that populations in Moffat County had declined by about 80 percent between 1978 and 1998. There was a 50 percent decline in the number of active strutting grounds. So there are some things in there that I knew were incorrect.

The authors are pretty correct, though, where they're concerned about what's going on. But I look at that, and there's just lots of things that were unsaid—things that were presented in a very rosy picture, even though the authors think that they've presented the worst-case scenarios. Some of the stuff is just downright misleading, very likely incorrect.

The effects of I-70 and I-80 (the Interstate system) really are downplayed. I-70 isn't mentioned. My remembrance is that 800-and-some-odd leks along I-80 have gone belly up, probably as a result of noise—truck noise—on I-80.

And so if you look at Interstate 80, for example, which goes through Wyoming clear to Reno, it's like driving a silver stake through the heart of the sage-grouse range. It's alluded to in the assessment, but it's really not. The human footprint on the landscape really isn't clarified because they want to publish those data someplace else. Consequently, it's not really summarized in here.

The genetics people did a good job, but they didn't want to be scooped on things, and so they didn't tell everything they know. There is one more paper coming out on the greater sage-grouse genetics issue which will clarify a lot of things, which could have been helpful in here, especially if you start thinking about trying to re-establish populations, or transplanting birds to increase genetic vigor. Look at the populations in Washington, for example, where you have only one or two or three haplotypes left.[29] They've gone through a population bottleneck. And so they have almost no chance of surviving the next catastrophic event because they are genetically depauperate. There's nothing left, so to speak, in terms of variability.

I can put my finger on almost any chapter in the assessment and can find things in it that I disagree with. I disagree strongly with their estimate of population size. I think it's too liberal. I think that some of the birds are "paper birds."[30] I don't think those birds really exist. And there are some good examples for that.

There are a few other glitches. The disease and parasites part of it is very, very poor. It wasn't prepared by anybody who knew anything about diseases and parasites.

I'll be the first to admit that the last several years have been pretty positive for sage-grouse. We've actually halted the decline in certain areas. And production has been better. There has been improvement in some populations, but not across the range.

If people think that sage-grouse are really doing well, I say, "Show me an area where they have expanded their range." No one can do that. There is no range expansion. There's range contraction.

The Fish and Wildlife Service is going to lean very heavily on this report, and by the end of September [2004], Fish and Wildlife biologists will have submitted their reports to higher-ups as to their recommendations on the

26 / *Governmental Personnel*

listing of greater sage-grouse. That listing very likely won't come out until early January [2005] at the earliest.

I've provided them a detailed review of this document.[31] I've also provided them a detailed review of the Western Governors' Association two-volume set of what they believe is positive about what they're doing for sage-grouse in general.[32]

It's easy to look at the WAFWA report and poke holes in it, but again I want to reiterate that it's a great step forward. It brings together a lot of information, and it provides a basis on which we can either build or we can shoot holes in. And some holes need to be shot in it.

CHAPTER 12

Predictions about the listing of sage-grouse

My intuition tells me that the biologists that work for the Fish and Wildlife Service will find that the petition to list the greater sage-grouse has merit and it should be listed.

My understanding of politics is that that won't happen—that the political masters of the agencies will instead find that either it will be warranted, but precluded because of other things, or listing is not warranted.[33]

If it's not warranted, there will be a rash of lawsuits.[34]

If it's warranted, but precluded, that puts lawsuits off for at least twelve months.

I don't know which way they'll go. I don't have a clue.

I think I know which way the biologists will go. There's enough evidence in there to indicate that things aren't wonderful. And it is downward, even if you count only nine of eleven states as being in a slide.[35]

But what will happen will be a political decision. It won't be a decision made by professional biologists whatsoever in this case.

I also anticipate that sometime next year the finding will come out that the sage-grouse in the state of Washington are listed as threatened. I think that will happen. It should happen.

I see several things happening in 2005 and all of them are related. The political people will be looking for the best way to go so that economic development of the West isn't stymied.

When you look at the West, the sage-grouse range, which is 60 percent federal and 40 percent private, except in a few states, there's great concern about grazing and mining. Those are the two dominant and historical uses of western rangelands. More recently, and second fiddle to both of those are recreation and watershed. Those are very minor even though they are widely trumpeted.

Sometimes you hear, "We're saving this land for watershed." Well, if you graze cows over it and you do serious damage with grazing, it's not worth much in terms of watershed—at least in terms of water-holding capacity. And, of course, livestock grazing can be negative for certain types of recreation.

So all these uses don't really fit together very well even though it's "land of many uses," or "multiple use" depending on whether it's the Forest Service or BLM. But really it all boils down to the livestock industry and the mining industry. And mining can be fluid minerals, oil and gas, which is a big deal right now—a very big deal, which is very negative for sage-grouse.

CHAPTER 13

West Nile virus kills sage-grouse
The bottom line on the West Nile story is that the disease has been reported from Wyoming, Montana, Alberta, California, and Colorado—one bird in Colorado. Several reports from Campbell County, Wyoming, which is near Gillette—mostly from areas where there are a lot of recent water containment ponds as a result of coal bed methane—where they had to pump water in order to extract the gas. The oil and gas industry is very much aware of this, and they're now starting to treat their water to insure that mosquitoes don't live there.

The West Nile outbreak last year was serious. The data collected by Dave Naugle and his students[36] at the University of Montana, indicates that most of the birds which got West Nile, maybe all of them, died. If you go back in and look at lek counts this spring in those areas where they had West Nile last year, the counts are almost zero—local extirpation.

This year the effect of West Nile on the sage-grouse population is very small. There have been fewer birds affected. Even though there are more radio-marked birds out there, there has been little evidence of a continuing, major die-off.

West Nile, my professional view of it: lots of birds have died, people have died, and horses have died. But people and horses are dead ends. For the virus to pass on, it's got to live in birds and be passed on by mosquitoes. If it doesn't live in birds, it's not successful. It can't be passed on if the bird dies. So I look at West Nile as a minor blip. It will ravage some local populations very likely impacted by the amount of standing water that's out there. That's different from the traditional way of irrigating hay fields or whatever, where there was no shortage of water moving across the landscape. But the key word is "moving," because moving water doesn't attract mosquitoes as much as standing water that's trapped in ponds.

And so it's very possible that the coal bed methane extraction process that left water in dugouts or pits or reservoirs, helped certain kinds of mosquito expand their range. And as West Nile came through, whether it was by a crow or a raven or whatever, it got to mosquitoes. And the mosquitoes got to the sage-grouse. And the sage-grouse died.

Temporary. Minor thing.

But it's a major impact where it took place because essentially all the birds died, which is not good.

Some people, a year or so ago, wanted me to jump on a bandwagon and say that because of West Nile, listing of greater sage-grouse is critically needed. I refused to do that. And some people were very unhappy with me. Other people were pleased. The reason that I didn't jump on that bandwagon was that listing should be based on a habitat decision, not on a short-term disease issue. When I looked at West Nile a year or so ago, when it first came up, my prediction then was that this would be a short-term thing. I also said in a newspaper someplace that a lot of birds would die. I think I was right.

I don't think that it will be the demise of the sage-grouse. If you have quality habitat and you have a genetically diverse sage-grouse population that can still expand, and it's not at some critical bottleneck, those populations can be replenished. But if birds have low genetic diversity, such as we have in the state of Washington, or such as we have in the Gunnison sage-grouse issue, West Nile could be pretty negative. But over the whole range of the sage-grouse, no.

CHAPTER 14

Negative impacts of livestock production on sage-grouse habitat
The livestock industry has had more negative impact on sage-grouse than any other single factor because western landscapes have been managed for livestock.

The addition of more water pipelines—more water tanks—tends to dry out the landscape, which is negative for grasses and forbs. Overgrazing leads to more sagebrush. Fencing divides up the habitat into smaller patches. There is more efficient use by livestock of the entire habitat. All these things are negative for sage-grouse. If I had to list the leading factor for putting sage-grouse where they are at the present time, the livestock-grazing industry is right at the top of the major issues. Right at the top. It's not the only one, but it's certainly one of the most important ones because almost all of western landscapes, public and private, have been grazed by livestock. It's rare to find any place that hasn't been grazed. Consequently, there really isn't any natural system left. If we're going to have sage-grouse, we'll have to manage habitats for them.

Some people take the simplistic approach of saying that if we just take away all these oil wells, all these roads or whatever, that sage-grouse will recover. There will be some recovery, but it won't be adequate to restore populations to the healthy state where they once were in this country. Consequently, we will have to manage landscapes. We have to make some wise decisions. Right now, we know enough to do that. We do not have the political will to do that.

Leon Fager

An early interest in hunting and fishing led Leon Fager to pursue a career in natural resources management. Fager began that career by earning a bachelor of science degree in range and wildlife management at the University of Arizona in 1965. Shortly thereafter, he became a wildlife biologist with the Nevada Game and Fish Department (1966–68), and then a biologist with the US Soil Conservation Service (1968–76). In 1976 Fager began work with the US Forest Service as a wildlife biologist on Arizona's Apache-Sitgreaves NF. He transferred to the Black Hills NF (South Dakota) in 1978, and subsequently completed an MS degree in public land policy at Michigan State University. Fager went on to serve in the Rocky Mountain Region as a regional fisheries biologist, and in the Southwestern Region as the wildlife program manager before becoming the region's endangered species program manager.

Shortly after his retirement at the end of 1997, Fager wrote a much-publicized letter to then Forest Service Chief Mike Dombeck in which Fager criticized several policies of high-level managers within the Southwestern Region.

Leon Fager made his remarks on the 28th of September 2004 in Albuquerque, New Mexico.

Chapters
1. Fager's letter to Forest Service Chief Mike Dombeck (23 February 1998)
2. Changes in Forest Service culture (1976–97)
3. Treatment of female employees by Forest Service managers (late 1990s)
4. Forest Service pursues uneconomical litigation (1994–97)
5. Livestock grazing degrades streams on the Apache-Sitgreaves NF
6. Economics of grazing livestock on national forests
7. Ungrazed areas of national forests
8. Forest Service personnel vilify environmental organizations
9. Forest Service interprets multiple use
10. Differences in ranching communities throughout the West
11. Cattle grazing changes natural fire regimes in national forests
12. Elk made scapegoats for cattle
13. Politics of cattle/elk management on Apache-Sitgreaves NF (late 1980s–early 1990s)
14. The Savory grazing method
15. Forest Service ignores research
16. Predator control on the national forests
17. Manifest Destiny appears in new form

CHAPTER 1

Fager's letter to Forest Service Chief Mike Dombeck (23 February 1998)

At the time I retired, a fisheries biologist[1] also took early retirement, and a botanist[2] resigned because of what was happening in the Forest Service. For example, my superiors were changing documents such as biological assessments—changing them to make them fit into the timber and range program. They would falsify information by changing words and data in the documents. And they would also delete items that would prove to be against livestock grazing on a particular area. So we were left basically powerless. And it was so upsetting to some of us that we rebelled and left the Forest Service.

Soon afterward I wrote a letter to the chief of the Forest Service, Mike Dombeck, because I was trying to expose a lot of the things that were happening in Region 3 (Southwestern Region of the Forest Service). This letter got wide distribution.[3] Consequently, I was called back to Washington, DC, with a couple other folks that were just out of the Forest Service to talk with Senate and Congressional staffs and a number of environmental organizations like the Wilderness Society. And I was involved in twenty-two call-in talk shows about the Forest Service throughout the United States.

We tried our best to get the word across. But the Forest Service protects its own. And they defend themselves very well. They've got very good attorneys. They choose to live in denial, I feel. They're very subject to political whims of administrations that come into the organization. Obviously the congressman has the purse strings. And many congressmen have interest in national forests.

For example, here in New Mexico, Joe Skeen, a former congressman who just passed away last year, actually had a grazing allotment for sheep in southern New Mexico.

There's also now, for example, his replacement, Steve Pearce,[4] who was elected a couple years ago. He's another anti-environment congressperson,[5] and he fights to the death to protect mining activities in southeastern and southwestern New Mexico. He used to own a mining drilling business.[6]

It goes on and on like that. We always felt that was a conflict of interest. But throughout the West you'll find a lot of the senators and congressmen have special interests in uses on the national forests, whether it be grazing allotments, mining claims, timber companies, or whatever. And in turn, because of the political clout that they have, they put a lot of pressure on forest supervisors and district rangers and regional foresters to get their will done. Plus they have a lot of money. And many of the environmental organizations don't have the funds that many of these large lobbying groups like the Cattleman's Association and the timber industry have.

The only thing the public can do that's been really effective is to sue the agencies. You tell the Forest Service, the BLM, the Park Service that they're not following the law. And then you take them to court. Environmental groups have won about 80 to 90 percent of their cases, which means the Forest Service is not following the law. They break the law in terms of the Endangered Species Act. They don't follow what they say they're gonna do in their environmental documents.

They've also broken the law in mining operations and in timber operations. And, believe me, this is very costly to the United States taxpayer because these lawsuits do not come cheap.

CHAPTER 2

Changes in Forest Service culture (1976-97)

I've seen the Forest Service evolve from a "can do" organization to more of one that's lost its sense of mission. For example, back when I started in the agency the emphasis was on getting things done on the land. There was little regard sometimes for resources other than timber and grazing. We tried to get things done, and we mitigated for wildlife and fisheries and endangered plants. But that wasn't a big emphasis at all.

When the Forest Service began issuing forest plans in the early 1980s, it became strongly evident to the public what the Forest Service was doing. And these forest plans also become documents that can be litigated against because they have standards and guidelines, and actually talk about the law, including the Clean Water Act, Endangered Species Act, and others.

So when the Forest Service started to implement the forest plans, they came up against lawsuits claiming that they weren't doing what they said they were gonna do. The Forest Service then began to view wildlife and fish and endangered species as obstacles to carrying out the grazing of livestock and the cutting of timber—activities that it had long considered the basis of its core mission.

The problem during the latter years of my career in the Forest Service was that the agency actually did not listen to any of us, its biologists and botanists, about the resources that we were paid to advise staff about. We all told them that they'd be in a heap of trouble if they didn't start harvesting timber in a way that protects the environment and the habitat for the Mexican spotted owl, for example.[7] Well, it ended up in court.[8] And then the court cases started flowing to all kind of issues.

I think that in the old days the only interest the Forest Service had in wildlife was in hunting and fishing species (elk, deer, turkey, trout). And to the nongame species and the endangered species, nobody paid attention—not even the biologists. Not even myself. Only through the lawsuits, brought because of acts that came into place, like NEPA and the Endangered Species Act, did attention start to be focused on those species.

Another difference between then and now is the change in the leadership of the Forest Service. In those days there was a very strong leadership. People—forest supervisors, district rangers, and regional foresters—were not scared to make a decision. They made it. They stood by it. And they expected their staffs to hang in there with them. And we all did.

I've seen the Forest Service go from a "decision-making organization" to a "consensus-building organization." Consensus-building to me means there's no leadership. They don't build that strong loyalty to our supervisors that we used to have in the Forest Service. The Forest Service, believe it or not, back in the early days—'60s, '70s, and '80s—was counted among the Fortune 500 companies[9] in the United States. And it's not there now.

When the Endangered Species Act came into place, it was basically ignored. It was given lip service only. And only because of lawsuits has the Forest Service now really paid attention. They have to or they're gonna be in court again. So they actually have put a lot of emphasis on endangered species. But when I say "emphasis" I don't want to say that too strongly, because the core values are still timber harvesting and livestock grazing. The emphasis is on how to get around the obstacle. Behind the scenes you'll hear rangers and forest supervisors laughing about how they conned everybody into thinking they're gonna obey the law. Sometimes they get caught and sometimes they don't.

In the deep depths of the organization they're still not really interested in threatened and endangered species. For example, when I was the endangered species program manager[10] here in this region, I actually drafted a letter[11] that was finalized by the regional forester here at the time and sent out to all the district rangers, saying that endangered species and the enactment of the Endangered Species Act and its rules, regulations, and policies would be followed as a number one priority. Basically, he said that. Basically, it was ignored.

There's a very loose connection in the Forest Service between levels of leadership nowadays. There's a loose connection between the Forest Service chief in Washington, DC, and a regional forester. Many things that the chief says, result in no action at the regional level.

Same between the regional level and the district ranger level. At district ranger level they act very autonomously. They're free to act the way they feel with little or no accountability to their forest supervisor or their regional forester.

During my last days in the Forest Service, I used to describe district rangers as having little fiefdoms. There would be no consistency in terms of following any of the laws or acts. You'd have one ranger, for example, that was strongly interested in, and put a lot of emphasis on, positive things such as protecting endangered species on his or her district. Then right next door, you'd have another ranger who had no interest—who thought it was a joke.

And so if you're working on a wildlife program at the forest level with seven or eight ranger districts under you, like I was, there's hardly any consistency in program development and program implementation.

CHAPTER 3

Treatment of female employees by Forest Service managers (late 1990s)

At the time, many of our technical people were female biologists, including a botanist and others. And because of the wildlife director's feelings, it caused problems with the women. But it leads again to the symptoms of a bigger problem—that the Forest Service has put on leadership that's no longer capable of leading. They're capable of a lot of politics, a lot of back slapping and hand shaking. But as far as strong resource management background, a large percentage of them do not have that. They've come up through the ranks in the Washington Office where there is a lot of politics going on. And when they send these people back to the field, they come with that baggage of politics and back slapping but without any leadership skills.

In the old days[12] the people were trained, and were field people before promotion to leadership positions. I can think of no one that wasn't a strong leader at the time. In fact, oftentimes I wouldn't agree with some of the things that my forest supervisor would do. But I'd back him a hundred percent. That was the kind of loyalty that we had towards our supervisors.

When I quit, I had no loyalty to any of these people. I had no respect for them at all because of the things they were doing that were unlawful:

changing documents, not defending our programs, ignoring us, and so on. It was very frustrating for me to work in that kind of an environment. So I retired early, along with some other folks. But we wrote this letter about our supervisor and the women. The whole staff signed this letter.[13] Of course, nothing was done about it. Same old story.

CHAPTER 4

Forest Service pursues uneconomical litigation (1994–97)
When I was working as an endangered species biologist[14] in the regional office in Albuquerque here, my job was entirely focused on litigation. I was on a team that reviewed all the lawsuits that came in that dealt with wildlife and plants. And I'll tell you, I know a number of cases where the Forest Service would spend a million-and-a-half, two million dollars defending a patch of timber that was probably worth twenty thousand dollars.

If they would have just left it alone and walked away from it, the lawsuit would have been settled and the taxpayer would have saved all that money. But they chose to defend themselves against this. And this happened time and time again.

The time while I was doing all the litigation, I felt that what the region really needed (probably all the regions in the Forest Service) was a biologist that does nothing but recovery work for endangered species. There was a rich field of endeavor there, but the Forest Service chose not to put that position together. They chose to keep on the defensive. And that's where they stand today.

CHAPTER 5

Livestock grazing degrades streams on the Apache-Sitgreaves NF
One issue that has really interested me is the livestock-grazing aspect of forest management. We've got many streams in the Southwest, but they're very isolated. And the streams frequently have threatened and endangered fish in them. What's happened over the decades is that the main rivers in the Southwest, like the Colorado River, and the Gila River, and the Salt River have all been dammed to the point where little natural flow is actually occurring. It's just lakes behind dams. And so what native fishes lived in those rivers many years ago, now occur only in the headwaters of small streams.

These small streams also have water and forage in them for livestock. And while I was in the Forest Service, these areas were being absolutely trampled to death. Livestock would trample the banks. They would eat all the vegetation. And many of these areas have a variety of endangered and threatened plant species in them, along with fish and other species.

When I was on the Apache-Sitgreaves National Forest, which is in eastern Arizona, we had the largest mileage of trout streams in the Southwest Region (Arizona and New Mexico). We put together a team of fisheries biologists, and they actually went out and intensively surveyed close to six hundred miles of cold water streams in the Apache-Sitgreaves.

A very interesting finding of that study[15] is that due to livestock trampling, grazing, rubbing, and bedding, the streams were becoming full of silt. The vegetation on the sides was being replaced by short annual grasses and weeds, which did not provide any shade or cover for fish. And the analysis showed that our streams on the Apache-Sitgreaves National Forest, because of livestock grazing, had 80 percent reduction on the capacity to sustain a fishery for native species.

So what the public essentially was lacking, in an opportunity cost, if you want to call it that, was that they were getting a $1.35 at the time for an animal unit month, which is one cow on the range for one month. And at the same time, they were losing 80 percent of their fishery capacity, which impacted not only threatened and endangered fish, but all cold water fish, including trout. This was published and brought up at a number of meetings and in papers. And brought up to our line officers in the Forest Service who chose to ignore that.

CHAPTER 6

Economics of grazing livestock on national forests
If you look at pure economics, it cost the Forest Service, we figured at the time, $10 to administer every cow because of salaries, overhead, and other things. At the same time, the Forest Service was bringing in $1.35.[16] Well, you look at the cost-benefit ratio and you see that the American taxpayer is losing a tremendous amount of money—strictly dollars.[17] But that doesn't include the amount of money that's valued in terms of ecosystem benefits, both to the environment and to all of us. Not only bird watching and fishing and endangered species, but the things that ecosystems do, such as filtering water, purifying air, stabilizing soil, and other things. We've lost those.

Recently some scientists have put dollar values on what ecosystems supply in terms of dollars, looked at from a global standpoint. And it's in the trillions and trillions of dollars that we're consciously, and sometimes unconsciously, giving up to graze livestock, to heavily cut timber, and to do other extractive things like mining.[18]

The Forest Service, every ten years, produces a national document under the Resources Planning Act[19] that talks about future projections of uses on national forests by region. The one that was published for this decade said that over the next five- to ten-year period, that monies that would come from livestock grazing to counties and to the government treasury would decrease down to 3 percent of the total amount of money that comes to local communities from national forests. About 70 to 80 percent was from wildlife and recreation.

As a company, if 3 percent of your market is selling one product and 80 percent is selling another product, why would you sell the product that only gets 3 percent of the market, especially if it hindered your ability to sell the other one? It's not good business. Economics plays a very low role in the Forest Service. They hardly ever look at the economics of their actions.

There have been many, many workshops that I've attended on economics. And they always point out that the best use for some of our lands is no use. Consider one case in point on the Apache-Sitgreaves on the West Fork of the Little Colorado River—a study area on grazing. The Forest Service had come up with several alternatives for managing it, such as 2-pasture rotation and 4-pasture rotation. All dealing with livestock use. They never came up with a No Grazing Alternative.

Well, one of our friends, who was a deputy forest supervisor[20] at the time, was very supportive of environmental issues. So he forced the issue—"Let's analyze what a No Grazing thing would be. Look at it only from a cost/benefit ratio in this analysis."

It won out based on the numbers. Figure the costs of grazing—people who have to administer it, the fencing, the water, the salting. All the things that go into livestock grazing. You get a $1.35 back, but it costs you $10 a head.

If you don't graze it, then other values come into play. You have better water. You have better fisheries. You have better wildlife habitat. You have better soil protection. It goes down the line. It's hard to put a value on all of those things, but they're there.

But these, again, are ignored. And you'll see today when EAs (environmental assessments) come out, they'll never have a No Grazing Alternative in it. Or, if it does, it's ignored.

When a No Grazing Alternative was put in there, the analysis would always come out skewed towards livestock grazing, because livestock grazing produces actual dollars that flow through the community—the local feed store, the cafés. Whereas *No Grazing* or *No Action* obviously doesn't produce a solid dollar. It produces intrinsic values which are hard to measure. But you know that people who use that area spend money in the local community for cafés, for sporting goods, hardware, gasoline, whatever. They do the same things, but it's just harder to measure, because you can't get the multiplier effect as easily as you can with that dollar of hard money.

So I never saw a No Grazing Alternative that was selected during my time in the Forest Service. It was always ignored.

An interesting thing that happened during the forest planning days, which is still with us, was the "guiding document" for the national forests. It sets the policies, standards and guidelines, limits in production, projections, and all that for the forest for a ten-year period of time.

The whole plan for the United States was developed using a computer model called "Forest Plan." It's a big cost/benefit tool that has all the resources in conjunction with each other. The problem with that model, though, is that when we put wildlife and fisheries and recreation values in there, there's no set fee. There's not a federal user fee for hunting or fishing or recreation like there is for grazing or timber board feet. But we did have national standards that were developed by economists for what a visitor day would be worth in terms of viewing wildlife, hunting, fishing, and so on.

When you put those values in there, it would always result in wildlife recreation being the number one alternative. Always. And so what the Forest Service did in those days was to take it out of the model. They took it out of the Forest Plan model, which was the basis for the whole planning process. Then they did the work outside of the model, so that it wouldn't influence the solution that the model came to. Somehow the Forest Service got away with that. And they're still getting away with not putting those values in there.

And here's another example of office politics influencing the formulation of forest plans. At the time I was on the Black Hills National Forest,[21] one thing I put in our forest plan, and it was accepted by our leadership and our regional leadership too, was a very, very simple statement. It said, "Any conflict that occurs between livestock and wildlife/fish will be resolved in favor of wildlife/fish." Pretty simple, straightforward direction.

I tried to use that same statement on the forest in this region, the Apache-Sitgreaves, and I was verbally reprimanded by the deputy regional forester for putting it in there. And I had to take it out. Everything was fuzzy in those forest plans. Nothing was direct. They told me that's a red flag to ranchers, and they don't want it in the plans. So that's what went on.

CHAPTER 7

Ungrazed areas of national forests
There are a few areas in this region that haven't been grazed by livestock for fifty years or more. Large areas, many thousands of acres, in fact. One is called the "Three Bar Wildlife Area" on the Tonto National Forest. Hasn't been grazed since 1943. And where it has not been grazed you're waist deep in desert shrubs, grasses, and other plants. Each canyon has a little spring or seep in it. Consequently, there's lots of plants and animals that rely upon those habitats.

South of the boundary fence, it's strictly bare desert and, what we call in the West, creosote bush.[22] Some people call it greasewood. And it's absolutely unpalatable for anything except the one lizard that survives on it. Other than that there's no vegetation out there. It's gone. It's gone to livestock.

CHAPTER 8

Forest Service personnel vilify environmental organizations
In a general way the Forest Service wouldn't dare criticize any organization. They endorsed them. But behind the scenes, they were called "the enemy."[23] For example, the supervisor I worked under at the time, director of wildlife now, many times was criticized by his fellow staff directors at the regional office for calling environmental groups the enemy.

There were times that there have been forest supervisors that I've worked with in the Southwest Region that have been highly respectful, and actually welcomed the litigation, because they knew that was the only way they could untrap themselves from something they were politically put into.

I often felt, though, that the Forest Service was in no position to criticize any group out there, because they're all taxpayers and support the Forest Service in our country. And no one's the enemy; no one's a friend. They're just people that have ideas and opinions. But the Forest Service never saw it that way.

CHAPTER 9

Forest Service interprets multiple use
The Forest Service leadership sees "multiple use" as "all uses on one acre," for example. They don't see that certain areas should receive only one use. Maybe two uses. They seem to feel they have to use every acre.

An example of that is one time in 1991 when I was on the Apache-Sitgreaves National Forest in Arizona. We were looking at a place called Eagle Creek, which is down on the Clifton Ranger District in the southern part of our forest. Desert country. And Eagle Creek is actually a riparian area with just a tremendous grove of cottonwood trees up and down that river. Heavily grazed over the years. There was no reproduction of cottonwoods, willows, or any other plant, because cows had eaten everything underneath.

The Forest Service, due to a land trade, acquired a large section of that river bottom. Clay Baxter, the ranger at the time, was very supportive of environmental issues. Basically, he said he wanted to set that area aside and not have any grazing on it until it recovered.

We were in the field in front of the other people. The range director in the region actually reprimanded Clay for saying that. He said, "We've got to graze it. Everything we've got to graze. And we'll manage it through livestock management."

That does not work.

I don't know why this mentality occurs with multiple use. For example, "no use" of an area, because of environmental reasons (water, plants, animals), is not seen as a multiple-use tool in the Forest Service.

The typical view in the Forest Service is that it's not multiple use unless you're out there managing the area—cutting timber, grazing livestock, mining, whatever. Then you're actually practicing multiple use.

CHAPTER 10

Differences in ranching communities throughout the West

When I was in Wyoming and South Dakota,[24] I don't remember noticing the presence and the verbal stuff and written documents coming out from the cattle growers up there. Even though I'm sure there were some. I don't remember that.

Our forest supervisor in the Black Hills, Jim Mather—a very strong leader, was once out in the field with some ranchers looking at overgrazing on some of the allotments. And one of the ranchers said to him, "We're really scared that these environmentalists are gonna take away our cows and permits." Jim's answer to them was, "Yeah, they probably will, if you don't start taking care of your allotments."

In the Southwest Region you never got that kind of response, I'll tell you that. I have noticed that in this region, the Arizona Cattle Growers and the New Mexico Cattle Growers are very politically strong—stronger than I've noticed anywhere else, probably because the country they graze is so harsh. It's really not livestock environment at all. There's no water. The soils are shallow. The plant species are not abundant. I know many, many allotments throughout the region on all the national forests where the topsoil has actually washed off the hillsides into the arroyos and washes. When you go onto the hillsides you're almost on bedrock. The plants that grow there can just be picked out with your hand. There's no root system to them. It's even hard to climb up these hills because of the loose soil. But they're still grazing these areas.

CHAPTER 11

Cattle grazing changes natural fire regimes in national forests
There are many photos throughout the West of how national forests looked prior to white settlement. For example, there's a very interesting document that was published by the South Dakota State University in the Black Hills where I was once stationed. When General Custer[25] rode through the Black Hills in the 1870s, he had a photographer with him. And the photographer, fortunately, took hundreds of photos of the Black Hills during Custer's campaign there.

In photographs of the forest at that time, you'll see streams with really dense willow populations. The forest was more open in those days because of fire. There were a lot more aspen than there are today.

The effects of cattle grazing were brought to the attention of forest managers by ecologists in this region,[26] and it was ignored. Livestock grazing removes grasses that carry fires, which actually kill seedlings of pine trees and obnoxious things like weeds. So when fires did occur, they were in trees rather than in grass. And they weren't allowed.

In Mexico, in northern Baja, the forests are extremely park-like. They're open ponderosa pine forest with grass up to your waist, because they're not grazed like we do here. But when you have livestock grazing denuding the ground, there's the possibility, over the decades, of forming huge, dense stands of small-diameter trees, which really cause the problem of fire throughout the West.[27]

Thinning, like the Bush administration wants to do, called the "Healthy Forests Initiative,"[28] will help, but it's not gonna be a large-scale operation. The smaller-diameter timber is worthless to the timber companies. So the Forest

Service has to also offer the larger-diameter trees in order to make it profitable for a company to operate within those areas. Consequently, the forests will be pretty well denuded. There's a sacrifice there. And as long as you don't control livestock grazing, in a few years you're going to have the same situation right back on your hands.[29]

The best thing that can happen on a national forest in terms of fire, either by prescribed or natural, is to let it burn. You've got to protect communities, but otherwise let it burn.[30] And what you end up with, which historically happened throughout the West in all mountain ranges, is that after a fire the first plant that comes in is aspen. And then you get a patchwork of aspen and pine and spruce in which fire cannot have a single track of pine trees to crown out and go for miles and miles. It can't burn an aspen grove because it's such a wet plant.

But in order to do that with the large elk populations that we have, and the livestock, you'd have to totally fence those areas to get that established. Otherwise, the aspen seedlings are prime forage for elk and livestock.

CHAPTER 12

Elk made scapegoats for cattle

Elk graze differently than livestock. Livestock tend to concentrate in riparian areas along stream courses. Elk will come down to use the water and forage, but then will return to the ridge tops and the hillsides. Elk, if there are too many, can cause the same damage as livestock. And ranchers are always saying that the elk are causing the overgrazing, not the cows. But, I'll tell you, I have never seen damage on a large scale by elk. It's always been on a small scale. Just certain small areas that happen to be attractive for elk wallows.

A good example of that is where the Apache-Sitgreaves National Forest in Arizona abuts up against the Fort Apache Indian Reservation. On the reservation side, livestock grazing is very, very light. But there are tremendous elk herds on both sides of the fence. On the reservation side, near a place called Reservation Lake, there are elk wallows all over in the meadow. Tremendous amount of elk use, but hardly any livestock use. When you look out over the meadow all you'll see is grass. And if you walk through the meadow you'll fall into that creek, because it's hidden there in all the grass. But as soon as you cross the fence into the national forest, the creek, rather than being two feet wide and two feet deep, is now fifteen feet wide and one inch deep. It's so apparent what's going on there. And it's the same elk use on both sides of the fence. The only difference is livestock use.[31]

CHAPTER 13

Politics of cattle/elk management on Apache-Sitgreaves NF (late 1980s–early 1990s)

Initially, when I was working closely with the Arizona Game and Fish Department, we had a forest supervisor who was very supportive.[32] We put together elk management zones on the forest. And within each zone we not only planned for elk, but also for fish, endangered species, water quality, and other things. And so the elk herds actually helped us to set the parameters for those areas. At that time we did cooperate with the ranchers. We'd go out together as a team: Forest Service, Arizona Game and Fish, and the ranchers to look at exclosures and talk about elk and livestock issues. It was working very well.

Then in early 1990 the Forest Service changed leadership at the forest level. A new forest supervisor[33] came in absolutely opposed to the Arizona Game and Fish Department because it was felt, not only by him, but also by the deputy regional forester, that the Arizona Game and Fish Department was out to take over the management of the national forest. And they would not stand for that.[34]

The Forest Service's line to the Arizona Game and Fish Department was basically, "We want your elk herds thinned out, big time."[35]

At the same time the Forest Service would tell ranchers, "Don't worry about the elk, you'll maintain your grazing permits out there."

I'd sit in meeting after meeting where the Forest Service leadership at the time was developing strategies to deal with Arizona Game and Fish, and to keep them out.

The forest supervisor actually was having secret meetings with the governor of Arizona. A lot of it was against the Arizona Game and Fish Department, and the Game and Fish Commission. The commission is very strong in Arizona, as it is in New Mexico, and it was stacked with people that would, because they are political appointees of the governor, oppose certain things that the Arizona Game and Fish Department wanted to do that the Forest Service also opposed. The Game and Fish Department gets its support through their commission.[36]

I was part of that selection team for the commission sometimes, so I was well aware of what the Forest Service was trying to do. I always lost the battle trying to select people I thought were well-rounded environmentally. Instead they'd choose somebody as a commissioner that was either a rancher, timber person, or whatever.

CHAPTER 14

The Savory grazing method

At one time I was the range/wildlife staff person on the Lakeside Ranger District on the Apache-Sitgreaves National Forest in Arizona.[37] And on our district we had one rancher that went to the Savory school and was implementing the Savory system[38] on his allotment. We found that under his system of high concentrations of livestock, with many movements of the livestock, that the range was actually degrading in its condition classes.

I've never attended a Savory school, but I know a lot about it through my experiences in talking with other people and reading some of Allan Savory's books. The system is based a lot on large herding animals that live in Africa.

You can't apply that situation, in my feeling, to desert and semi-arid environments like we have in the Southwest—down here in New Mexico and Arizona—because the soils are different and rainfall is much less. So you end up having a breakdown in the system.

Savory talks a lot about the effect of grazing and hoof action on the forage, of breaking it up, allowing it to plant seed, and so on. That might be fine for the natural habitat of the buffalo, which never occurred in our region, by the way. Buffalo, as we all know, moved in tremendously large herds in the old days. They'd go through an area, graze the heck out of it, and then move on, perhaps not coming back for years.

And after the buffalo would graze an area down to the ground there'd be a sequence of things that would happen. First, prairie dogs would come in. Then would come a lot of species associated with prairie dogs: black-footed ferrets, burrowing owls, and others.

And eventually, and this has been proven through studies from the Rocky Mountain Research Station in the Badlands of South Dakota, what happens is really surprising. That's very arid country. But if you stand on a vista overlooking the basin out in the Badlands, you'll see little green "postage stamps" scattered throughout the area. And what these green postage-stamp-looking things are, are exclosures that keep livestock out. And guess what? There are no prairie dogs in them. Prairie dogs are all out where it's grazed heavily, because they need a site where they can stand up and see predators at a far distance. When the vegetation grows so heavy that they can't see, they simply move. But if you have continual heavy grazing, the prairie dogs will stay there. You'll always have prairie dogs.

Well, here in the Southwest, where Savory systems are being applied, I have experienced the same thing. With high densities of livestock moved rapidly through several pastures, I began to see prairie dogs. There's nothing wrong with having prairie dogs. They're part of the natural system, but you'll see them as a symptom of overgrazing.

The Savory system may be workable in the Northern Plains, which is more like Africa, but I think if you try to apply it to the Southwestern ecosystems, it's a disaster.

CHAPTER 15

Forest Service ignores research

I worked with a fellow who was the head biologist at the Rocky Mountain Research Station in Rapid City, South Dakota. Dan Uresk was his name, and he was doing all the prairie dog studies in the Badlands.

In fact, I saw him on a TV special one night when they were talking pros and cons of the prairie dog in the West. His research basically disputed all the old stories you hear like, if you have prairie dogs, the horses and cows will break legs, and the prairie dogs will take the grass away. It's just the opposite. The cows take the grass away. And he said there's never been a documented case of a cow or horse breaking a leg in a prairie dog hole.

I was at a meeting with him when I worked in the Black Hills. At the time, the Forest Service was financing prairie dog control. People would go out in four-wheelers and put poison in their holes. And at this meeting, Dan got up and presented data from his research that it wasn't the prairie dogs, it was the overgrazing of livestock that was causing the problem. He was ignored completely. Later, when he was sitting behind me, I heard him tell another fellow, "Boy, so much for all you spend on research."[39] And that's a good story. All the money spent on research is basically ignored by the Forest Service.

There's been many, many stacks of reports written from meetings, studies, and committees about riparian areas in the Southwest. There are literally hundreds of them. And they all point to the need to get livestock off these creeks. Yet the only reason they come off, many of them, is because of lawsuits from environmental organizations. That just proves the point. The Forest Service does not care about facts and data, or expertise. What they care about is politics and core values: grazing livestock and cutting timber.

CHAPTER 16

Predator control on the national forests

I was part of what they call the Animal Control Committee, as were all the biologists in this region.[40] And there's actually a part of the Forest Service manual that talks about animal damage control. Animal damage control basically means controlling coyotes and wolves, but mostly coyotes. And most actions deal with coyote predation on sheep and calves.

What happens on a forest all depends on your leadership. When I worked under Nick McDonough, the forest supervisor on the Apache-Sitgreaves, who was strong on environmental issues, I had the wonderful responsibility of calling the director of the animal damage services of this region in Denver, Colorado, and telling him that if his trappers were found out on the Apache-Sitgreaves National Forest that we'd have federal marshals arrest them.

It also went to the other end of the spectrum, where we welcomed every person they'd send out to shoot coyotes—to preemptively kill coyotes with guns or poison before the herds of sheep would come in. They'd do that on a smaller scale with wolves. The Mexican gray wolf was reintroduced a number of years ago in the Southwest. When a wolf starts eating livestock or gets outside the zone he or she is supposed to be in, they either trap and remove them to another site, or they kill 'em.

To me, the public lands of this country (Forest Service, Bureau of Land Management, and National Park Service) are needed in the year 2004 and beyond for recreation, and the preservation of rare and unique gene pools. I don't believe that cutting timber on a large scale, or grazing cows to the point where you have degradation of the environment, is a proper use of the national forests in this century. I think it's way outdated and that the Forest Service is hanging onto a past that should have been buried a long time ago.

But that's what's going on. And it will continue until there's enough pressure put on. I don't know where this pressure will come from though, because I know that the Roadless Area Rule that the Clinton administration issued[41] has now been significantly weakened. That was endorsed by, I believe, 80 percent of the people that talked at over 600 public meetings.

"No, we're gonna give you cows. We're gonna give you mining." And it just doesn't make sense. It seems our democracy is going crazy sometimes—that our elected officials aren't really representing the people. They're representing special interest groups and corporations. That's a large-scale thing, but it seems that's where we're heading. And it's the wrong way to go, because I can see the end point. There's always been talk of privatizing the national forest and BLM lands. And you see steps towards that right now where they're giving the timber companies "stewardship permits." They let the timber company actually go out and select the timber they're going to cut and then do all the work. They make the selection of how they're gonna treat the land rather than the Forest Service.

You see it too on the mining claims. They give these mining claims to oil and mineral people. They let them go out and develop how they're gonna reclaim the site rather than the professional people that the government has hired to set those standards. It's like the "foxes in the hen house" approach to a lot of our programs now.

I see the Bush administration is reducing budgets. When the national budget came out, the budget that increased was for the Department of Defense. All the domestic budgets, including the environmental organizations (Forest Service, Fish and Wildlife Service), were reduced. And when you keep reducing your budget so that all you can afford to do is pay your people, you're gonna have an organization that's inefficient and ineffective. And I think that's exactly what they want, so they can say, "See, we told you this wouldn't work. So therefore we're gonna replace you with XYZ Corporation that will work." Maybe I'm wrong, but that's how I see things happening in the future if we're not real careful in this country. And when it happens, it will be too late to take it back.

CHAPTER 17

Manifest Destiny appears in new form
When we started developing this country towards the West, we had this thing called Manifest Destiny. Basically, what that means to me is that anything in our way was gonna get trampled. They first encountered the American Indian tribes in the West. They stopped our so-called progress. They tried to defend their lands. We simply stomped them into the ground by killing their food supply, the buffalo. We killed their people, their women, their children as they tried to defend their lands. We put them on reservations to get them out of the way. Then we traveled on.

Today, instead of American Indians, we're encountering rare habitats, endangered species, clean water. Manifest Destiny is still in the heart and soul of America. And, if we have our way, we're gonna put 'em in an aquarium. Put 'em in a botanical area. Or just forget about 'em. They're expendable. Let 'em go.

This will bring a tremendous cost to our future generations, not only in terms of natural beauty and knowing that things are no longer there. But how many things from the ocean are picked up as medicinal cures for people? Or something that can be used as a fabrication for industry—dyes, for example.

Right now in this region, the desert topminnow[42] that lives in little hot pools in Arizona is under study for treatment of malignant melanoma—skin cancer. Because it's exposed to the sun all the time, it has chemicals in its body that prevent that from happening.

If we let those things become extinct, that gene pool will be gone—a possible cure will be gone. And that's what scares me, because my kids and my kid's kids may not have that medicine.

The national forest system lands are probably the most beautiful, highly naturally productive lands in the United States. And they are owned by the American public. They're not owned by the ranchers. They're not owned by timber people. They're owned by the people of the United States. And they are indeed *national* forests.

Theodore Roosevelt, a Republican, set up public lands—national forests.[43] And succeeding Republican administrations have been the most beneficial to the environment. The Endangered Species Act, the Clean Water Act all came under Republicans, up to the time of Reagan.[44] And then things started to change to the other way around. Now, the ideals of Roosevelt are gone.

The current administration, on the record anyway, is very anti environment. Today there are no wildlife staff officers on national forests in the Southwest. They're all working again on timber or range staff like they did in the old days. And where you sit in an organization says a lot about how the organization feels about your input into their decision making. So what they're telling us is that we don't really count. What really counts is a range program and a timber program just like they've always counted.

From speaking with biologists that are still on board, I know the Forest Service today is actually worse than when I retired. One friend of mine just recently resigned because he couldn't stand the pressure any longer from his supervisors. He's a wildlife biologist with a PhD who worked under a range staff person. And here's a guy who's a real smart, intelligent person that cannot take the heat from the Forest Service because they're making him change documents. He wasn't allowed to talk to wildlife biologists in ranger districts, because they didn't want him to influence them to write certain things in the documents that are based on data.

And I know that the Bush administration is trying to push through many things in the Forest Service and BLM that are contrary to the historical functioning of those organizations. For example, the leadership of those organizations is actually getting direction from the Energy Task Force, the undersecretaries of agriculture, and undersecretaries of interior, instead of getting it from where it used to come—the chief of the Forest Service and the national director of the Bureau of Land Management. Politically they're being put under the knuckle from a Washington standpoint. And it's due to the administration's friendship with the timber industry and the livestock and mining industries.

Renee Galeano-Popp

Renee Galeano-Popp began working short-term jobs with the US Forest Service while still an undergraduate at Northern Arizona University (NAU). Upon receiving her BS degree from NAU in 1978, she embarked upon what would become a twenty-year career with the US Forest Service—a career that included roles both as technician and professional, with wide experience in timber and range. In her professional capacity, she held positions as range conservationist, regional botanist, and manager of the Wildlife, Fish, and Rare Plants Program on the Lincoln NF of New Mexico.

Galeano-Popp resigned from the Forest Service in April 1998 following a six-month-long dispute over the evaluation of grazing allotments containing federally listed species. Since leaving the Forest Service, she has privately consulted on energy-related projects in addition to performing land management planning for the BLM.

Renee Galeano-Popp made her remarks on the 4th of August 2004 in Livermore, Colorado.

Chapters
1. Galeano-Popp's early life and education
2. Overview of Galeano-Popp's career
3. Forest Service avoids gathering data about rare plants (mid-1980s)
4. Congress degrades Forest Service management
5. Social pressure on land management personnel
6. Animal Damage Control harms wildlife
7. A personal anecdote about Animal Damage Control
8. Attitude of APHIS about the killing of coyotes
9. Implementing Integrated Resource Management on the Lincoln NF
10. Treatment of endangered species on the Lincoln NF
11. Management agencies fail to support their personnel
12. Galeano-Popp resigns from the Forest Service
13. Allotment management at BLM declines in rigor
14. Experience with the BLM's Great Basin Initiative

CHAPTER 1

Galeano-Popp's early life and education

I was raised in the suburbs of New York City in Westchester County and attended a private high school in Scarborough up by Ossining, New York.

My family is now out West because when I was looking at colleges, I looked into Prescott College and my mother followed me. Although I didn't get into Prescott College, she bought a former ranch near there, and my family moved to Arizona the year I graduated high school. They've been there ever since.

My brother married an Indian and lives on the Yavapai Reservation. He runs the cattle that they manage there, and on some of the holdings that the tribe has around the Prescott area.

Early on, as far back as I can remember, I've always revered nature. Henry David Thoreau[1] was probably my first idol. And then along came Rachel Carson,[2] who really stirred one to action. I didn't know exactly what I was going to do with this interest in nature, though.

I did a semester at Boston University—went from a high school where I graduated with twenty-eight people to classes with seven hundred. And it was the biggest culture shock you could imagine. I didn't last a semester. I dropped out. Hitch-hiked around the country. Floated around. Worked in co-ops.

But then I went to Friends World College in British Columbia, one of the few Quaker colleges where you don't have to be a Quaker. At that time they were forming a campus on Vancouver Island.

For our educational program there, six of us built our own kayaks and kayaked the Inside Passage for the summer. We were living off of the land. And I was learning the plants and seeing their diversity. Because I had not had high school biology, I didn't even know there was such a thing as a genus and a species. So I started learning from some resident botanists and foresters. That was really my awakening.

It was the year in British Columbia, foraging for food and learning the botany of the rainforest, that started to bring it together for me. Then I came back to the states, and I went to Prescott. My father had died. He had always been on my case about what I was I going to do. I had tried typing. I had tried waitressing. I had tried everything. I didn't know how to manifest my interest. But I took that interest that had sparked in British Columbia, and I took a botany course at Prescott College. I looked through a microscope, saw the parts of a flower in there and I just went berserk.

Taking that botany course was the biggest turning point for me. After that I got involved in academia, finding out what the Forest Service is, what public land management is. That's a western thing. You don't get exposed to that in New York.

CHAPTER 2

Overview of Galeano-Popp's career

While I was in college, I had a couple of short-term jobs for the Forest Service with the research branch at the experimental forest outside of Flagstaff. Then I became a full-fledged summer seasonal in '78, the year that I got my bachelor's degree.

I started out as a timber marker for two years. The first year it was great. I thought, "Oh, my God, if somebody's got to pick the trees to be cut, it might as well be me." You get to walk through the woods all day. You get raptors yellin' at you. You get turkeys scurrying by.

By the second season the novelty started wearing off. My eyes started opening, and I'd say things like, "Why are we in this timber sale? Why are we cutting trees at all? This is nothing like what they taught me in the textbooks."

And I would get answers like, "Well, there's one tree per acre out here, and we're gonna cut some of them."

And then I'd ask, "Why are we doing this? These are the only trees out here."

The answer being something like, "Well, we have to cut to get the money to do more planting."

That kind of circular logic they didn't teach me in school.

Eventually, I became curious about the history of stands. I started asking questions about how long ago a stand had been logged.

I was told, "We don't keep those kinds of records."

"What do you mean you don't keep—? I just went to four years of school that was all about rotation, ages, and histories of stands."

Now we cut trees when it's economically appropriate—when we "need to." It's all just seat of the pants.

That's when I said, "Okay, time out, I can't do this anymore."

So if a summer seasonal can resign, I did that my second season. I wasn't going to do that anymore.

The third year, the wildlife biologists had picked up on me and said, "We want her to work for us."

So I went to work for the wildlife biologists as a summer seasonal, '80 through '84, and I went to graduate school, '80 to '83, again at NAU.

Altogether I spent twenty years in the Forest Service. The first ten were essentially technician work. And the second ten were professional.

As a technician I worked in timber for two years, as I said, then wildlife for four or five. I worked on a soils crew, mapping ecosystems. I was a range con for two years in Arizona on the Springerville District of the Apache-Sitgreaves.

From there I filled in as a botanist here and there, mostly doing rare species work. And then I became regional botanist in Albuquerque, which is a whole story unto itself.

You take a field person, who is used to wearing jeans and flannel shirts and knowing what's behind each rock, and you put them on the eighth floor of the Forest Service regional office with people running around in suits and ties, and it's another culture shock. I spent two-and-a-half years there in the world of politics. I did not like what I was involved in there either. So I worked a deal where I could take a demotion, and go to the Lincoln[3] to manage the Wildlife, Fish, and Rare Plants Program.

Essentially there's a hierarchy in the Forest Service. Within the biology shop itself, you have the "totem pole" with the wildlife biologists, the fish biologists, and the botanists. The botanists being on the bottom was very frustrating for me, because I felt that I could run the program as well as anyone. So it was a big deal for me to take over an entire wildlife, fish, and rare plants program, because I had been labeled as a botanist.

CHAPTER 3

Forest Service avoids gathering data about rare plants (mid-1980s)
One of the bigger issues I had to deal with as a range con[4] was the "what I don't know won't hurt me" attitude. There were times when the Nature Conservancy pleaded with the forest supervisor and the district ranger to conduct botanical work on some rare plants. And they refused to let me do it even though the conservancy in Tucson kept name requesting me to do it. My supervisors would make up every excuse in the world why I couldn't do the work—"She's pregnant. She can't do it."

So what if I'm pregnant? I'm working. I'm walking around. I can go look for rare plants.

It was just very clearly, "What we don't know won't hurt us," and "We don't want to know that it's out there." They would do everything to not gain information.

For the most part, though, I think we were doing a fairly good job at that time. But in those days we did the management plans by ourselves. We didn't go through a NEPA process. Or, if we did, it was on paper. There was not a lot of controversy.

CHAPTER 4

Congress degrades Forest Service management

I firmly believe that Congress is a major player here. I learned in school that politics were part of the equation, but I sit here now and will tell you that politics *are* the equation. That was hard to handle—a very difficult thing to acknowledge that politics are running this thing, and that science has very little to do with anything.

The pressures that are on the decision makers, whether it be a district ranger, forest supervisor, or regional forester, are such that they can't take the political fallout for what they need to do. And they've been far and few between, the ones that have done that—Gloria Flora[5] and a few others. And they took the hit for it.

But what I mean about Congress is this: these systems out West are long term, they're here for eternity, and yet we manage them on an annual whim. It's not going to work that way.

In the 1970s, Congress passed the Renewable Resources Planning Act,[6] they passed NFMA, they passed FLPMA. And all of those laws said the same thing—that we would plan out midterm, long term, whatever you want to call it—that we would not do any more of this seat of the pants management, where one minute it's one way, the next minute it's another way. Instead we would go through a plan. Call it a "forest plan" for the Forest Service. Call it an "RMP" for the BLM. Whatever you want to call it, it's a ten-, fifteen-, twenty-year plan that gives direction.

I was mandated under law to do those plans. But the agencies are budgeted on an annual appropriation. And those annual appropriation bills do not necessarily conform to, or are consistent with, those plans that are contracts with the public.

You can have a forest plan that says, "I'm going to take as a major goal that we will have 50 percent of the 'fair,' or the 'poor' condition land brought into 'good' condition in fifteen or twenty years." And yet there isn't an iota of progress, nor is there any kind of urgency because, I think, it's conflicting for everyone. On the one hand, you have the forest plan that tells you what to do. But then you get the annual budget that says, "Don't do it."

I know that it's a power game. Congress wants the power and they don't want to give it up. I know the environmentalists also don't want to give the Forest Service too much power by having five-year appropriation bills. But I think that you have to somehow address the fact that this "dance" isn't working. As a person who tried to implement these things, that was the big issue for me.

I'm very sympathetic to the "mission impossible" that the agencies are given. And one consequence is that it causes stress for some people. When people are under stress, all kinds of things happen. For example, it forces people to be creative on how to tell a boss, "Yes, I did this," knowing that you did something else.

Congress's power trips, and Congress's whimsical changes, and shifting priorities are bad news—it's not just bad news—it's virtually impossible to manage a long-term resource within that context.

CHAPTER 5

Social pressure on land management personnel
In the early 1990s, when I was T and E program manager, we gave an award to people in the region for doing award-worthy work for T and E species. One year, we unanimously decided to give it to a ranger from Reserve[7] on the Gila National Forest.

Why did we do that for him? He didn't do anything special. But he had the most contentious timber sales. His children were being black-listed in school. His wife couldn't go to the store.[8]

Forest Service people, who live in these rural communities, were being black-listed there, because they did what the law said. Yet this ranger went through with the protections for the spotted owl the way they were written.

We took a lot of flak because some people didn't understand why we gave that award to him.

Hey, putting your family through that just to implement the plain and simple rules, I think, is award worthy.

CHAPTER 6

Animal Damage Control harms wildlife

What hurts wildlife is Animal Damage Control—the predator control done for the ranching industry. That is what affects wildlife on a big scale, especially some of the canids—foxes. Coyotes, I'm not so worried about.

But you know what's interesting about that? In my understanding, the ranchers "double dip." Number one, they get lower grazing fees on public lands[9] because they have to tolerate predators.

Ranchers don't tolerate predators. They get the agency to wipe them out.

So that's an impact on wildlife that would essentially go away if we were not putting livestock on public lands.

When you look further into that whole subject, you'll find that there are animal husbandry practices that can minimize losses to predators. Some ranchers will do it, but a lot of them have the view, "I have ADC to take care of me. Why should I get all my animals bred at the same time, so that they'll drop at the same time?"

It's pretty predictable when the golden eagles are going to migrate through, and when your lambing season, or calving season is. I'm not saying you could eliminate losses to predators, but a lot of the ranchers have no interest in taking control of matters before they call Animal Damage Control. It's just a right—it's just a God-given right of a rancher to have Animal Damage Control.

CHAPTER 7

A personal anecdote about Animal Damage Control

The first assignment I had on the Lincoln[10] was my big immersion into wildlife and Animal Damage Control. I did the environmental assessment and worked with APHIS and other people to come up with a plan for how the APHIS would operate on the national forest.

ADC uses several devices to kill predators. There's something called an "M-44" with sodium cyanide in it, which is regulated by the EPA. Then you have the steel traps, and the snares, and some other things.

The M-44 was the most controversial. They had not been used on the Lincoln for at least ten years, but I had historic data as to how many animals had died, target and nontarget, from the M-44s.

The M-44 is scented with something that attracts canids, so it's somewhat humane, because it is specific to the canid and is very quick. They die within three feet of getting it. But it attracts canids indiscriminately, so it also kills the non-targeted gray foxes.[11] And because they're elusive there are very little data about them. The Game and Fish Department doesn't have data, nor does hardly anybody else as to their densities or their abundance.

It was very apparent that for every target coyote, there was an inordinate number of foxes that were being taken. There's a lot of literature on what happens when you take out coyotes—that they have density-dependent reproduction, and that they can rebound. We do not have that kind of data about foxes.

So I said, "I have a problem here."

The way we decided to resolve it, was that we gave APHIS what we called the "threshold for more information." It basically said, "Under the Lincoln National Forest Plan, if APHIS takes more than this many foxes in this period of time, they will cease use of M-44s until they can produce information that shows that the fox population in that area is okay."

When this decision went out the door, both sides attacked it. Environmentalists from the Predator Project felt that I hadn't gone far enough.

And guess what? The New Mexico Department of Agriculture said, "You don't control the populations of wildlife. You're the habitat people. You're stepping on the state's toes."

The Forest Service didn't want to hear that. They backed off. You know what they asked me to do? They said, "Listen, Renee, we have to withdraw this decision, and you know it better than anybody else, so could you find a flaw in your own work that we could use as a basis to withdraw it?"

I said, "You've got to be kidding. I'm not doing that."

To make a long story short: The shit was hitting the fan in DC because we had done this. The Director of Wildlife in DC called me up and said, "Renee, when we told you to do NEPA, we didn't mean for you to analyze the wildlife population."

"Oh, well, where's it say that? Could you get that in writing for me? That I do NEPA, but I don't do the wildlife populations because that's the state."

The end of the story is that they withdrew the agreement between the two agencies, and gave NEPA responsibility all to APHIS. The Forest Service backed out of the whole thing. They took my EA and my program and threw it in the garbage. And now they have M-44s anywhere, anytime they want on the Lincoln. No limits.

I was proud of my little threshold. I thought it was great. I'd let them kill all the coyotes they wanted, but I didn't have the data from the foxes to say that they can rebound. I've been to the state. I've been to NMSU. I asked them all what is the sustainable harvest. None of them had that data.

Again, that program is for the benefit of the rancher. And I'm sympathetic to the rancher, but he is double dipping.

CHAPTER 8

Attitude of APHIS about the killing of coyotes

The Lincoln National Forest is surrounded on three sides by sheep and goat ranches. It's the biggest part of the economy down there. Well, sheep and goats are very vulnerable to predators, more so than cattle. And the attitude was that we have to shoot these coyotes before they leave the Guadalupe Mountains.

I'm not buying that. We want them there. They're part of the system. And so it was a big deal for me to work ad nauseam with them to confine them[12] to within two miles of the national forest boundary. In my view, they can come in, in a hot pursuit, or in certain other cases. But not to just have at 'em.

And, of course, these are just varmints in the state's eyes.

APHIS does break their program into *preventive* and *corrective* components. When I had the program, it was supposed to emphasize the corrective as opposed to the preventive. But APHIS wants to do preventive everywhere. That was the story of the Guadalupes.

If you need it for hot pursuit, fine. But they should not be coming into the national forest just for the sake of wiping out a coyote before he does something. It's especially hard to accept when many ranchers won't do something to reduce their own losses by changes in animal husbandry.

CHAPTER 9

Implementing Integrated Resource Management on the Lincoln NF

Integrated Resource Management (IRM) is, or was, Region 3's[13] little package for how to implement NEPA. You put together a proposed action. You scope it internally. You see if there are issues. You work to change it a little bit to accommodate these issues. Then you go to the public. And then you write an environmental analysis. It's the whole package of how you do NEPA. The essential message there is that the entire spirit of NEPA is *integration*.

It's one thing to put a team together and say, "We need to do a timber sale. What's the best way to do it? I want to take care of wildlife. I want to take care of soils. So let's develop a timber sale with a holistic kind of concept."

Or you can let the timber guys or the grazing guys come up with their proposed action, and then everybody else has their alternative. There's the *Wildlife Alternative*, and there's the *Grazing Alternative*. Basically, you're saying, "Here's the timber sale the way he wants it. And here's the way I want it. Now, decision maker, pick between us."

Well, that's not what the NEPA is. That's not the spirit of integration. That's called "combat biology."

The Integrated Resource Management process was literally portrayed and promulgated as an interdisciplinary exercise to land management. But yet we ended up just being document writers instead of being full-fledged professionals participating in a real process. This is a farce. And it's important because it is at the core of business for the Forest Service and the BLM. The NEPA is the core of it.

In the Forest Service, timber used to be king. And there's a "totem pole." But it's not an *integrated* totem pole. Timber was king and everybody else had to fight for consideration. But that's not the way NEPA reads.

That particular point is so central to Forest Service business that agency officials will argue with me that what I've just said is not true. They are in total denial that I'm correct because it's so core. In fact, I've seen remarks stricken from records when we said that things are not integrated enough, or that there isn't enough internal coordination. Well, the reason for the denial is because you have a combat culture here.

CHAPTER 10

Treatment of endangered species on the Lincoln NF

When a species is listed as endangered, the way the system is supposed to work is that a federal agency then looks at all of its programs, and checks whether any are affecting the species and, if so, go through the legal consultation procedures.

That isn't the way it's worked historically, though. What has happened is that nobody looks around at anything, and then one day the Forest Service will do an allotment management plan or a timber sale, and they'll say, "Oh, guess what, there's going to be an effect." And then they go into consultation.

I was taking a couple of allotments into consultation through on-going grazing—it was the biggest thing in the region. When I hit the Sacramento Allotment[14] I said, "You're not up for decision,[15] but you have major effects here on T and E species, plants mainly, in addition to the Mexican spotted owl." And I needed to take them into consultation.

The deputy regional forester said, "Over my dead body are you going to take an on-going allotment with no fresh decision into consultation with the Fish and Wildlife Service. What are you trying to do?"

So we had a big pow wow. I convinced him that it wasn't going to be World War Three—that the cows would not have to come off, that we would work through it by studying the effects of the grazing as they happened.

And I finally got it through. But that was a major issue to bring any kind of scrutiny to an allotment that was not under formal review.

CHAPTER 11

Management agencies fail to support their personnel

The most contentious part of the Lincoln's grazing issue right now, and has been for the past fifteen years, are these plants: prickly poppy[16] and the Sacramento Mountains thistle.[17]

First of all, with plants you don't have the protections that you have with animals. You need to have "jeopardy" to get a mandatory action for a plant. Otherwise they're "conservation recommendations."[18]

Anyway, Fish and Wildlife Service is always on the Forest Service's case about heavy herbivory on these two plants. Prickly poppy being even worse than the thistle.

But come to a head. Okay, let's get the Fish and Wildlife Service, the Paragon,[19] the county commissioners, and the permittee down here. I want to talk to them.

Fish and Wildlife Service goes down there and says, "If you need to put the cows on, put the cows on."

They totally back down on everything they had been saying before. You get people up in front of the opposition and they melt.

But let me also say this. I used to blame people. I now blame the system.

I believe that Pete Domenici[20] is more responsible for this than the forest supervisor. I believe that when Pete Domenici, or any congressman, is out there on the opposite side of the federal government, the federal government has no backing. They can't do anything.

What do you think a forest supervisor's going to do? Cut an allotment and have Domenici on his ass? Or just say, "I don't think I'll make a decision today. I don't think I'll make a decision today. No, I don't want to cut 'em."

The political pressures, on the individuals that are involved, are incredible. The Fish and Wildlife Service comes out into the field and they buckle under. They just crumble before your eyes. But it's not them. It was the fact that this person knew that when they went back to Albuquerque, they weren't going to get support.

Leon Fager[21] spent a lot of time going after individuals. And I never got on board with that because I felt like, if we get rid of *that* guy, we'll get five more like him in his place. Granted, there are some outrageous ones that Leon targeted, that should have been targeted. But that was not my approach. My approach was never to target individuals, because individuals are not the problem.

It's a culture. It's a culture in this country. It's a culture in Congress. It's a culture in the Forest Service.

CHAPTER 12

Galeano-Popp resigns from the Forest Service

It would have been mid-'90s, 95ish when the lawsuit came down that accused the Forest Service of being in violation of the Endangered Species Act, because of all these on-going grazing allotments that had never been evaluated.[22]

What's the Forest Service going to do?

"Well, we're gonna do biological assessments on every allotment. Okay. Renee's the best one to do this. Renee, your calendar is cleared of everything else. You're gonna evaluate every one of these allotments in a few months and you're gonna tell, in a biological assessment, whether or not they're affecting T and E species."

In a nutshell, what happened has to do with "allowable use monitoring," which is something that the Forest Service made policy years prior as part of the goshawk policy. The goshawk guidelines[23] said, "You will use allowable use." That means that instead of saying, "Because this land is in X condition, we're going to do this and this and this to get it to the new condition that we want," we're going to get more aggressive. Now we're going to look at how much is grazed—what is the utilization level that's allowed? And they adopted that as policy.

Okay, I'm supposed to evaluate all these allotments. And my proposed action is to implement regional policy of allowable use on every one of them.

So how are you going to implement it?

Well, I need to know what the condition is, and what the species composition is here today. Then I need to know what the desired community is. I also need to know what tolerance the species has for grazing. And I need to know how fast I want to get it to this new condition.

To do that requires that you have a clue as to what the current conditions are, so you can *prescribe*. The second thing it requires is *monitoring*, to see if you're getting the use that you prescribe.

So I'm working with the range cons to get those allowable uses that had never been assigned to allotments. A "poor" allotment, for example, should have a lower allowable use than a "good" allotment. There should be some kind of correlation. Some kind of logic. Some kind of system that's not arbitrary.

This work is going on for months and months. First of all, the only condition information I have is thirty years old. Trying to get current information is like pulling teeth. And I'm starting to get a little irritable about this.

These guys are pressuring me that they want the analysis completed by a certain date, but yet I can't do it because they won't "play" with me.

They give me data. And I say, "Well, Rick, how come if this allotment's in such bad condition, you're giving it this allowable use? Could you help me understand this? Because I'm the one that's going to have to defend it to the Fish and Wildlife Service."

I got very little cooperation. And it was all so contentious.

In the end, I must have asked the range cons one too many times to explain to me the relationship between the condition of the allotment and the assigned allowable use, because I was called into the forest supervisor's office.[24] And I got called in when my main supporter wasn't there. Funny thing about that. But the range staff was there. And they sat me down and they said, "Look Renee, number one, when the range con says that it's *this* allowable use, you don't have to understand what it's based on. You just take it."

Well, that's not how I was raised as a professional. To just swallow something. I'm the one who has to defend it.

So they're saying, "No." They're saying, "Renee, you're out of control. You will buy it hook, line, and sinker if the range con tells you this. And, Renee, the reason why one day it's 20 percent here, and another day it's 40 percent there, and there's no consistency, and the boys aren't giving you the attention that you think that this lawsuit should be getting is because they know we're not really going to implement the allowable use monitoring."

"You're having me spend six months of my life writing biological assessments for the Fish and Wildlife Service, a legal document on a proposed action that you have no intention of implementing, and I'm supposed to be quiet? I'm supposed to not say anything to anybody about that?"

"That's right, Renee."

I was stupefied. I had never been spoken to that way. Not nasty, but "This is the way it is, baby. Take it or leave it."

And when I walked out of there I said, "I think they just intimidated me. I think they just did something they're not supposed to do." So how could I go on with this whole effort knowing that they had no intention of implementing it?

It wasn't, "We're not gonna have the money for this" or "We're scrambling to find the money to get all the monitoring people out there." It wasn't that. It was, "We have no intention of doing this."

But implementing it was mandate. Forest plan to me was "Bible." I took the forest plan very seriously. It's a contract with the public—a huge big deal.

So, ultimately, I left the Forest Service. And I don't know what happened. I turned my back. And the fact that I had a butterfly listed in that area, right as I was leaving, didn't make it easy for me to continue communicating with people. But that's my story.

CHAPTER 13

Allotment management at BLM declines in rigor
When I first started out in range with the Forest Service, we worked a lot with species composition and range condition. We classified the range as "very poor," "poor," "fair," "good," et cetera. We made goals for ourselves in our allotment management plans such as, that we would bring so many acres up to the next condition category within a specified period of time. We had permanent transects, and we would check whether the condition and trend had changed. Then we would adjust from there on a decadal basis.

Then I got involved with this allowable-use-monitoring concept, which established a monthly or annual basis for meeting a goal. It was a more intensive approach.

But then an interesting thing happened when I worked with BLM for two years on a resource management plan in Nevada.[25] There I saw that they were going in an entirely different direction. They don't want to do allotment management plans anymore. They want to jump from the resource management plan (RMP) to the grazing permit. And they want to use the permit as the vehicle.

They want to cut out all of this environmental analysis. Not only do they want to do this, but they have this thing called the "RAC Standards."

The RAC Standards are now a big deal. They say things like "The stream width should be appropriate for the site, and it should have adequate cover." And then they have a nebulous, abstract, very undetailed desired future condition.

That's scary that the processes they were putting into place seemed so unimplementable, so difficult to measure. It's the "trust us" approach. I can't think of anything that is less timely.

CHAPTER 14

Experience with the BLM's Great Basin Initiative
My experience in Nevada with the BLM gave me a number of observations that the agency was going in a direction that would make livestock grazing management even harder than it has been. By that, what I mean is this. If you compare the dollars per acre that the BLM and Forest Service each get from Congress, BLM has always had mega-land and mini-dollars, and Forest Service has always had the opposite. The Forest Service has always been perceived by the government as having timber resources that are much more valuable. So we've always had this disparity between the two agencies on their ability to intensively manage the land.

There's been a push in the BLM recently to consolidate districts, which are enormous to begin with. They take three five-million-acre districts and make one BLM unit out of it that's fifteen million acres. They couldn't deal with the management of the livestock to begin with. Now we're spreading the management even thinner.

My involvement with BLM was in connection with what's called the "GBRI"—the Great Basin Restoration Initiative. It's the big thing that BLM has bought into in the Great Basin. And it's very pie in the sky in terms of how they're going to remove and eradicate cheatgrass, how they're going to restore fire regimes, and so on.

Because there's been the unified policy to use watersheds as a consistent unit of analysis and unit implementation, I was required to analyze an RMP for thirteen million acres with the assumption that they could do a watershed analysis on all sixty-two of the area's watersheds within a ten-year period.

They have not produced a single watershed analysis in several years. Not one!

Yes, use the watersheds rather than grazing allotments as the basis of the management plans. I'm all behind that, ecologically. But from a simple administrative standpoint, if the restoration of the Great Basin is a real goal, the last thing I want to do is dilute the effort.

The Forest Service and the BLM have problems. And a lot of those problems are funding and cultural. It's the culture of their thinking. It's the old "can do" thing. "We can do anything." But they can't, and the land is suffering.

Their goals are extremely unrealistic, and I think that they should be called on the carpet for it. If we were serious about keeping livestock on public land, given what we know, we need more intensive management. BLM is going in the opposite direction.

They're very complacent at the BLM about that. I think we're all very complacent about the conditions of the Great Basin. The average person does not see the impacts of grazing. It's not like Napalm. It's not like a hydrogen bomb. It's not a like a clearcut, which is very striking to an individual.

Instead it's like a warm and furry animal. You just show it and the hearts bleed. You show the grasslands with all the irises and the sneezeweed,[26] or the rabbitbrush, and people think it's gorgeous.[27] They don't understand the insidious nature of the impacts of grazing and how widespread they are. And they may never understand them until *their* area turns into gullies and rills and is eroding away.

Steve Gallizioli

An early interest in hunting and fishing inspired Steve Gallizioli to pursue a BS degree in fish and wildlife management at Oregon State University. Soon after graduation he joined the Arizona Game and Fish Department as a regional biologist, a position that afforded him the opportunity to observe the negative effects of abusive livestock grazing on wildlife populations throughout Arizona. In the 1970s Gallizioli gained notoriety for his articles and presentations about environmental damage inflicted by livestock overgrazing.

Gallizioli retired from the department, as chief of its Wildlife Management Division, in 1983 after thirty-two years of service. Since that time he has served on the board of the Arizona Wildlife Federation and as the editor of its newsletter.

Steve Gallizioli made his remarks on the 24th of September 2004 in Forest Hills, Arizona.

Chapters
1. Gallizioli's youth, war experiences, and formal education
2. Gallizioli joins Arizona Game and Fish Department (1951)
3. Politics influences livestock management on the Crook NF (1951)
4. Politics thwarts management proposals of Arizona Game and Fish Department
5. Predator control
6. Rancher attitudes about predators
7. Livestock grazing depresses Mearns' quail populations
8. Gallizioli goes to Washington (1976)
9. Gallizioli's experiences with holistic resource management

CHAPTER 1

Gallizioli's youth, war experiences, and formal education

I was born in Italy, northern Italy. And moved to the United States in 1932 when I was eight years old. Lived in northern Michigan for twelve years until I graduated from high school in 1943.

My first interest in wildlife came from my dad being an avid hunter and fisherman. I followed him around for a few years without a gun, but immediately with a fishing rod. When I reached the age of eleven, he bought me a .410 shotgun, and I started hunting with him with my own gun.

Right after I graduated from high school, I was drafted into the Navy. And after going through boot camp at Farragut Naval Training Station in Idaho and then Quartermaster's School in San Diego, I was assigned to the USS Jet, a patrol boat based in Pearl Harbor. I spent the next two years on the Jet, mostly patrolling around the Hawaiian Islands and escorting cargo ships to islands as far west as Midway. Never got into any action, fortunately—my regret at the time, though. Like most young guys, I was looking for action, and maybe I might have had my head blown off if I'd got into it, especially on a small patrol craft like that.

The only interesting thing that happened in those long months at sea was my learning about a possible career in fish and game management. It happened one day when I reported to the bridge a few minutes before the start of my Quartermaster watch. There on the chart table was a copy of either *Outdoor Life* or *Field and Stream*. And near the back of the magazine an article discussed a new career. I read the article and learned that some colleges were giving degrees in fish and wildlife management, and that work for graduates was available with state wildlife agencies, and with federal agencies.

And on the spot, I decided that's for me. I decided right then and there, I was going to college.

And so I did. I was discharged from the Navy in 1946, and I spent a year at San José State College. Then I went up to Oregon State where I graduated in 1950 with a bachelor of science degree in fish and wildlife management.

During my last term, I began contacting various state wildlife agencies around the West. The Oregon Fish and Game Department offered me a temporary job as a creel clerk with the promise of a permanent job by the end of summer. I accepted, and began looking forward to a pleasant summer in the Cascade Mountains of Oregon.

I had also applied to Arizona Game and Fish Department. And before the end of the term, I got a call from O. N. Arrington, who was a division chief there to whom I had been recommended by a friend. And on the strength of that recommendation, he was willing to give me a job—a permanent job. I talked it over with my wife and decided we'd come to Arizona.

CHAPTER 2

Gallizioli joins Arizona Game and Fish Department (1951)

So I came to Arizona, and I spent a summer on a white wing dove nesting study. There were several study areas—one just south of Phoenix and one near

Tucson. All of them were in mesquite thickets, and I hated the heat. I knew I could never live in this country, but I decided to stick it out for the summer just so it couldn't be said that I had been whipped by the weather down here.

But by the time the summer ended and the white wings had flown to Mexico, I was sent to Fort Huachuca, which had been turned over by the military to the state. It was like moving from hell to heaven.

Unfortunately, our stay at the fort was short-lived. We moved down in September and had to move out by the end of January, because the military took over the fort again. But I loved that place. There was so much wildlife. Right behind the house we were living in, we would frequently see white-tailed deer and coatimundis. Walk down the sidewalk and there'd be coveys of Mearns' quail. Just a wonderful place. I thought, God, I could spend the rest of my life here. I was really unhappy when they moved the Game and Fish people out of there.

About the same time, the department reorganized to a certain extent. Most of the field personnel were game rangers at that time. There were only three or four regional biologists. I became the eastern Arizona regional biologist living south of Pima in eastern Arizona at Cluff Ranch, a facility owned by the Game and Fish Department in the foothills of the Graham Mountains.

I had all the country from Pima to the New Mexico border south to the Mexican border—about 10,000 square miles or more—and I was suppose to manage it as the only wildlife biologist in the region.

CHAPTER 3

Politics influences livestock management on the Crook NF (1951)

One of the first things that struck me, as the regional biologist, was the condition of the range. It was shocking to me that it was in such bad condition. The area, including a unit of the Crook National Forest,[1] was in terrible condition.

Once I actually found an allotment where the cattle had been left to literally starve to death. Two range cows had already died from starvation. And the condition of the range in that area was so incredibly bad, you could understand how they would starve. Even such things as manzanita and turbinella oak, a scrub oak, had been chewed down to stems the size of my thumb.

As soon as I could, I stormed into the office of the local district forest ranger and demanded to know "What in the hell are you doing up there? How can you allow that range to be so badly treated? Why don't you do something about it? Or are you doing something about it?"

He sat through my tirade half smiling. Then he said, "I can appreciate how you feel, but the permittee for that allotment is a man name of Jim Smith. Jim Smith is a state senator and is running for governor of the state at this moment."[2] And he said, "Jim Smith is a big man, figuratively speaking. And there's nothing I can do with somebody like that." He said, "If I try, word will get to the regional forester in Albuquerque, and it probably won't get any farther than that. And the word will come back to me, 'You lay off Jim Smith. He's too big a man to mess around with.'"

And that turned out to be the case with other situations that I became familiar with. One of the big problems with proper range management turned out to be that even if the agency personnel were concerned and wanted to do something about it, they were blocked by politics. And, unfortunately, I think that's still the situation even today.

CHAPTER 4

Politics thwarts management proposals of Arizona Game and Fish Department
Livestock reduction proposals have always been an issue. And it isn't one we can address by going to the rancher. We would do so by going to the agency, which is usually the Forest Service but also BLM.

Most of the time, though, the federal agency would even refuse to consider our complaint. Because the minute they would start saying to the rancher, "Hey, we got to cut back on your permit" for this or that reason, the rancher is gonna scream like a scalded cat. He can't afford it. He's gonna go broke. Or he probably gets on the phone and calls his congressman and senator and says, "You know what they're doing to me again? They're trying to get me out of the cattle business."

It's difficult—virtually impossible for a member of a state agency to persuade a federal agency that they ought to do something, especially on behalf of wildlife. They've always catered to the rancher. That's their first consideration. They might pay lip service to the needs of wildlife, but they don't do much more than that.

CHAPTER 5

Predator control

I think a lot of ranchers would like to see the availability of Compound 1080 again, but I don't think that's likely to happen. And without something like that, you'll never get the wide-scale killing that we had for many years. Of course, coyotes were poisoned even before the advent of 1080. Ranchers have been having the Fish and Wildlife Service do the killing for them. Before 1080 they used strychnine and thallium, and I don't know what else. They also use "coyote getters," a trigger device that shoots a load of cyanide into the mouth of the coyote when he pulls on it.[3]

But back when they had to use toxins other than 1080, the coyotes were never really reduced in significant numbers. It's an amazing animal.

Coyotes respond to a reduction in population in two ways according to studies that have been done in California. One is by producing larger litters. The other is by having a better survival of pups. So to really reduce the coyote population you've got to stay in there.

Aerial shooting is pretty effective, but it's extremely expensive. It requires a helicopter and charter rates run five, six hundred dollars, or more an hour. So every coyote that's killed represents a lot of money out of the Game and Fish Department budget.

Of course, Game and Fish gets criticized every time they do any coyote control using aerial gunning, but they feel the cost is justified to improve pronghorn fawn survival in places like Anderson Mesa.

Compound 1080 wasn't supposed to kill animals besides coyotes, because the experimental work that had been done showed that the dog family was extremely susceptible to this poison. The coyote being a pretty large animal and being so susceptible could be killed with a very small amount of the poison, such that other animals were not affected.

That worked in theory, but in practice some predator control guys felt like some people do about medicine—that if a teaspoonful is good, a cup ought to be even better.[4] Then they would shoot horses and burros, pump 1080 into the animal and use that as a bait for the coyotes.

I've heard guys say that this is what happened. They weren't supposed to do it, but they did anyway. Maybe that's why it was so effective, because in addition to the coyotes being so susceptible, they got a much heavier dose.

CHAPTER 6

Rancher attitudes about predators

Lions don't kill very many cattle, but I don't think you could ever get a rancher to consider that that's an aspect of doing business as a rancher.

A sheep herder loses many sheep to disease every year. More than he loses to predation.[5] But yet he'll scream to beat hell about the predators taking his sheep, and not worry about how many are being killed by diseases. Because, I suppose, it's a lot easier to kill the predators than to treat the sheep, or find out what's killing them and do something about it. It all boils down to the fact that if he loses animals by any means it represents a loss of income to him.

CHAPTER 7

Livestock grazing depresses Mearns' quail populations

After we closed out a study on Gambel and scaled quail in 1965, we went to work on the Mearns' quail farther south in Arizona, in the higher elevations, about four to five thousand feet up in the oak woodland. There we found that livestock had a much more severe impact on the population level than the grazing did on the other two desert species.

We were surprised to learn that the impact from livestock was almost entirely due to overgrazing eliminating escape and nesting cover. Heavy grazing was actually beneficial from the standpoint of the food that the birds fed on, which was different from the other two quail. They feed mostly on seed; the Mearns' quail feeds on underground bulbs and tubers. They scratch down and dig for them. A heavily grazed area was conducive to the production of bulbs and tubers, but that didn't necessarily mean an abundance of the bird during the winter when hunters were seeking them.

One area in particular illustrates that situation. Richard Brown, the biologist assigned to the study, was trying to find a study area with an abundance of quail. He called me up about the end of September, and he said, "I think we've got the area that we want to study. Come on down and look at it."

So I showed up with my Brittany spaniel. Richard and his dog had found fifteen coveys on one-half of a section—one half of a square mile. He had marked the location of each covey he had found with the dog just a couple of days before on an aerial map. He suggested we hunt the area again using my Brittany.

He said, "Let's go around to where I found these coveys before." We found thirteen of them with my Brittany, who was not nearly as good as his dog.

But the reason I brought up this anecdote is that before the season opened on the first of December, all those quail had disappeared. Apparently, there were so many quail in the area that they had exhausted the food supply, and they had to move out. And so here was a situation where we had good grass cover. And that's why there was a really good hatch and good food supply for a while. But then there were so many quail that they actually ate themselves out of house and home.

One of the main things that Richard learned on this study was that if you've got a lightly grazed area or an ungrazed area, you had a lot of birds during the hunting season. And good hunting. If you had an overgrazed area you wouldn't have any birds, or practically none. And then a moderately grazed area would be in between. The number of birds was directly related to the intensity of grazing and to the resulting cover conditions.

So that wasn't the kind of thing the rancher was happy to hear. But that was a good example of where proper range management was beneficial to this particular species.

CHAPTER 8

Gallizioli goes to Washington (1976)
In 1976 I got excited enough to do something about Arizona's overgrazed rangelands. I presented a paper at a symposium in Washington, DC.[6] When I returned from the symposium, I sent copies of my paper to a number of newspapers, including to Bob Thomas, who was the outdoor columnist for the *Arizona Republic*.

Bob apparently liked my paper because he devoted an entire weekly column to it. His story then got picked up by other newspapers and pretty quick it was down in the Arizona legislature.

Then John Hays saw a copy of the article. Hays was not only a state representative, but also had a ranch in Peeples Valley. He wasn't very happy with my remarks, and decided he would have to have a chat with me and the Game and Fish director, Bob Jantzen.[7] Hays called the Game and Fish office and demanded that Bob and I both show up at his office—pronto. Bob happened to be out of town, so Roger Gruenewald, the assistant director, called Hays and offered to fill in for him.

He agreed to that, but he demanded that we come down to his office. Roger said, "Okay, when would you like to have us down there?" And he said a couple of days after that.

So I said, "Okay, we'll be there."

Then I called Bob Thomas. And I said, "Hey, you know what's happening? I have to go down and defend my position on grazing, apparently, to John Hays. And Roger Gruenewald is coming with me. He wanted the director, Bob Jantzen, but Bob is not here. Maybe you'd like to come on down and listen in on it."

He said, "You bet. Not only that, but I'm gonna see that other people are there."

Bob spread the word. And pretty quick all the radio stations and the TV stations were calling John Hays' office. They want to know what time this meeting is 'cause they want to be there. The next thing you know, Hays backed off of it. We never heard another word from him.

The governor's committee
Soon after this, back when Bruce Babbitt was governor,[8] he created a governor's committee on range conditions. There were about a dozen people on that committee. Every one of them a rancher!

The next thing I know, I'm invited to make a command appearance down on the ninth floor in the governor's office. He had a conference room up there. And Roger Gruenewald came along, the assistant director.

Anyway, we went down there. Babbitt didn't put in an appearance. It was just the committee. And the committee chairman introduced everybody to me, and said there's some concern on the part of some of the committee members about this paper I gave back in Washington, DC. And they had some questions for me, or words to that effect. Anyway, "What the hell you have to say for yourself?" was the gist of it.

So I said the same thing in fewer words to them that I had said in that paper back in Washington. And they would ask me questions. And bluntly I didn't back off of anything. For one thing, I didn't say that every square foot of rangeland in Arizona was overgrazed. I said, "I was talking about overgrazed rangeland. And there's a lot of it." Some ranches are in pretty good shape.

One thing I didn't expect, and I really kick myself for what happened. One of the ranchers finally said, "You work for the Game and Fish Department?"

"Yeah."

"How long have you worked for them?" And I told him.

He said, "You ever work for anybody else? Ever have any other job?"

I said, "Well, as a matter of fact, except for three years I spent in the Navy, I never had any reason to want to leave the department. So, no, I guess I haven't worked for anybody but the state."

"So you never had to get a real job?"

He was implying that because I was one of these guys feeding off the taxpayers, I had a lot of nerve complaining about how ranchers treated their leased public lands.

I should have responded by pointing out the various ways ranchers fed at the public trough. Various subsidies and tax breaks. Grazing fees much lower than those for private lands. Predator control paid for by the state.

It wasn't until I got out of the elevator down below and I thought, "Why didn't I think of pointing that out? I wonder what he would have said then?"

So, anyway, that started my short career as a grazing activist.

I then wrote a number of papers, mostly at the request of agencies—the Forest Service; the BLM asked me to give a talk; New Mexico Fish and Game asked me to come over and give a talk. The California Division of the Wildlife Society[9]—they had their meeting on the Queen Mary docked at Long Beach. Everybody stayed right on the big boat.

Almost always I would use a bunch of slides to illustrate my talks. And it was kind of hard to argue with those slides, and claim that it's all bullshit.

CHAPTER 9

Gallizioli's experiences with holistic resource management

At that time the word was going around that Steve Gallizioli was really mad at the ranchers. In fact, I wound up getting the nickname "Overgrazioli."

Allan Savory[10] heard about me and, like he's done with other people, I was offered a free tuition to one of his courses in Albuquerque. I still had to pay my other expenses, but Game and Fish paid for that.

Before I took the course, I had heard Allan Savory speak a few times on what at that time he called the "Savory grazing method." It became "holistic resource management"[11] shortly thereafter.

Having watched him in action, I knew Savory was a fantastic speaker. He had the kind of charisma that would allow him to sell ice boxes to Eskimos in the middle of winter. And I knew going in that I would have to watch myself to not be brainwashed.

And he brainwashed me anyway—to such an extent that when I came back, I even wrote a letter to a number of people who knew how I felt about grazing. I told them that I thought I'd found a solution.

One of my concerns at that point was—can we ever possibly get together with livestock people and reconcile our differences, so that we can have both livestock and wildlife without one impacting the other? And I had about concluded that it's impossible. It was never going to happen. And then here comes a guy who proceeds to demonstrate how you can actually have better wildlife habitat with livestock on the range. And I bought it. For a while.

In fact, I even gave a talk at a meeting of the Arizona Cattlemen's Association[12] up at Prescott, where I stuck my neck out and said, "I think this is something that all ranchers ought to be considering. I think it's the way to go. It's the way to get away from this fight between ranchers and environmentalists."

Then I sat back and waited to see if anything was gonna happen that would change my mind. And before long, I realized that Savory had been preaching this for quite a while in Arizona. He came from Rhodesia, what is now Zimbabwe, and had developed his grazing system there.

In 1987 I went to Africa for seven weeks. I went to Kenya and Tanzania, and down to southern Africa into South Africa and Botswana, and got up to Zimbabwe for several days. And before I went over, I called Allan Savory and I told him what I was gonna do—that I was planning on going into Zimbabwe, and I'd like the names of a couple of ranchers that had been practicing holistic resource management, so I could see how good the habitat was after being subject to that kind of management for a long time.

Much to my surprise he said, "You know, after I left there, those people just gave it up."[13]

I said, "What! You're preaching that this is an easy system to implement, and that it's gonna do all these wonderful things for the rancher, for the range, and for wildlife, and now you're telling me that the people involved in it weren't impressed enough to stay with it when you weren't there to hammer on them?"

He said, "I'm sorry, but that's the way it is."

Well, that did it. I figured that this has got to be just nothing but a bunch of bullshit.

But I continued to correspond with Savory for a while and ask questions, most of which he couldn't answer.

One time I accompanied several Forest Service people from the Tonto National Forest on a hike up to Dutchwoman Butte,[14] an area that had never been grazed.[15] That was the roughest climb I ever made. And I was amazed. Here was a place that had never been grazed, because it was too hard for the cattle to get up there. But grasses were abundant. I forget how many different species of perennial grasses we counted. I think it was something like nineteen or twenty-nine.

And this is in Tonto Basin, where off of this butte the cattle had just completely destroyed the habitat. If you read Croxen's report from the Forest Service,[16] you'll learn what the Tonto Basin was like at one time. It must have been a fantastic place. The grass used to be so high on these mesas that they would cut it for winter hay. You look at these mesas now and they're nothing but scrubby mesquite, cat claw, and prickly pear. If you can find a perennial grass, you're doing damn good.

And yet about a thousand feet higher, we had all these fantastic grasses. And it's not as if when you get to that elevation off of Dutchwoman Butte that you'll find these same species. You won't.

I took a bunch of pictures on the butte. And I sent half a dozen of them to Allan Savory. I told him that this is an area that's never been grazed by livestock. Never! Cattle can't get up there. And there's practically no large animal up there. A couple of deer tracks and a couple of javelina tracks are about all I could find.

He replied, "Oh no, there must be a big deer herd up there for that kind of a situation to develop." He's talking about the grasses that I showed in the pictures. According to Savory's theory, in the absence of large grazing animals, grasses do not develop.

He as much as told me that I was lying. I felt like really telling him what I thought about him at that point, but I didn't.

David Gilman

David Gilman was born and raised in northeastern Vermont, where his early life centered around farm work and domestic animals. Seeking to help Vermont farmers, Gilman attended Utah State University, where he focused his coursework on soils and meteorology. After receiving his bachelor of science degree in 1968, he mapped soils for the USDA Soil Conservation Service in Woodstock, Vermont. Gilman relocated to the Targhee National Forest (Idaho) as a soil scientist in 1974. The following year he became the zone soil scientist for the Challis and the Sawtooth National Forests, a position that gave him the responsibility for insuring protection of the soil against any activity that might disturb it. When every national forest was subsequently assigned its own soil scientist, Gilman remained in that position on the Sawtooth until his retirement in 1994.

David Gilman made his remarks on the 23rd of August 2003 in Twin Falls, Idaho.

Chapters
1. Gilman's position with the Forest Service
2. Cattle compact soil and strip vegetation
3. Soil compaction reduces plant productivity
4. Forest plans fail to account for changes in vegetation
5. Allotment evaluations not performed
6. Differences between livestock-grazed and ungrazed riparian areas
7. Soil is essential to life

CHAPTER 1

Gilman's position with the Forest Service

My position with the Forest Service was mostly as a consultant in an advisory capacity. I could ask the rangers if they could do something to mitigate impacts on soil. For example, we might find in doing my soil survey work that there could be a 100- to 200-acre meadow at a high elevation that would be suitable to graze a band of sheep for, say, five days. But for the sheep to go from the valley bottom to this high meadow, they might have to traverse up to a mile of undesirable, unstable slope. The slope did not have sufficient vegetation to protect it from erosion and compaction by the passing livestock. But the district rangers, who were under the gun by the permittees, as well as by the political machine, oftentimes couldn't make any decisions to protect those fragile hillsides. The viable tradeoff would have been to protect those fragile areas, and forget about grazing that higher meadow. But they couldn't do it because of the political machines.

It's not necessarily the ranger's fault. He's trying to do a good job, but he just can't do it because he is being informed by his superiors that those sheep will go up there, and they will stay on that allotment for so many days, regardless of the consequences to the resources.

CHAPTER 2

Cattle compact soil and strip vegetation

Undisturbed soil in a natural condition is a very friable, permeable natural resource material. And a soil that has developed in place, where you've got ample organic accumulation and decomposition, builds up a very fertile, productive soil. Oftentimes I related to people that a good condition of the soil would be when it's like a jar of marbles. The soil can have a "crumb" structure, but it also allows the water to infiltrate into the soil profile. It's also a good medium to exchange the soil gasses.

When you have this type of condition, the root development of the plants is ideal because they are able to expand as they develop. They can take up the water that they need for their purposes. It's a beautiful thing.

But when you have weight impacting on the soil from livestock, particularly if it's concentrated livestock year after year in the same area, the soil will become compacted. A good mental idea of that is when the soil becomes compacted, it becomes "platy" in the top six to eight inches. Then, instead of a crumb structure, it changes to a platy structure. An example of that would be like a deck of cards lying down on a table. This becomes an impervious layer to the downward movement of water—to the development of plant roots. It also shuts off the exchange of gasses in the soil, which is important to the growth of roots and other organic organisms. Carbon dioxide is released and oxygen enters the soil. When soil compaction closes the soil pores and the worm holes, this stops the exchange of gases.

It's very easy to see this compaction and the platiness. Any time there's a high-intensity rain, instead of the soil soaking up the water like a sponge, it just runs off the soil surface like a parking lot. What little friable soil, or loose soil, there is on the surface is eroded away. And then this erosion causes plant pedestalling, whereas the only soil that's held is that which is held by the roots of the plants. And so you get these little plants sticking up above the soil surface. This is called "sheet erosion."

This is an indication that the soil has lost its infiltration capacity. The excess erosion then reduces the potential productivity of the soil. Soil also runs into drainageways and creates problems with water quality. And then, of course, that can get involved with the fish spawning because it suffocates the eggs. A lot of problems are created when the soil does not stay in place.

The vegetation is basically the only protection that the soil has from raindrop impact or from flowing water in a stream channel. And when the vegetation has been removed by livestock, the soil is vulnerable to rapid erosion. That's why we have a lot of this plant pedestalling, and why we have a lot of our gullies developing and getting larger in our present-day watersheds.

CHAPTER 3

Soil compaction reduces plant productivity

I did a soil compaction study,[1] which indicates that soil compaction does have a direct detrimental effect on plant productivity. Of course, when you don't have any plant productivity, you also don't have any available forage for livestock.

There are many indications of soil loss. One that I mentioned is plant pedestalling, but another is the loss of vegetative cover on the hillsides. You can look at rilling, which are small, minor gullies just developing. That's an indication of rapid water movement over the soil surface.

There are a lot of things out there that you look for that will give you an indication that the soils are compacted. And when they are compacted, we have a great reduction in our vegetative production potential. That applies not only to the grasses and the forbs, but also to the trees.

I worked with a guy at the research station out of Boise in timbered areas where they'd selective cut the trees. Then they'd go back and try to hand-plant two-year-old seedlings. In the areas where they've had a higher degree of soil compaction—either from livestock or mechanical means during the harvest methods—the roots of the seedlings will go down until they hit this compacted layer. And then they will form a "J"—they will just go down and make a right angle and follow the compaction layer. They call that "J-roots."

This can also be found in the rangeland areas where the soils are compacted. Roots do not develop. They can't penetrate down through this compacted soil layer, so their productivity is reduced because they can't access the water. Because the roots don't develop, they can't take up the minerals that they need for the plant functions. Then you get stunted plants. And, for example, instead of two thousand pounds to the acre of dry forage for livestock use, you may only have five hundred pounds.

Once those soils have become compacted, it takes a very long time for them to be removed, because the only natural way for the compaction to be reduced is from frost action. But if the water can't penetrate into those compacted layers, we don't get the frost action. There needs to be seasons when there is ample rainfall in the fall to get the soil saturated to the point where this frost action can take place.

CHAPTER 4

Forest plans fail to account for changes in vegetation
When we were working on the Sawtooth National Forest Plan in 1985, the allotment management plans, and the forage production for livestock stocking rates had been done anywhere from twenty to thirty years earlier. And, to my knowledge, since that original vegetative inventory was done, there has been no update of that vegetative information.

CHAPTER 5

Allotment evaluations not performed
The district range conservationists are supposed to go out, as I recall, before the grazing season begins and do an allotment evaluation. But with the small number of personnel in these district offices, and the large amount of time it takes to check every allotment on a district, it's not being done.

Another thing that's kind of sad—nobody is counting the livestock on. Some of these permittees could be putting on more than their permitted head (cow-calf pairs). If that's the case, they are still in their same rotation schedule, but there's maybe a third more livestock than what the allotment plan initially stated. The amount of impact is going to greatly increase. And this happens year after year.

Then when you hear the words, "We are getting an improvement in the watershed" or "We are getting an improvement in the range conditions," it is very hard to believe. When you go out there and look at the range conditions, you can see for yourself that we're not getting an improvement. I have walked four hundred miles of riparian myself. And so when you look at the condition of these riparians, where they have grazed and where they haven't grazed, it is a remarkable difference.

CHAPTER 6

Differences between livestock-grazed and ungrazed riparian areas
One of the most striking differences that you'll notice on a severely degraded riparian, versus one that's in good condition, is the absence of willows and some other aquatic species—alders and birch. Some of these species depend on the geologic type and elevation. Aspect also has an influence on what shrub species will be present. But when you have a good riparian condition, both sides of the streambank will be lined with willows of different species. The willows will be overhanging the stream. The stream will be deep and narrow. Willows hanging over the stream keep the water temperature much cooler, which is necessary for the fisheries, along with other aquatic biota. The riparian vegetation, Carex sedges, in some cases can be nearly knee high.

Floods should be occurring every year. When the water comes up in the spring from a rapid snow melt or a high-intensity rain, it will just flow over the streambank, and go out through the vegetation. The vegetation filters out the sediment, which also re-fertilizes the floodplain. And also, we can have a tremendous flood, and streambanks will remain stable because the vegetation is protecting the soil.

Those channels that are lined with the willows, and oftentimes some of the taller species, such as cottonwood, are the main wildlife areas that birds use for nesting, breeding, and rearing their young. They're also important for larger big game for shading and hiding cover, as well as for rearing their young.

When you get flooding, and you have good vegetation on the floodplain, that water slows down, because it has to go through the vegetation. That also gives the water time to be absorbed by the soil, which recharges the ground water table. During the summer, when you have little rain, that stored water in the floodplain will recharge your stream to maintain a flow. That way you will not have a dry streambed as we commonly see around here. That vegetation in the riparian area is extremely important, not just in the watershed, but for people and activities downstream. That little strip of riparian area is probably the most important piece of real estate in the watershed.

A very undesirable riparian condition, the most common that you'll see—the stream will be wide and shallow. Streambanks will be sloughed in. Instead of nearly vertical streambanks, they will be laid back at a low angle from the livestock walking into and out of the stream channel. You won't have vegetation that is lush and knee-high. You'll be lucky if you find *any* vegetation. Oftentimes it looks like a feedlot. Willows will be absent or nearly so.

Also, the streambank or the stream channel, when it's wide and shallow, has no shading. The water will be very warm and mossy. There will be no fish. Aquatic species will be almost absent.

CHAPTER 7

Soil is essential to life

Soil is one of our most important basic resources. When you don't have soil, you don't have much of anything. You don't have any water. You don't have any vegetation. You don't have any large wildlife. You have a sterile environment. And so when the soil is lost, everything is lost.

When the soil is not protected because of political greed, for whatever reason it may be, not only does the Forest Service lose, the whole community and the state loses. All life in the watershed is directly impacted and influenced by the loss of soil. As the streams go down into the bigger streams, the major drainages are also directly influenced. And it's all negative.

So this is my main concern, that the soils are not sufficiently protected. Very few people understand what the value of soil is, because they can't see it. They see the top of it, but they don't see the whole picture—what it does to our life. Without the soils we have no life.

Martha Hahn

Martha Hahn traces her love for the outdoors to her childhood fishing trips with her grandfather, along with summers spent at her grandparents' high desert California home. She furthered her interest in the natural world by attending Utah State University (USU), where she obtained a bachelor of science degree in forestry and outdoor recreation. Then as a cooperative education student with the Bureau of Land Management (BLM), she obtained a master of science degree in outdoor recreation behavior, also at USU. Hahn spent twenty-six years working in the natural resources management field with BLM, National Park Service, and Grand Canyon Trust. Her career at BLM included four years as the Colorado associate state director, and seven years as Idaho state director. Since leaving federal employment in 2002, Hahn has operated her own consulting business, The Sage Project, which is dedicated to professional coaching and teaching leadership skills.

Martha Hahn made her remarks on the 23rd of August 2004 in Boise, Idaho.

Chapters
1. Hahn's early life
2. Overview of Hahn's career in natural resource management
3. Hahn's experiences with grazing impacts on public lands (1970s–80s)
4. Frequency of lawsuits brought against BLM over grazing impacts
5. FLPMA and SVIM change how BLM does business (mid-1970s)
6. BLM's notoriety increases during the Clinton administration
7. Jim Baca's tenure as BLM director (1993–94)
8. Interior Secretary Bruce Babbitt pushes rangeland reform (1994)
9. Policy changes at BLM resulting from Rangeland Reform '94 (1995–98)
10. Lack of planning rewarded by BLM (1996–99)
11. Idaho Watersheds Project influences BLM's range management
12. Owyhee ranchers benefit from an Air Force bombing range
13. George W. Bush administration changes BLM management (2001)
14. Hahn's removal from BLM Idaho state directorship (March 2002)
15. Suspected influence of the livestock industry in Hahn's reassignment

CHAPTER 1

Hahn's early life

I grew up in Southern California, where my interest in the outdoor side of things came from my draw to the ocean and the beach. My grandparents lived in the California high desert. Complete open space. And I spent a lot of time in the summers with them, as well as fishing with my grandfather.

I think my love for the outdoors really was born from those types of experiences. I had the opportunity to attend Utah State on an athletic scholarship, which was really a springboard for my interests in natural resources. I had been thinking how being in the city would restrict me to things like community recreation. But now I had an opportunity to go to a school that provided natural resources training. And that was the premier boost for me in getting my career going—getting two degrees in natural resources, and then beginning to work.

So that's really what brought me into it—the background with my grandparents, my whole passion for the outdoors, and then having that ability to go to Utah State on an athletic scholarship.

CHAPTER 2

Overview of Hahn's career in natural resource management

I started out as a GS 3 working summers in fire and recreation with the Bureau of Land Management. During graduate school at Utah State University, I got on as a cooperative education student working with the BLM. In that situation, I was hired by the agency and split my time going to school and working. I did that for a year, completed my thesis work, and was offered a position without competition.

I began as a recreation planner in the San Rafael Swell in central southern Utah, where I worked for about four years. My main focus was the original wilderness reviews in Utah—establishing wilderness study areas. That would have been in the late '70s or early 1980s.

From there I moved to the National Park Service, where I was a resource management specialist at Grand Canyon National Park. I got involved with the Glen Canyon Dam environmental studies, and the decision on how to operate the dam to best provide for the river environment. I also established a lot of research and studies around the impact of use in the river corridor as it related to recreation, and commercial and private outfitters. That work helped establish the user capacity and the number of permits that could be allotted to each group. I worked on that for three years before moving back to the BLM as the resource area manager in Kanab, Utah. It was the mid-1980s that I worked there overseeing an area of more than two million acres with a staff of ten people.

I was one of the first women to become a line manager for the BLM in Utah. It was quite the experience, not only for me, but, I think, for the people of Utah, especially southern Utah. I felt that quite heavily in terms of how much impact my decisions and my style had on that community and its future.

I worked there for two-and-a-half years and then left government altogether to work for a nonprofit organization, Grand Canyon Trust, in Flagstaff, Arizona. I understood the government side of natural resource management, and I wanted to understand what it meant in terms of the private side. I saw Grand Canyon Trust as a group that really looked at the people's side of natural resources and conservation. It was a new organization. And as the vice president for conservation, I was able to go in and actually influence the structure and creation of their programs. I stayed with that for two-and-a-half years.

I always thought I'd go back into government work with the natural resource agencies, though. I had a criteria, however, and that was I would do it only if I could be in charge. I really found, throughout my career, that the more authority you had, the more influence you had. And I really wanted to make a difference. I applied for and was selected to be the associate state director in Colorado for the BLM.[1] That's where I began to really understand what it was like to motivate employees and other people. I was able to observe the effect of that on relationships, decisions, and programs.

With that under my belt, I realized I really wanted to be a director of a state BLM operation. Within four years I was selected to be the state director for BLM in Idaho.

CHAPTER 3

Hahn's experiences with grazing impacts on public lands (1970s–80s)

BLM is a very small organization; you don't have a lot of specialists. Everyone has to work together. So it's really important that everyone understand each other's programs. When I was working on recreation, I was concerned about how other uses of the land would impact it, and conversely, how recreation could impact other resources. So I worked very closely with all the different specialists. For example, if I was out doing some kind of inventory work in the field, and I saw cows in a place I knew they weren't suppose to be, I'd immediately get a hold of the range conservationist and say, "Hey, there's twenty cows out there. Are they supposed to be here? Here's the tag number." We would help each other in that fashion. Same thing occurred when someone was out doing their work and they would say to me, "There's a bunch of 4-wheelers rippin' up this hill. Did you know that they're out here?"

We also had to work as a team when we wrote environmental assessments and environmental impact statements. We were very aware of each other's programs and what that meant in a written sense. It wasn't unusual that I'd be assigned to do an environmental assessment on something to do with grazing.

That was really my experience early on. I understood what was happening in grazing, oil and gas, hard rock mining. All of those issues.

In the early '80s, I sensed conflict between recreation and grazing. But there wasn't a lot of recreation going on out in the middle of the desert—in the canyonlands. It wasn't a trend. It wasn't something people clamored for as they do now.

My feeling was more of "Hmm, you know, I have a hunch that this could be bad in the long run." For example, a water hole might be beat out by cattle use, and I'd think, "Hmm, I wonder how that will impact people or this area in the future?" And then, "What effect will that have on the recreation or wilderness use?"

I had a sensitivity and I had a hunch, but I didn't necessarily have the feedback, because the use was not there. And people just didn't express their objections like they do now.

In those days, when you put out an environmental assessment, you rarely got public comment. When we had meetings on wilderness, we had very little participation. It was amazing. That would have been in the late '70s and early '80s.

CHAPTER 4

Frequency of lawsuits brought against BLM over grazing impacts
You didn't have people really using the legal system at that time.[2] You might have someone come in and complain, but rarely did you see lawsuits. It wasn't

until the inventory work in wilderness, and when BLM started to come out with their final recommendation to Congress, that coalitions started being formed. There were groups beginning to come along and use the legal system. That would have been right around 1981, '82. That's when they started to gear up. And they could see how effective they could be. I believe "wilderness" was the place they cut their teeth, so to speak.

It was the small, local groups that were forming. You didn't hear from the Sierra Club. You just didn't see their effects. During that time period, however, SUWA was becoming active, along with the Wilderness Society.

I would say it probably was around '84, '85 that their momentum started to pick up. But nothing like I experienced when I got to Idaho in '95. Ten years of experience really catapulted those groups into action. Even when I was an associate state director in Colorado in the early '90s, we didn't have nearly the challenges, in a legal sense, that I saw when I got to Idaho.

Some of my fellow state directors, when I started in Idaho, said, "Boy, being a state director isn't what it used to be even a year ago." So it really, I think, took on a whole different character. The public became engaged, along with the national and local groups, in starting to use the legal system.

CHAPTER 5

FLPMA and SVIM change how BLM does business (mid-1970s)

The most significant change at BLM was in 1976 to '77 when it got its organic act, FLPMA. Everyone was running around trying to figure out what to do with this thing.

At the same time, grazing was taking on its own studies in what they called the "SVIM inventory." And that was basically looking at the vegetation. What is the vegetation that we have out here? Based on that, how will we manage?

This was huge for the agency. It hadn't been done before. Grazing had been set, based on what people knew about the countryside—what the ranchers knew about it, and what they wanted. Now we had the organic act, which basically said, "You will look at this. You will do it in this fashion." And so these inventories aided in the assessment.

Actually, the SVIM inventory was an outgrowth of the lawsuit that came out in the early '70s, where BLM was sued in terms of their whole grazing administration.[3]

CHAPTER 6

BLM's notoriety increases during the Clinton administration

When the Clinton administration came in,[4] the emphasis changed completely in terms of grazing. Secretary of Interior Babbitt took that to the next step by saying, we need to really be looking at this together.[5] I believe that was probably the first time that that type of public involvement, and rancher involvement, had been solicited.

What the secretary saw (and was right about) is that there were tremendous conflicts between what the general public wanted out of their public lands, and what the ranching community had had over many, many generations. So he took it and went in that direction.

He also, I believe, put BLM on the map. Previously, we were never really recognized as an agency that had valuable land and resources to provide to the public. It was more like we were a machine that produced oil, gas, minerals, grazing, and some recreation scattered here and there. BLM wasn't a household name.

And it was really surprising to see the change. I did little tests on airplanes. Rather than saying, "Bureau of Land Management," I'd say, "BLM." In the early '80s, and even into the '90s, people wouldn't recognize the term. But once we got into 2000, everybody started to understand what BLM was. I believe that the Clinton administration, as well as Babbitt, but mostly Babbitt, deserves credit for bringing the agency to the public.

CHAPTER 7

Jim Baca's tenure as BLM director (1993–94)

I was the Colorado associate state director when Jim Baca was appointed BLM director in spring 1993. When he came in, it was one of those situations where he picked people he trusted. Basically, everyone else he didn't trust. This is a common mistake that a lot of political directors make. They don't recognize the value of the employee, the seasoned civil servant, who's been there for ages. And rather than just being calm about it, and noticing and valuing what they have to bring, some of these political types will come in having it all figured out and take sides. They know who's good and who's bad, immediately. I don't know if it's more of a protection for themselves (because this is so new to them), or whether they do it because they only have a short amount of time.

Jim Baca came in pretty fast and furious picking sides and really driving his agenda. As a result, those that were not on that side, he was enemies with. He tended to create quite a few enemies rather than using his relationships to his benefit.

He had an opportunity to fill some state director positions, and really wanted to make a difference doing so. But because he had polarized so many people, in a political sense, he was removed before he had the opportunity to select people to be directors.[6]

I believe, in his situation it was a miscalculation on how to operate. He had the right stuff going, he just didn't know how to use it. He tended to polarize a lot of the ranching community.

CHAPTER 8

Interior Secretary Bruce Babbitt pushes rangeland reform (1994)

Jim Baca left, and Mike Dombeck came in as acting director. At that point, Secretary Babbitt took on range reform. He came to Colorado. In fact, I was the associate state director when he came. He went to Gunnison because he was curious about a group of people in the Crested Butte area who wanted to get together—the ranchers and the environmental group and so forth.

He'd say to them, "Well, tell me what it is you're thinking. Tell me what it is we should do."

He then committed to coming back once a month until they came up with something he could work with. That's when they produced the whole structure of range reform. It was very moving to watch him work in that atmosphere.

Now, that was Colorado. And the Colorado ranching community, in general, at that time was not characteristic of some of the other states, such as Idaho and Utah. It was, in a sense, ahead of its time. The ranchers were willing to look at this as a business, and to do something that would work for them, as well as the environmental conservation community. They saw the good in what the environmental side wanted. And they wanted to do well—they wanted to have a business that prospered. So they put that energy together.

That was a little harder to swallow for folks in Idaho, Utah, and other places. The secretary launched it from Colorado, and then came up against resistance from other communities. However, he never backed off from his personal commitment to meet with them.

Now, contrast Babbitt with the new secretary, Gale Norton,[7] and her range reform, or her changes in grazing regulations. She's absent. She's invisible. Babbitt was there front and center at every meeting. He'd go to a small community. It didn't matter. He'd take whatever flak they would throw at him. And at the end of the meeting they would be friends. It was amazing to watch him.

And then you look at Gale Norton and how she's attempted to make changes in range reform—how she's done it invisibly and tried to push it through for the favor of just the rancher. It's a huge contrast.

And, of course, we don't have any changes yet. Public comment ended, and we haven't heard a peep since.[8] Whereas you constantly heard Secretary Babbitt, and you heard the progress, and you heard the resistance. It was all out on the table.[9]

CHAPTER 9

Policy changes at BLM resulting from Rangeland Reform '94 (1995–98)

The very first thing we did out of the chute was create the resource advisory boards.[10] Their first task was to work on standards and guidelines. As a state director, I was required, within a time frame, to have those boards work together and to come up with standards and guidelines for my state operation. As soon as those were created, I had to implement them in terms of the new permits that were coming on line.

In Idaho almost 80 percent of the permits were up for renewal. I now had to bring into the fold the new regulations and the new standards and guides which were, I would say, the cornerstone for the whole permitting process and the whole operation.

The hardest part was having no increase in money or personnel to help support that effort. So I had to draw from within to accomplish this daunting task. I succeeded, but yet it could have been much better if, as an agency, we had received the support we needed to get that done.

I also knew that times change, and that there was an election coming up[11] that could very easily give us a new administration with a whole new direction. So I wanted to make sure that what we had created out of Rangeland Reform[12] was implemented as much as possible. Our efforts might still be eroded by changes, such as what the secretary is doing now. But the purpose was to have something solid for the Idaho range program to work from that couldn't easily be shot down because of a new ideology.

CHAPTER 10

Lack of planning rewarded by BLM (1996–99)

I was trying to avoid a backlog in evaluating grazing allotments, so I shifted the emphasis in my budget to get that done. Some other states didn't do that. I don't know what their situations were. But I knew what my goal was. And as a result, I met all of my requirements. When it came around for a new budget, however, I had more to do, but I got zero funding to do it. All of the money went to other states that hadn't planned ahead. Oregon and Idaho were in the same situation. They had met all of their requirements, and then received a zero budget in the rangeland permit renewal process. The incentive was to not succeed. My incentive, however, was to get this done because it made sense, and it was the right thing to do.

CHAPTER 11

Idaho Watersheds Project[13] influences BLM's range management

Idaho Watersheds Project was starting to get noisy. They didn't have a mechanism to make noise with yet. But when Rangeland Reform happened they got their "hammer." And the hammer was the standards and guides, and renewing of the permit.

Once the standards and guides are in place, BLM has one year to show progress in rangeland conditions. If an allotment was not making progress towards meeting those standards, then that was IWP's hammer.

So they just waited. They were poised. And when the year came around, if an allotment wasn't showing improvements towards those standards, they started filing the lawsuits. In fact, you could probably go back and see that most of the lawsuits began in '96, a year after Rangeland Reform was implemented. That was really when IWP began their quest.

The reality was we cheered them on, in a sense. We looked at it as, great, now we have someone else being able to hold our feet to the fire. Because before, a rancher could come in and challenge what we were trying to restrict them on, and no one would stand up for us. And we didn't have any true basis to be able to say, "Well, you have to be meeting this."

So now we had the standards. Now we had a commitment. When the rancher didn't make it, then the Watersheds Project was right in there with the lawsuit. And the judge was saying to us, "You have it right here. When are you going to do this?"

That's when the lawsuits started coming in. And it gave BLM the opportunity to turn to the rancher and say, "The judge is telling us we have to meet this."

And it's sad, but that's exactly how the process began to work.

CHAPTER 12

Owyhee ranchers benefit from an Air Force bombing range

The Air Force was trying to push through a proposed site withdrawal in the Owyhees that wasn't tasteful to anybody.[14] This was also an issue with Director Baca. The last straw for him, actually, was when he publicly opposed the proposal to then Governor Andrus.[15] They had quite a disagreement. That display pretty much did Baca in as a director.

I came into the matter in '95 as Idaho state director knowing that one national director had gone down, two EISs had failed, and this was still a very hot issue. Consequently, we started to consider another area—a place where the Air Force could get their training range and everyone could live happily ever after. We were working on an area far, far east of where they began, which was in the heart of the Canyonlands. Yet even though we were getting farther away from the Canyonlands, the flight approach to this training area would still have caused low-level flights over these significant canyons.

We had quite the negotiations on altitude restrictions. Consequently, we came up with a much more palatable solution for a bombing range compared to where the Air Force began. We eventually got a plan that restricted flights during certain time periods of the year when people were recreating, ranchers were operating, and wildlife would be most affected—avoiding the bighorn sheep lambing period, for example.

At the same time we were negotiating, the Air Force also wanted to insure that the ranchers, who were most affected, would buy into this agreement. So they sweetened the deal by providing them compensation for what they would "lose in terms of grazing in the area." And I say that in quotes because they really weren't going to lose anything except for the footprint of the bombing area. For example, it was an 11,000-acre withdrawal, and they worked on it as if the rancher was losing all 11,000 acres. But in reality he was going to be able to graze in most of that area. The bombing area really only takes up maybe 1,000 acres. The other 10,000 acres is surrounding. And grazing can continue on that.

So they paid him for losing 11,000 acres of grazing. And they paid him for all of the structures that he had developed, which was quite the intricate water system that came from the Jarbidge Mountains clear across the desert.[16] Those parts of the water system that were in the path of bombing were removed and diverted to other areas.

The Air Force also was providing for displacement of his cattle from the 11,000 acres even though he was going to continue grazing. They were buying him grazing permits from another rancher. As a result, they paid that rancher for his displacement.

It all ended up at a tune of over a million dollars that went basically to two ranchers. They continued to graze as before. So it was quite a sweet deal. And the interesting thing is that the daughter of the main rancher worked for Senator Craig.[17] Of course, they all deny she had anything to do with it.

This was done under the guise of the Air Force, because the land was withdrawn to them. But BLM still had to do all of the environmental work, as well as adjust the permit. What's interesting about that is that no money was given to BLM to make those changes. And when I requested our effort be paid for by the Air Force, Senator Craig came in and scolded me over it, and put in the appropriations that I would take it out of my existing budget. Basically, BLM got to "eat" well over a million dollars just for that to happen. I'm not sure where we took it from, but Idaho BLM had to scrounge it up.

And another footnote to that is that this happened right at the time Senator Craig was ready to vote on a new director for BLM, Pat Shea.[18] Senator Craig was on the committee to vote for Shea's confirmation. I always laughed at how much it cost to buy a new director. It was basically, "What can DOI, BLM give Senator Craig so that he will vote for this new director?"

The way it worked is that Senator Craig would want something, and the acting national director would then tell Idaho BLM to give it to him out of its budget. Typically, it was something like developing a committee and paying for their activities, such as roads or improvements.

CHAPTER 13

George W. Bush administration changes BLM management (2001)

Immediately what I saw change, being a state director, was my authority. There were a lot of things that I had the authority to do, such as put out a federal register notice and announce that we were going to have a meeting. That authority was taken away. In fact, all federal register notices were to be approved by the secretary of interior. Actually, it went through one individual in the assistant secretary's office, who made judgments on whether I should be holding certain meetings or making the decisions. I saw that right away. And I saw it as it related to significant actions on the public lands.

A good example would be the mining withdrawal in Bruneau Canyon, which had been established in the first Bush administration.[19] The withdrawal didn't impact miners, because there are very few minerals of interest in that canyon. It's a very scenic, beautiful area that's very rich in wildlife and other cultural resources.

It was just the simple act of issuing a federal register notice to initiate public meetings for discussing whether we should extend the withdrawal. And that simple little process was totally aborted by taking away my authority. The federal register notice needing approval by the secretary went into a deep, black hole long enough for the withdrawal to expire and the place to be opened to mineral entry with no public involvement whatsoever.

I saw public involvement in all arenas starting to be taken away. The proposed changes in the grazing regulations are a good example. If you look at them, you'll see that the public is recognized as "the rancher"—the person with the permit. The ability for the rest of the public to have a say in how that permit will be issued and what will be in the permit and how that person will operate on public lands has been removed. That was the most significant policy change that I noticed.[20]

The other major change was decisions, in and of themselves, being made at the secretarial level and taken away from me, as a director, at the state level. It was usually spurred on by someone who was politically connected—a local person who went to their congressman or senator in Idaho, who then went directly to the secretary's office. Someone from there turned around and gave opposite directions to whatever kind of activity I was emphasizing.

Those were the biggest changes that I saw. And I lasted a year within that administration.

CHAPTER 14

Hahn's removal from BLM Idaho state directorship (March 2002)

There was a lot of turbulence regarding my tenure, right from the beginning. Just my coming to Idaho as BLM state director in '95 started it. I believe most of it was because I was a woman coming to run the BLM in Idaho. That concept was new and different for the very conservative types of people who lived there. They weren't used to that. Nor was the congressional delegation used to that.

There seemed to be a lot of suspicion when I came, and it continued throughout my whole time in Idaho. I had to really work to prove myself, in a sense. And even then, I wasn't going to be accepted because I came during the Clinton administration.

So even though I was a career employee—had been working for BLM since 1977—that wasn't taken into account. I was a Clinton appointee. That's how I was regarded. I had to constantly fight that image.

The other factor, too, is that I had to run the operation the way that I was directed by the administration.

So just starting out, I had three strikes against me. When the administration changed in 2001, I saw where it was going in terms of authority, and the different ways of doing business. I saw how I was being circumvented in terms of contact by departmental people.

Also, I saw how departmental people just happened to be from Idaho. For example, Governor Kempthorne,[21] who was a senator before he became governor, had a lot of his staff working for the secretary of interior's office.

In addition, Tom Fulton, who was in the assistant secretary's office that oversaw BLM, had been campaign manager and staffer for Senator Craig.

As a result, I started to see a lot of "Idaho agendas" created in the secretary's office, along with circumventing of my authority by these individuals. And the picture became pretty clear to me that something was going to happen. Either I would not be able to withstand this type of treatment in terms of losing my authority, or I would be moved somewhere else. My expectation was to be given an assignment within BLM to another state.

When I received my directed reassignment to New York, to a position that didn't even exist, in another agency,[22] it was very clear to me that that action was on purpose. The purpose was for me to decline and to leave the agency—to leave government work altogether. I only had two options: take the job or resign.[23]

I also saw the fingerprints of Senator Craig on my reassignment. During a lot of my time interacting with him as state director, there was a lot of tension. And he very, very much tried to keep me under control, as if I had my own personal agenda, and I was out doing something other than what the

Clinton administration wanted. But I also think he saw an opportunity, with his friends now in the DOI, to get whatever he wanted. Even before I announced I was resigning, he was already speaking to the new national director of BLM about my replacement.[24] And yet, supposedly, no one knew about me being reassigned. So that's what gave me the indication that he very much had a lot to do with me being removed from Idaho.

CHAPTER 15

Suspected influence of the livestock industry in Hahn's reassignment
I think that what really got people in DOI to create a job in New York for me had to do with the grazing community. We were in the midst of issuing our decisions in relation to the lawsuit that Western Watersheds Project brought several years before. We were to produce decisions on these grazing permits, and how they would be operated in line with standards and guidelines. I was basically directed by the courts to issue a decision, and then I'm having a senator go to the media and object to my decision. The timing was interesting in that it happened just prior to my directed reassignment.

The other thing you can bring into the fold is that the individual who took my place happens to be a rancher, who had worked for the BLM ten years prior. Had retired from the BLM. Came back to Idaho. Went back to his family ranch. Was a very active rancher. And lo and behold was the choice of the senator as the state director to replace me.[25] And again, the senator discussing this choice took place before people knew that I was leaving.

I think if you put all of those things together, you can say that the impetus for my transfer probably came from the ranching community, most specifically from the Owyhee ranchers who were up against the court-ordered decision that I was implementing.

David A. Koehler

Born in Chicago, then raised in Illinois and Texas, Dave Koehler went on to graduate from Texas A & M University with a bachelor's degree in range and forestry. After two years of service with the US Army Military Police Corps, he began his career in natural resources management as a supervisory forestry technician (range) on the Sierra NF (California). Koehler subsequently held positions as range conservationist on the Deschutes NF (Oregon), research biologist with the Oregon Wildlife Commission, range conservationist on the Santa Fe NF (New Mexico), supervisory range conservationist on the Rio Puerco Resource Area of the BLM (New Mexico), area supervisory range conservationist and area natural resources manager for the Bureau of Indian Affairs (Navajo Reservation), and finally, resource area manager, and rangeland ecologist with the BLM in Idaho. Interspersed with his work assignments, Koehler completed an MS degree in ecosystem ecology at the University of New Mexico, where his research focused on the environmental impacts of feral burros in Bandelier National Monument (New Mexico). In 1985, Koehler received a PhD in range ecology from Colorado State University, where he conducted research into the restoration of plant communities on sites disturbed by oil shale development. Since retiring from the BLM in 1999, Koehler has worked as an independent range consultant.

Dave Koehler made his remarks on the 2nd of September 2004 in Delta, Colorado.

Chapters
1. Koehler's youth and early professional career
2. Koehler serves as a range conservationist on the Deschutes NF (1968–69)
3. Koehler studies feral burros at Bandelier National Monument (1973–74)
4. A conflict with Hispanic ranchers in New Mexico (1979)
5. A rancher who subleased and overstocked his BLM allotment (1985)
6. Koehler's experiences as a resource area manager, Idaho BLM (1993–96)
7. Koehler's experiences as a rangeland ecologist, Idaho BLM (1996–99)
8. Idaho Watersheds Project pressures BLM (1990s)
9. Poor planning by the management agencies harms the environment
10. Mismanagement of BLM
11. Wild horses on western public lands

CHAPTER 1

Koehler's youth and early professional career

My father was a native of Texas, but he came north to find work, and he settled in the Chicago area. On one of his trips back to Texas, he met my mother and they were married. I was born in Chicago.

My mother, my brothers, and I would go to Texas for summers where my grandmother and uncle had a ranch. They raised beef cattle and some agricultural crops. I got to ride horses a lot, and mules. So I knew that when I grew up, I wanted to do something like that.

Meanwhile, when I was back in Skokie,[1] we lived in a little development that was surrounded by dozens of square miles of empty prairie. There were some willows and woods close by, where I could go out with a bow and arrow, and hunt pheasant and cottontails. Occasionally we'd get some ducks on the wetlands there. I could lose an arrow trying to nail one of them too.

When I was about fourteen, my mother wanted to go home. So we packed it up, and went to Texas where I finished high school.

I matriculated at Texas A & M University, which at that time didn't have a forestry program, but they had a curriculum called "range and forestry." I thought, well, that's about as close as I'm gonna get. So I enrolled in it. On the downside, A & M didn't have any female students at the time. We would have to go a long way for a date. I look back now and I think, how did we stand it? It went co-ed in about 1970, but I had already graduated.

In the early 1960s, I was drafted into the Army for two years. Vietnam was cooking, but it hadn't really become huge. I was fortunate enough to go to Military Police School in Fort Gordon, Georgia, and then go to Germany. I want to make a distinction here between the Military Police that do town patrol, and the Military Police that we see as prison guards on TV lately. When I was in there, there was separate jurisdiction over American and German nationals. We couldn't touch theirs; they couldn't touch ours. We did patrols downtown and at all the bars.

In early 1964 I returned to the US, applied for some jobs, and took one in the Sierra National Forest of California as a supervisory forestry technician (range). That's where my career began forty years ago.

When I look back at that time, what strikes me in contrast to today is that it was literally possible to disappear into the mountains for weeks or months at a time. I had a three-man crew working on streambank repair and restoration, and on eradicating alien forbs that had invaded some of the high-country meadows. We were probably working at about 10,500 feet, 11,000 perhaps. We were above timber line. During the entire summer up there, we saw only three people in a group that came backpacking through. That was kind of my anchor point because I've seen, throughout my career, the public lands and national forests become more and more crowded. Along with that, I believe that the agencies have attempted to accommodate more and more people, more and more activities, more and more programs than we ever had before. Years ago, we never anticipated that we'd have anything like off-road vehicles out here tearing across the landscape, rippin' through meadows.

I guess, if there's a point to be made here, looking back over forty years, it's that we just didn't have all the stuff that we have now. We didn't have snowmobiles. We didn't have ORVs. We didn't have mountain bikes. We didn't have a lot of the things that have come to be so popular today. And it seems to me, we also had managers who were able to say no. "No, you can't do that." "No, we don't have the capacity or resources to accommodate that."

As my career has progressed, I have seen fewer and fewer strong managers—people who could say no.

That's a shame because, looking back over forty years, we've given most of it away. Every time there's a Republican administration, it seems like there's a rush to rape and pillage our natural resources. And every time we get a Democratic administration the concentration turns to being nice and not hurting anybody's feelings—being sensitive to people's needs. And that hasn't been helpful either.

Of course, I'm being a little bit facetious. On the other hand, I have some clear data in my own mind to base those opinions on. The present intensive concentration on oil and gas extraction in western Colorado is an example. Extractive industries get what they want and can circumvent previous environmental law. I couldn't have imagined that happening twenty years ago.

CHAPTER 2

Koehler serves as a range conservationist on the Deschutes NF (1968–69)
In 1965 I transferred to the Southwest Experimental Station in Berkeley, California, which assigned me to research studies on range/wildlife interactions in central California and Eastern Oregon. In 1968–69 I became a range conservationist on the Deschutes National Forest in Bend, Oregon, with

primary responsibilities for range allotment and wilderness management in the Cascade Mountains.

One of the best jobs I ever had. I had only eight grazing permittees, and some of those were inactive. My range management duties actually only required counting livestock on in the spring, counting them off in the fall, and doing an occasional utilization check out there on the allotment. The rest of the time my supervisor would tell me, "Why don't you go up in the Three Sisters Wilderness and do a little public relations and cleanup."

I had a Forest Service guard who worked up there in the summers picking up trash. He'd bag it and stash it. Then I'd come around with a pack string and take it out. At that time, in the late '60s, it was still possible to go up into the mountains and get lost. You'd see somebody once in a while, but there weren't a lot of people up there in the wilderness area.

Within the Forest Service itself there was a lot of resistance to the idea of "wilderness" because the Forest Service people had been indoctrinated with the multiple-use philosophy. They were not inclined to accept protected areas—wilderness areas. So there was a bit of conflict within the agency at that time. But I was able to ignore it and take my horse and my pack string into the wilderness area and bliss out.

In order to make wilderness as a singular use palatable to some agency personnel, the agency undertook an intensive public relations campaign, which accomplished the objective of public support, but at the cost of more popular use. There was a trade-off of acceptability and internal support for the experience of absolute solitude. I also think the first Earth Day in 1970 was a landmark for wilderness popularity. After that, I would typically pass dozens of people on the trail.

CHAPTER 3

Koehler studies feral burros at Bandelier National Monument (1973–74)

I moved in the early 1970s from Oregon to New Mexico, where I worked on the Santa Fe National Forest as a range conservationist for a while. All but a few of the ranchers out there were Hispanic—they were all small operators belonging to grazing associations. They worked together, and they were wonderful to work with. If you wanted to get all these guys together on a weekend and put in a spring box, an underground water line, and a drinker, all you had to do was call them. And they'd all be there on Saturday morning, early. They all loved to go out on horseback and look at stuff or show me things. I really enjoyed working with those guys.

I had an opportunity then, at the beginning of 1973, to go back to school. The National Park Service (NPS) was having a problem at that time with feral burros in Bandelier National Monument near Los Alamos. The Park Service wanted to remove them, but the Fund for Animals did not. NPS offered me a research grant, and I was still eligible for the GI Bill. So for almost two years, I worked in there collecting data, mostly from horseback in the backcountry. The biology department at the University of New Mexico (UNM), where I completed my MS program,[2] was a terrific academic venue. The faculty was involved, helpful, interested, and professional. My advisor there was Dr. Loren Potter, who was a good friend to me and is still widely revered by his former students. I eventually completed my master's degree at UNM in ecosystem ecology with my thesis based on that work. It was called "The Ecological Impacts of Feral Burros on Bandelier National Monument, New Mexico," and was subsequently published by NPS and widely circulated.[3] Over the next ten years or so, I was called to testify at several hearings when the Park Service was trying to get the burros out of Bandelier and other monuments and national parks in the Southwest.

As my career has progressed, I've come to realize that any kind of introduced or feral organism in this country is bound to have, is destined to have, an extremely detrimental impact on the landscape and on the ecosystem. That was the case with the burros. They had grubbed out every possible vegetative source of food and were pawing some of the woodland species so that they could get down to the roots and eat them. The impact extended not only to the soils and vegetation, but to the native wildlife that should have been in there, like the deer and elk and smaller mammals. I think that some of those have recovered somewhat since then.

CHAPTER 4

A conflict with Hispanic ranchers in New Mexico (1979)

During the period 1979 to 1986, I became the supervisory range conservationist for the Rio Puerco Resource Area, Bureau of Land Management, headquartered in Albuquerque, New Mexico. The resource area abutted the Navajo Reservation on the west, the Jicarilla Apache Reservation on the north, and the Santa Fe National Forest on the east. Our permittees were primarily Hispanic, and most were seriously involved in the land-grant movement that was pervasive at the time. *La Alianza de las Razas*[4] contended that all public and national forest lands belonged to the descendants of Spanish colonists by virtue of a land grant from the King of Spain in the 17th century. One point of dissension was that active members refused to submit to the authority of federal agencies, which made for openly hostile relations. My employees were literally shot at and threatened at gunpoint. When we directed complaints to the State Office, we were told, "Aw, you boys just need tougher hides."

I remember one occasion, while we were attempting to impound trespass livestock, when the situation really turned ugly, and our crew was threatened with assault and death. They were standing around the corral pelting us with rocks. We were about eighty miles by road from Albuquerque, but had one law enforcement agent with us, which probably prevented more serious trouble. On the way out, the patriarch of the local villagers fired four shots at me from about 400 yards out while I was on horseback. Charges were filed with the US Attorney, and the old man and three sons were charged with aggravated assault on federal officers. The offenders were brought to trial and given probated sentences for three years.

Sometime later, the family invited me to their house in the village of Casa Salazar for coffee and conversation. The old man explained that he had fired all his bullets when I was a long way out, so he wouldn't have any more when I got closer. That made perfect sense to me at the time.

The mother of the family wanted to show me a new chapel (*iglesia*) they had constructed on a little hill behind their home. She told me that she had promised God that if her men were not sent to prison, they would build the chapel. So they did. It was a beautiful little place.

Another issue of that period was the Sagebrush Rebellion, essentially a movement to privatize public lands. Even President Reagan[5] proclaimed himself a "Sagebrush Rebel." This created even more hostility toward our agency and created an atmosphere of antipathy toward the federal government. It was difficult to enforce rules and regulations that the public regarded as superfluous and purposeless.

CHAPTER 5

A rancher who subleased and overstocked his BLM allotment (1985)

Once, in 1985, I learned that a BLM permittee north of Albuquerque, with a total permitted authorization of 208 cows was subleasing to other people, who actually had a total of 880 cows out there. They were grubbin' that place out.

I had worked up there eleven, twelve years earlier for the Forest Service, which had allotments contiguous to the BLM ones. So I knew a lot of these people. And a lot of them I called friends.

Anyway, I was able to document that this guy had 880 cows up there. And they belonged to twenty-eight individuals, as best that I could determine. I had known the permittee for twelve years. I had gone up there and had lunch with him. Ridden with him on horseback. I thought he was a pretty good guy. But he was profiteering off the lands. He clearly had violated the terms of his permit. So I cancelled it.

Subleasing of a permit is illegal. In other words, if you have a permit for twenty cows. And you only have fifteen cows to put out there, you can't go to your neighbor and say, "Hey, I have excess forage on the BLM allotment that I'll let you rent from me for fifteen dollars a month." It's not legal. The livestock that you have on your allotment need to belong to you.

The permittee had a partner, who had been acquitted of two murders. So they really weren't people that I wanted to fool with. But I cancelled his permit and ordered him to remove all livestock from the place.

He failed to do that. He had about a 120 cows left on the place when I executed an order of impoundment. I went out there with a livestock truck and gathered all his cattle in a corral. Just before we loaded them, he arrived with a trio of attorneys and an injunction to keep us from impounding his livestock. Basically, our hands were tied, and we had to go home.

I thought I was doing my job. But one day, the deputy state director arrived at my office and announced that they were going to investigate my activities—that they thought that my activities had been illegal and inappropriate. And they were going to audit my program. So they interviewed all my employees. They went through all my files. This went on for thirty-seven days.

Fortuitously, for me, the adjacent ranch belonged to the man who was the former and future governor of New Mexico—a man named Bruce King.[6] After thirty-seven days of searching for something dirty that they could put on me, the state brand inspector, whose name was Mel Sedillo, caught these two guys in the process of changing the brands on sixteen of Governor King's yearlings.

Saved my butt. You know, that deputy state director couldn't wait to drop the matter.

My question then, and my question now is, "What the hell was that all about?" Was it routine in New Mexico to persecute employees that did their job if it inconvenienced a rancher? That seemed to be the case. And if

Governor King's yearlings had not been stolen, and if the chief brand inspector had not happened to go out there on an inspection and found them in the act of changing the brands, they could have fired me on some kind of trumped-up, fictional charges. I have seen people either fired or harassed until they resigned from federal agencies.

You know what? That was nineteen years ago, and I'm still mad about it. I always wondered if they hadn't caught those two guys changing the brand on the governor's yearlings, what would have happened to me?

What did happen was that they sent me a letter signed by the state director stating that they had determined that I had been involved in "serious improprieties." But they would not specify what those were. I retained an attorney. We sent a series of letters to the state director asking what the serious improprieties were, and how they thought they could have been handled differently. All I ever received was that letter saying I was found to have participated in serious improprieties. No other explanation. I think the whole thing happened as a result of an over-officious bureaucrat.

I was given an hour interview with my state director, but he had other business to take care of during that time. So I was never able to have a dialog with him about this incident.

My conclusion is that these agencies "play games." And I don't feel that game playing is appropriate in the context of federal land management. It happens a lot. You have people trying to "get" each other. You have these little "dog fights" and "cat fights" going on all the time, where people are trying to discredit and embarrass some other person.

My argument would be that we don't have time for this stuff. We don't have the time, and we don't have the energy. We don't have the resources to put into personal feuds when there's so much to do out here on the ground.

CHAPTER 6

Koehler's experiences as a resource area manager, Idaho BLM (1993–96)

In 1993 I transferred back to Bureau of Land Management in the position of area manager in Shoshone, Idaho. It was pretty much a major disappointment, and I lived to regret my decision. Opportunities to go to the field—opportunities to do real things—were pretty rare. I don't want to be too harsh here, but it seems to me that I spent most of my time listening to a lot of narcissistic blather from my immediate supervisor. Unfortunately, she had an associate who was just as enthusiastic about her own assets as my boss was. So it wasn't a very gratifying or fulfilling experience. I didn't get to the field very often. I wasn't able to do the things that I would have liked to have done. I'd say that 75 percent of my time was either in meetings, or was spent waiting for a meeting to begin, or for the opportunity to participate in a meeting. I don't feel that was a productive time. Especially not, because it didn't seem relevant to our mission to be forced to participate in the naked ambitions of a couple of self-promoting supervisors.

From talking to my staff, I had a pretty good indication of what the problems were out there. But I was really not able to participate actively in the field with my staff on resolving some of these problems, because I was required to attend these meetings in which nothing was ever said; nothing was ever done. It was the most unfulfilling three years of my career.

CHAPTER 7

Koehler's experiences as a rangeland ecologist, Idaho BLM (1996–99)
I became more of a consultant to the field when I was transferred to Boise. Actually, because I lived in Twin Falls, we worked out a deal where I was stationed in Boise, but I was co-located in Twin Falls. That way I didn't have to commute to Boise except once or twice every couple weeks. That worked out pretty well for me. I don't have any complaints about those times. I worked on a lot of special projects. I worked on mine land reclamation plans, and on an annotated riparian bibliography that was published in 2000.[7]

I also worked on environmental analyses and some range studies that allowed me to get out in the field—in the Jarbidge area. I was the program manager for the wild horse program in Idaho, as well as the program manager for the weeds program there.

CHAPTER 8

Idaho Watersheds Project pressures BLM (1990s)
We especially received a lot of pressure from a man named Jon Marvel, who resided up in Hailey, Idaho, which was inside my jurisdiction. He formed an organization called the "Idaho Watersheds Project," which later became "Western Watersheds Project." And he's expanded to other states.

Jon could be very aggressive and very critical. And, in fact, I attended some meetings with ranchers at night in remote locations, like schools in small towns, where I actually feared for his safety. I was afraid ranchers would just take him outside and hang him. But he didn't seem to have any fear at all.

Jon Marvel is an interesting man, and he has an interesting agenda. He has, apparently, a pretty good funding source and a pretty good following. And he's able to raise the red flag every time something happens out there. He knows what's going on, on the grazing allotments a lot sooner and a lot faster than the BLM people do.

You might wonder why that's so. But it goes back to the fact that BLM people are tied up in a lot of meaningless exercises and meetings. They're not out there doing what they ought to be doing. And I can say that with absolute perfect personal certainty.

CHAPTER 9

Poor planning by the management agencies harms the environment
I think it's fair to say that a lot of the western wildfires that we're experiencing are the consequence of having exotic plant species occupying these sites instead of native ones. Specifically, one of these is *bromus tectorum* (cheatgrass); another is hoary cress.[8] Cheatgrass is a very competitive plant. It crowds out the primaries. It creates a very fine, flammable environment for fires to establish and move through very quickly. It's something that we didn't see coming.

A lot of the problems that we were trying to deal with earlier in my career we didn't see coming. We clear-cut lodgepole pine to create more grazing area, and now it won't reproduce. We eliminated large predators, and now we're trying to get them back. Hindsight is wonderful.

Up until the Yellowstone fires in 1988, we weren't thinking about fire having a natural role in the ecology of the West. All anybody was able to see, even for the first year after that, were blackened trees. Then, after a year or two, they started to see some green coming up. And it occurred to a lot of people that this was something that had probably been happening for hundreds of thousands of years before man started to manipulate it. Certainly the "Smokey Bear Policy" of putting out every fire has not been helpful. A lot of these fires should have been allowed to burn. They did a lot of good. They cleaned up a lot of undergrowth, underbrush, litter, and accumulated solids that provided a fuel base for fires when they did occur. Now we're having to pay the price for seventy-five years of Smokey the Bear—"Put out those fires." It almost had a religious intensity. We're at the point where we can look back now and say, "Boy, we really screwed up."

For another example, we spent millions of dollars out here in the West trying to extirpate sagebrush. And, unfortunately, we were pretty successful. We got rid of a lot of sagebrush, and seeded exotic wheatgrasses. These stands, in turn, have been invaded by cheatgrass. So now we have the potential for huge, destructive fires that are extremely difficult to stop. Every time we have them, they take out more of our sagebrush, and expand the communities of invaders.

Sagebrush—twenty years ago we thought it was a pest species that we had to eliminate. Today, it's fair to say, there's an absolute panic in the agencies that sagebrush will eventually disappear from the western landscape because of bad management, and because of the incursions of exotic plants. There are a lot of birds and animals that are associated with sagebrush that are also being placed in danger because without sagebrush habitat, they don't have a place to live.[9]

I'm sure you've seen in the news during the last few years that the BLM and US Fish and Wildlife were proposing listing sage-grouse as a candidate species. I can remember twenty-five years ago going into western Colorado and Wyoming and southern Montana, and not being able to walk fifty feet without kicking up sage-grouse. They were all over the place. We'd have to move pretty slowly through these stands because when they'd jump up, they'd scare the hell out of you. You practically have to step on them before they fly.

Now the populations are way down,[10] and I think it's truthful to say that a lot of our management practices have been based on misinformation. I think grazing allocations have been way too heavy. And I think that a lot of the agency managers didn't have the courage to say, "No, you can't have more cows out there."

CHAPTER 10

Mismanagement of BLM

I think it's fair to say that, at the field level, managers really don't have a whole lot of autonomy. They're kept on a leash. They're not allowed to exercise their own judgment in managing public lands and national forests.

For example, in the early days of the Reagan administration when Robert Burford was our director,[11] he had the BLM turned around in circles with one priority or another. One of the things we did was spend about six months "combining" the US Forest Service (from the Department of Agriculture) with the BLM (from the Department of the Interior). It eventually went away and hasn't come back. Everybody had to go to those meetings and write reports and get involved. And nobody could be out to the field and do their job.

The BLM, particularly, has a history of fixating on some kind of program or problem like that and throwing millions of dollars at it before deciding that it wasn't any good in the first place, and that they needed to move on to something else. Then somebody else in Washington (I have to assume that this stuff comes out of Washington) will come up with yet another fantastic project that we have to implement. And it's going to cost millions of dollars, and it's going to take hundreds of work-months away from the field.

The best example I can think of is what was called the "ALMRS Program"—the Automated Land and Mineral Record System—a program for accessing and storing all of the resource information in the United States. At the time, I was working in Idaho. And every year, 50 percent of our natural-resources budget was taken so that we could throw it into the ALMRS pot. Finally, in 1999, they realized that it was impractical. It was never going to work. And they abandoned it—four hundred million dollars and seventeen years later! It absolutely consumed the careers of hundreds of people over that time period. And you have to come back and say, for what?

CHAPTER 11

Wild horses on western public lands

When I was the manager for the wild horse program in Idaho during 1996–97, all the facilities for holding horses were overcrowded. The "pipeline" was clogged. The eastern states, which had been a primary source of adoptions, had decided not to take any more horses because the market had been saturated. Up 'til that time, they'd been accepting about 4,500 wild horses a year for adoption. Now, there were 4,500 horses surplus in the western states that had to be disbursed every year.

So BLM began a program of what they called "satellite adoptions," where we went out to every little town in Idaho that we thought might have a market for wild horses. That was in addition to having regularly scheduled adoptions.

The market for adoption has pretty much dried up. Ten years ago there were a lot of kids out there that wanted to adopt a wild horse. Now they want a motorcycle or an ATV. Areas are getting more and more populated and crowded. And people are just losing interest in the program, sorry to say.

Factors in stabilizing wild horse populations

Lions have a large range.[12] We keep on building in rural areas. We keep on throwing up subdivisions and strip malls, and we just expect these animals to go around them. It's not going to work that way. Nature has an equilibrium and a rhythm that we don't respect and we don't acknowledge. I know that more predators are not the solution to stabilizing horse populations.

Immunocontraception is another possibility that's been kicked around here for the last fifteen years, but it's being opposed by all of the wild horse groups because they don't want the horses gathered or bothered. And they don't want the natural birth rates interfered with.

Bringing horses in from the range and implanting the mares with contraceptives is going to be an extraordinarily expensive operation. At this point in time, I don't think there's a practical solution to it.

There was an equine species[13] here until the late Pleistocene.[14] We all know that, and all recognize that. But that doesn't mean that contemporary ecosystems have evolved to accommodate an equine species. It only means that there was some prehistoric ancestor that no longer inhabits this part of the world.

I don't have the answers to managing wild horses, but I can tell you there are two sides to it, and both sides are too extreme. And I'm not sure a compromise is possible.

Don Oman

Don Oman grew up on a farm in central Montana, then went on to earn a bachelor's degree in forest management from the University of Montana–Missoula. Over the course of his thirty-five-year career with the US Forest Service, Oman served on numerous ranger districts in Idaho, Montana, and Wyoming. But it was his nearly ten years as the district ranger on the Twin Falls Ranger District (Sawtooth NF) in Idaho that brought him to national attention. Arriving on the district in 1986, Oman found the land overgrazed, facilities that were in disrepair, and ranchers who showed little interest in abiding by the terms of their grazing permits. In striving to improve this situation in all respects, Oman was subjected to intense social and political pressures generated by ranching interests. Nonetheless, Oman prevailed, and much was accomplished for the health of the land.

In 1996 Oman was promoted to the position of ecosystems staff officer on the Sawtooth, where he remained until his retirement three years later.

For his efforts to protect public lands resources from poor grazing practices, Oman was awarded the Wilderness Society's Olaus and Margaret Murie Award in 1991.

Don Oman made his remarks on the 25th of August 2003 in Twin Falls, Idaho.

Chapters
1. Oman's early years, education, and career
2. Historical records of environmental conditions on the Sawtooth NF
3. Environmental conditions on the Twin Falls Ranger District (October 1986)
4. Oman stands up to ranchers on the Twin Falls Ranger District (fall 1986 & spring 1987)
5. Environmental recovery within a livestock exclosure on Trout Creek, Twin Falls Ranger District (summer 1987)
6. Additional conflicts between Oman and ranchers (1987–89)
7. Counting cattle at the fall 1989 roundup
8. Environmental recovery within a livestock exclosure on Dry Gulch, Twin Falls Ranger District (1989)
9. A water pipeline for ranchers poses environmental risks (summer 2001)
10. Political obstacles to achieving sustainable land management

CHAPTER 1

Oman's early years, education, and career

I grew up in Montana, actually on a farm in the little town of Manhattan west of Bozeman. Then I went to the university there at Bozeman for one year. Then the University of Montana–Missoula. That's where I graduated from with a degree in forest management. That was in 1965.

I got my first permanent job on the Hebgen Lake Ranger District at West Yellowstone on the Gallatin Forest in the fall of '65.[1]

From there I moved to north Idaho—Priest River country. And I was mostly in timber management.

From there I was transferred to the Madison District on the Beaverhead Forest where I had a good, well-rounded experience in recreation, timber, minerals, and special uses.

Then I transferred back to north Idaho to the Wallace District on the Panhandle Forest. There I was in charge of pretty much everything except timber—including grazing, which was a fairly small resource there, though.

Then I moved to the Madison District on the Beaverhead Forest. That was a big range district. There I was in charge of a number of things, but not grazing.

From there I transferred to Jackson, Wyoming, to the Hoback Ranger District on the Bridger-Teton Forest. And there I had timber, gas and oil development, and the fire program.

From there then I was moved to Vernal, Utah, where I was in charge of the Northern Utah Shared Services Timber Group—a new group that was formed to prepare timber sales for the Ashley, Uinta, and Wasatch-Cache National Forests.

So I had a pretty well-rounded experience, but not too much in grazing.

Then in October 1986, I was transferred here to the Sawtooth National Forest as district ranger on the Twin Falls Ranger District—one of the units south of Twin Falls. It's about a 320,000-acre district, and we had a big range workload there. And so this was the first opportunity that I had to really do something about some of the grazing problems that I'd seen everywhere else I'd worked.

CHAPTER 2

Historical records of environmental conditions on the Sawtooth NF
There's a history of the old Minidoka National Forest[2] that was written in the 1940s. It did a good job of documenting physical damage to the land by grazing. People that were here at the time, even though the attitudes were always strongly pro-grazing and pro-development, made comments on how they saw the soil eroding and the tremendous increase in sagebrush and reduction in grass. And they described how in the springtime the cattle and sheep were out there on the range all the time.

In the springtime the cattlemen and the sheepmen would all try to beat each other into the hills and reach any bit of grass that was starting to green up. Well, of course, not only were the sheepmen in competition with the cattlemen, but cattlemen were in competition with other cattlemen. And so everyone was racing to stay right behind the snow, gettin' their livestock up in the mountains out of the valley here—out of the desert country. So it never gave the grass a chance to even get started. Of course, we know that one of the things we need to do is let the grass get a good start and get some volume to it before you put any livestock out there.

So the country just eroded terribly in those days and that was documented by people on the ground clear back in the 1940s. But there weren't any detailed studies. We started some on watershed and erosion after I and my range guys got here.

CHAPTER 3

Environmental conditions on the Twin Falls Ranger District (October 1986)
I had two really excellent range conservationists working for me who were new here too. They had a career-full of experience in grazing, wildlife, watershed, and other resources. And we, working together, immediately recognized the very heavy grazing.

As we went around and looked at allotments, there were livestock, mainly cattle, still on the Forest two weeks after the two-week extension they'd been given. And yet there was damage—severe grazing and impact to streams all over the district.

So when we told ranchers they had to actually get their cattle off, it was the first time they'd ever been told that. They'd always been allowed to just let them drift with the snow, apparently. So that generated some of the first "congressionals"[3] that we got.

Probably two-thirds of the water developments were in disrepair. And, of course, those were vital to keep the cattle away from the creeks.

And fences were in such a state of disrepair that cattle could go anywhere they wanted.

So there was no control. No way of keeping cattle in a unit of an allotment for the time they were suppose to be there. In other words, cattle were on every stream on the district all the time—all summer long, and then allowed to stay for a month, month-and-a-half after the grazing season ended. It just resulted in serious damage to almost every stream and pretty much everywhere cattle grazed.

They were always near streams. Some of the streams that we saw that first fall were running just a small amount of water, and they were running green with manure. Just manure and mud. An absolutely terrible situation.

Areas away from the streams still had tall grass in places, because they just weren't getting the cattle to those locations.

And large numbers—hundreds of cattle were in the sheep allotments that didn't belong there.

So the fences were in such bad shape and permittee attitudes were such that they didn't see the need to actually get the cattle out and keep 'em out. They we're just always allowed to have 'em everywhere.

There were other problems, of course, with all the wildlife and watershed values and recreation. Very important recreation sites were being impacted severely by grazing. And we even had campgrounds and picnic areas where sheep were standing on the picnic tables, and the paved trails were covered with cow manure—as many as three hundred cattle in one picnic area that we counted. It took us several years to change that situation.

We also had sheep being driven up the main recreation highway—totally denuding the vegetation along the road and breaking down the banks.

So we had a lot of things we needed to start working on right away.

Livestock impacts on archaeological sites
That district is a very important area for archaeological values. The Indians of the day, probably the forerunners to the Shoshone and Bannock, used numerous sites up there. I think there were approximately fourteen hundred identified archaeological sites. They were being severely impacted because they're

mostly around water—around the springs. And so trampling and removal of the vegetation and compaction was very severe, and any artifacts that were there were being broken and destroyed.

CHAPTER 4

Oman stands up to ranchers on the Twin Falls Ranger District (fall 1986 & spring 1987)
We told them that they had to get their cattle off immediately, and that the way things would work in the future was that there'd be no extensions unless they were deserved, which included no large areas of real heavy use, and that they had followed the operating plan for the year—moved the cattle from unit to unit, cleaned the units,[4] and operated like they're supposed to.

Of course, we were informed it had never been done that way. They'd always been allowed to get most of the cattle off, and the rest would just drift with the snow. So they immediately started talking to my boss, the forest supervisor, and politicians—the congressmen and senators, county commissioners. And so the ranchers immediately applied pressure to try to avoid the limitations that next year.

The next spring things really hit the fan. For some of the allotments, we went out with the ranchers ahead of time and estimated when the cattle could go out in the spring—when the range would be ready. That first year the area had been hit so hard in the past that we told them that we felt they needed to be held off two weeks that spring—go on two weeks later, and we'd see about giving them extra time in the fall. But they needed to give the area a break and get it started better. It was a late spring.

Well, they just had a fit over that—the one allotment we were dealing with there. And so we compromised and went back to a one-week delay in the spring. We thought they'd agreed to that. But then we got congressionals in from one senator and two congressmen complaining about that and inquiring why we would do such a thing.[5] And as always, not a true inquiry, but a defense of the ranchers and a pressure tactic to get us to let 'em do what they wanted.

So that was the first round of actual official congressional letters that we got on their behalf for just trying to hold them off a week in the spring to give the range a break.

CHAPTER 5

Environmental recovery within a livestock exclosure on Trout Creek, Twin Falls Ranger District (summer 1987)

The first year my range people and I were here, that summer of '87, we put in the first of our exclosures over in Trout Creek—a three-hundred-foot length of the stream bottom. Our specialists in soils, watershed, wildlife, and fisheries were a big help to us. And that was really a key thing, because until we started putting in these exclosures there was no place on the district where you could compare what things could be like to what they were. And that exclosure, and the numerous ones to follow, really served to show the devastation that was taking place there. Of course, back in those days I wasn't allowed to say the word *devastation* or *destruction* or *plowed* or *roto-tilled*, but that's what the case was.

Trout Creek, for instance, is a creek that runs a lot of water, all from fresh springs. It has Yellowstone cutthroat trout in there—one of the few populations in the area. And here it was being just totally denuded from bank to bank. Vertical banks. No vegetation in the flats. Willows all highlined and all the sprouts eaten off. No regeneration.

These exclosures immediately began to show, in just one year, an unbelievable amount of recovery. The willow was there. The sedges were there. The watercress was there. And immediately the stream and the surrounding flats were filled with vegetation. So the exclosures were key to accomplishing some of the things we were able to do later.

The ranchers, many of them, had the attitude that that's just how things are—"We're taking good care of the land, and it can't be different. That's just the way it's always been."

Well, it probably was the way they always saw it.

CHAPTER 6

Additional conflicts between Oman and ranchers (1987–89)

We had continual problems with the Goose Creek permittees in getting them to fix their improvements and their fences. They'd have cattle in the wrong places—down in the sheep allotment. Of course, the permittees would always say it was the tourists, the hunters, and the recreationists driving the roads that left those gates open.

We knew they had too many cattle on, several times, from the reports of some of their own people. But there wasn't much we could do about it. We couldn't count 'em on in the spring, because they all just went on out of their own ranches in different parts of the Forest boundary—the BLM boundary first.

The first spring I was here,[6] one of the fellas from the Goose Creek Allotment called me up—and he wasn't even the president of the association.[7] He said they were going on that next Monday. Well, the on-date wasn't even for a while later than that. And so I told him on the phone, "No, you can't do that. You're not goin' on without us takin' a look to see if the range is ready for an early on."

Well, he'd never been talked to like that in his life, he said. And all I did was deny his request and made arrangements to meet him out there.

Then when we went out there to meet him that next Monday, we met with all of 'em. Well, he thought I called him a liar when I just disagreed with him once. And he jumps around wanting to fight in front of everybody. So crazy stuff like that goin' on.

Then there were always congressional aides along and county commissioners supporting the permittees in this effort to not comply with the rules and the permits they signed.

There were continual problems with several water developments being maintained. They'd call up and say that all the water developments were maintained in the first unit for the on-date.

So we went out and inspected, and they weren't maintained. In the first unit one year there were, I think, sixteen water developments and, I believe, only something like six of 'em were working; ten were not. Pipelines weren't even started, and one of 'em was the Owens Corral Pipeline.[8]

We told permittees they couldn't go on until those were fixed. Well, they'd never been talked to like that before. So every little thing you did just created an uproar with a congressman and congressional aides and county commissioners and anybody they could get to support 'em. This was building over and over for a number of years.

One of the biggest fusses was always in the fall of the year. Some of these allotments would hold back in their rotation—moving cattle from unit to unit. They'd hold back until the area was way overused—totally beat. And then they'd move on. Well, in the fall of the year then, they had held back so long that they did have grass left when their off-date was getting near. So, of course, they had to have their extension—extra time in the fall.

They'd call up and ask for their extension. And we'd have to tell 'em no, because other areas were abused. And they'd have tours, and we'd get people out there and they'd see all the tall grass. And we just made 'em come off.

We couldn't do that for a while because the administration—the authority for the range permits—was in the hands of the forest supervisors. And I was made to give the Goose Creek Allotment an extension one year by the forest supervisor, because it was within his authority. But the next year, nationwide, the authority was passed to the district rangers where it should be—to administer the permits and make those kind of decisions.

Then it was in my hands to make the decision with my range people as to when the permittees would go on and when they'd come off. And we never gave one more extension in the fall on any allotment, because there was never total compliance with the rules. And there was never an area that wasn't overused on these allotments.

But it really built when these guys would hold back and have the cows moved into the next-to-last unit, and the off-date's coming. And they got the entire last unit left with tall grass.

And we'd say, "No, you're gonna have to go."

So that would generate all the congressionals and all the tours and everything again. But we'd go ahead and make 'em get off.

A lot of the district rangers were afraid to do that. But the way I and my people looked at it is, it's within our authority. We're doing what we know is right, and if the permittees want to take exception to that, then the proper way is for them to appeal the decision to the forest supervisor.

And the forest supervisor then, if he wants to, can say, "No. I'm letting them stay under this appeal."

But they weren't willing to do that, because we always had good evidence. And so it never happened, but still we went through the fuss.

This kind of thing just built, year after year after year. So when we did the livestock count, in the fall of 1989 at Piney Cabin on the Goose Creek Allotment, the permittees went into orbit, because that was the last straw—to actually come out there with brand inspectors and count their cattle.

The reason we did that was because we'd been told by one of their own people that two permittees had too many cattle on. We had good evidence. We did it all properly with brand inspectors and everything. There was no abuse or threats or anything out there except from the ranchers.

The story got all blown out of proportion, but we had a very professional count. And it had to be done there because that's where they gather the cattle. They ship there on the Forest.

So that was the last straw—to actually count their livestock out there. And that's what created the big uproar that started the effort to remove me with congressmen involved, and the regional forester involved and his people, and the Forest, and the ranchers all working together. Which, of course, we ended up fighting with a whistleblower complaint, and won.

And we stayed here and did a great job of managing that piece of government land for six more years.

CHAPTER 7

Counting cattle at the fall 1989 roundup
What happened was that the permittees never gathered all the cattle in to that roundup. They still were not cleaning their units totally and they had cattle that had gone on ahead.

One of my range guys was up that morning flying in an airplane trying to get a count on all these scattered cattle. Well, that was really hard to do because of the long shadows and the aspen patches and all the places they could be. But he counted something like three hundred cattle that were scattered all over.

The count did come out short. But even then, the count for two of the permittees was a little bit over, or right on. Really close. So that indicated that two permittees were over when we applied that three hundred that weren't accounted for to the whole allotment. We tried to pro-ration that to the five permittees. And when we pro-rated that for those permittees that were really close, or right on, it put them over to the tune of twenty, thirty, forty animals. And they were the two guys that we were told that were over by their own people. So that's the way it worked.

Well, we were never allowed to apply that figure when it came time to take permit action because of the count. The forest supervisor's office said, "No, you can't apply that. That's too much of a guess applying that extra three hundred that isn't accounted for proportionally to each individual."

The other interesting thing was that one permittee that was over, had something like a 106 or 108 percent calf crop. Now, that means each cow had a calf and some had twins. The average for that count was something like a 87 percent calf crop. And sometimes it's even lower because all the cows don't have a calf in the spring. Some are barren. Some calves get killed out on the range for various reasons.

So for one permittee to have a 108 percent calf crop, after the cattle had gone through two-thirds of the grazing season, indicated that there were some other problems going on.

What probably happened, I'm told, was that the guy was somehow acquiring calves as new ones were being born, before the cattle went on the allotment. And he was mothering them up with cows and giving a lot of cows twins. So that was another aspect that nobody ever pursued. Law enforcement really should have gone after that one.

But that cattle count was the big last straw that caused them to get with officials and demand that I be moved. One permittee, old Winslow Whiteley, the fellow that threatened to cut my throat in the *New York Times* article,[9] told the range deputy and the forest supervisor right in front of me, "We don't want Don fired necessarily, he's a nice enough guy." He said, "Just move him back to north Idaho, where there's no grazing in that timber country, and then he won't be bothered by it."

So the answer to the heavy grazing problem and the resource destruction down here on the Sawtooth Forest was to get rid of people that were worried about it.

CHAPTER 8

Environmental recovery within a livestock exclosure on Dry Gulch, Twin Falls Ranger District (1989)

One of the exclosures that we put in, I believe it was in 1989, was on Dry Gulch down on the southeast corner of the Twin Falls District. It's a long gulch that comes down around the Goose Creek Allotment. Soil analysis and indications on the banks of this gully, which is thirty feet deep in the lower end, indicates that it was a stream on the surface before livestock grazing came to the country a hundred and forty years ago. But it's one of those areas where it was heavily grazed way back. And the combination of denuding the vegetation and compacting of the soil has created such tremendous sheet erosion, especially during summer thunderstorms and spring runoff, that this has turned into a vertical-sided gully thirty feet deep, and probably two, three miles long.

We wanted to fence the lower mile-and-a-half of it for a demonstration—a permanent exclosure just to see what a deep gully like that could do to recover. We based that on some of Wayne Elmore's[10] work in Oregon.

And so we wanted to fence the area from the BLM boundary upstream a mile-and-a-half.

Well, the ranchers saw no need for that. They tried to talk us out of it, because they said, walking down there, "Look at the green grass growing in the bottom. Shoot, the thing's healing itself."

But we saw fences hanging in space, where erosion had cut the sides of the gully far enough that fence posts were hanging out over.

Together with the SCS, we did a study based on our 1957 photos and our brand new 1987 photos. They were able to measure the depth, width, and length of that gully for a five-thousand-foot stretch for the last thirty years.

Well, to be brief, they confirmed and calculated that we had lost 317 acre-feet of soil from that five thousand feet of that gully in thirty years. Then we calculated that out in the form of ten-cubic-yard dump trucks, and it came to about four-and-a-half dump truck loads a day for thirty years that we had lost from that gully.

So we fenced it, and the results were dramatic. The first year without cattle, we grew tall grass and a little bit of sedge in the bottom. The gully was mostly dry. Just water here and there that was from the spring upstream. And we only had maybe a quarter mile of water in the whole gully. We had enough grass that first year to catch sediment from some summer thunderstorms.

And then, of course, after the next spring runoff, all that grass was pretty much buried. But it was buried in sediment that came down and drifted in there three, four, five inches deep. In places, sediment drifted eighteen inches deep.

Then that next year, sedges and grass and willow started showing up—grew up through that and laid a mass of vegetation eighteen inches, or so, high. And, of course, with every storm there was more sediment.

So that's the process of rebuilding those bottoms. It's a natural process. And we found that within the first three years, based on actual studies and cross sections in three different areas, we had rebuilt the bottom of that gully, on the average, eight inches. And we had places where it was much bigger than that.

After six years we'd refilled it, on average, something like eighteen inches.

And the last year I was on the district[11] there was an old water trough by the spring that had been sticking up out of the ground about, oh, probably twenty-four inches. And it had disappeared! Totally disappeared in the heavy sedge and grass and the sediment that had filled in there. And, of course, then that creates a big sponge that holds water in the spring, and from winter and summer thunderstorms. We ended up with live water in that stream for at least half of that mile-and-a-half. Maybe closer to a mile of it. So that was another dramatic exclosure with studies that really show the results of improved management.

There's another interesting thing in that draw—these are really fragile, volcanic-ash soils. And there's banks of volcanic ash that the cows chew on and eat—not that they really need it, because the ranchers give them salt other places. But in the course of doing that, the cattle would climb the banks that had sloughed in, and so they would totally churn anywhere from six- to ten-foot slopes of material that sloped into the gully bottom. Then, every storm, a lot of that sediment would wash out.

Well, after we stabilized that, the moisture in the bottom of the gully started seeping up into those wet areas and into those banks—those loose banks. Turned them wet. And it wasn't long until they were growing heavy grass and sedges that hung down into the draw and totally protected those banks probably five feet up above the bottom.

So an unbelievable turnaround. We went from losing four-and-a-half truckloads a day, on the average, to probably rebuilding something in that sort of a figure. Because in one storm you catch maybe a hundred truckloads of sediment in the tall grass and sedge and willow that was growing in there.

And we started to see a lot of deer activity. A lot of elk tracks down in there. And the elk were fairly rare in this country. Waterbirds, sandpipers, all kinds of wildlife moved into that area that we hadn't seen before when it was just absolutely bare.

CHAPTER 9

A water pipeline for ranchers poses environmental risks (summer 2001)
I believe it was two years ago that the country was pretty dry and a lot of these counties here got declared disaster areas, which they have been for the last two years also. So the permittees apparently applied for funding to put in a pipeline to pump water directly out of Goose Creek.

The Farm Services Agency, I believe it was, gave them something like $60,000 to build this pipeline. It's heavy-duty reinforced rubber pipe that's about three inches in diameter and is just rolled out on top of the ground—not even buried in most places. And put up through Coal Banks Creek, across the BLM, up on the Forest. It was really done without any permission or water rights from the state, without permission from BLM or Forest Service. The funding came, and they just did it. So that's how they've gotten used to operating again out there.

But the other big problem I saw was while I was on the district for ten years. Even though in the driest years we had water to those same areas coming from existing springs—the Owens Corral Spring out to the north and west up on the Forest runs good water, but it goes through a pipeline that's roughly three miles long. And it feeds four troughs out across the country. The permittees never did like to maintain that, because the pipeline goes down into some deep draws, gets air locks, and it takes some maintenance. That's when we required them to maintain that every year and get the troughs working. They were always able to do that. And in the driest years it ran water.

Well, we had word from two of the Forest's own employees that the spring was running sufficient water in that year 2001. So the ranchers finally found a way to put a pipeline right up the draw, pump out of Goose Creek, which they didn't have water rights for, and avoid having to maintain this pipeline with four troughs that probably, over the years, public money has been spent on, to the tune of many thousands of dollars—probably $20,000, $30,000, or more.

To me, this new pipeline was just a way of getting around that. They always wanted to do that when I was there, and we said, "No, we're going to make this pipeline work."

Anyway, since the pipeline went in, they've worked with the Forest Service, and they got the state to switch water rights to that area from a couple of other springs, which we discovered were dry. They've gotten the Forest Service to do a minimal categorical exclusion to approve it.

The first year it was put in, after some of us complained about it, they very strongly said this is going to be a temporary thing. I was even told by the forest supervisor, "It'll just be temporary for this year. Sure, everything wasn't quite done right, but it's temporary. It will be out this fall. They'll be back to using the old system. We're just getting them through this dry year."

It's still there two years later. Still not buried in most areas. On top of the ground. Pipe and debris lying everywhere.

And then there are environmental problems that it could create. Probably the biggest one is that that whole area of Goose Creek upstream, clear into Utah and Nevada, has been quarantined because of leafy spurge—a noxious weed. A very, very bad noxious weed that spreads very fast. Has a seed that's viable for twenty to thirty years.

And so here's all this leafy spurge growing along Goose Creek upstream. Even the hay crops are quarantined by the state and county. They can't be removed from those farms and ranches up there for use anywhere, because it would spread leafy spurge seed.

Yet these guys are pumping water out of Goose Creek way up onto the BLM and Forest. And this stream contains leafy spurge seed. There's no question. They've talked the forest supervisor, it sounds like, into believing that leafy spurge seed floats, and that any leafy spurge seed will float across the pools, while their intake pipe is in the bottom.

But you know what? I've never seen a seed, pinecone, anything that doesn't get waterlogged and sink. So I can't believe that some of that seed's not being pumped up onto the Forest and the BLM. And this year there's so little water in Goose Creek it's hard to tell what they're pumping up there. The flow is really way down. So it's a big problem.

Of course, the other problem is that Goose Creek doesn't run much water in the summer anyway, and they're removing probably most of the flow right there. So they should be using the old pipeline up on the Forest that's provided and paid for. And all they're doing is just finding a way to minimize their amount of work that it takes to get water up there.

CHAPTER 10

Political obstacles to achieving sustainable land management
I was put in a position as district ranger with the responsibility for taking care of this land. I could not continue to see it degraded like it was and feel I was doing my job.

The politics behind all these government agencies trying to do their job is a serious, serious problem. I don't think that's a secret. But anyone who thinks anything of their career and has ideas of advancing on to a forest supervisor position, or whatever, has to play the game and pretty much has to keep their nose clean—follow the rules.

Well, I was in a situation, where right from the top the ranchers had the attention of the politicians, the congressmen and their aides, the county commissioners, and the Forest Service people. This even included some Forest Service people who knew what needed to be done right, and were following a forest plan—were trying to enforce permit rules that are written on the back of the permit. Right on the back of the permit!

And these ranchers had signed that and had agreed to all these things. And yet you're really not supposed to do 'em, because it costs the permittee a little bit to do what's right, and it aggravates 'em. And anything that aggravates 'em, aggravates the politicians, and that comes down on the government officials.

So every line officer in the Forest Service and BLM knows this. If you're gonna advance, you pretty much keep your nose clean. You may try to do some things, but if you're told "lay off," then you lay off.

In our case, we said, "We can't. We've got a forest plan to abide by. I respect your authority, but we've got to do these things. The place is coming apart. Has been for a hundred and forty years."

If we'd had five thousand years do to this like they have in the Middle East, the country'd look just like that. We can see it happening right under our nose.

People in government agencies are under a tremendous amount of pressure. A number of 'em, I feel, want to do the right things. And a lot of 'em are. But if they want to advance, and have a career, and move ahead in the agencies, they know the "right answers." There's only so much they can do. They can make some progress, and there's been some progress made all around the country. Some people have really bit the bullet and taken some risks and done some great things. But it's all held back and tempered by the politicians and the administration that's in charge.

But even during an administration that is pretty pro-environment and wanting to solve problems for the environment—for the people of the United States—there's only so much the government agencies can do, because they're still under the gun of all the politicians. And most of them bow to the pressure from the ranchers.

Robert W. Phillips

Bob Phillips grew up in the agricultural community of Independence, Oregon, where he developed an early interest in fishing the nearby Willamette River and floodplain lakes. Although planning a career as a teacher, Phillips soon switched majors and went on to earn a bachelor of science degree in fisheries from Oregon State University in 1953. Upon graduation, Phillips joined the US Fish and Wildlife Service in Ketchikan, Alaska, where he studied the migration and survival of pink salmon. The following year he began a fifteen-year tenure at the Oregon Department of Fish and Wildlife, where he held positions as district fishery biologist in management, fish research biologist, and assistant chief in basin investigations (later renamed habitat management). During this period, Phillips continued his studies at Oregon State, earning a master of science degree in fisheries in 1962. He joined the US Forest Service in 1969 as a regional fisheries biologist working out of the regional office in Portland, Oregon. Phillips transferred to Washington, DC, in 1977, where for three years he worked in program planning and budget for fish and wildlife habitat management on the national forests. From 1980 to 1983, Phillips worked in Alaska at the Juneau Office, and then completed his career at the Portland Regional Office in 1988.

Phillips has served as president of the Pacific Fisheries Biologists, the Oregon Chapter of the American Fisheries Society, and the Portland Chapter of the Izaak Walton League.

Bob Phillips made his remarks on the 11th of August 2003 in Portland, Oregon.

Chapters
1. Eastern Oregon's arid landscapes are easily damaged by livestock
2. Difficulties of managing livestock on public lands
3. The genesis of destructive livestock grazing in Eastern Oregon
4. Autumn cattle grazing can damage streams
5. Recovery of the Bruneau River after termination of livestock grazing
6. Diminishing efforts to protect grazed streams in Eastern Oregon

CHAPTER 1

Eastern Oregon's arid landscapes are easily damaged by livestock
In the eastern part of Oregon, particularly in the arid and semi-arid areas of the state, it was obvious that livestock grazing, particularly cattle grazing, was having a significant adverse impact on fish habitat. The reason being that we have dry summers on the national forests in Eastern Oregon and Washington, and throughout the Intermountain West. From about mid-July until the fall rains in mid-September, or as late as mid-October, there is virtually no green forage for cattle except in the riparian areas—down along the streams and where there are the spring developments. And the cattle concentrate there, because the cattle we're working with originated from Europe where they're used to a moist climate. They need a drink of water at least once a day. And also, they like shade from the summer heat. So they basically live on the riparian area. And they overuse it. That's where our problems begin. Not only that, but this use occurs year after year after year.

Many times you'll hear ranchers say, "Well, we're providing a great service because we're reducing the grasses out there, which reduces the fire hazard."

There's a grain of truth in that statement, but basically we'd be a lot better off if the range burned every ten or fifteen years, rather than have it cow-burnt every year.

CHAPTER 2

Difficulties of managing livestock on public lands
One of the basic problems with grazing cattle is the variation in forage production in Eastern Oregon—those drier rangelands. And that's because there is a ten-fold variation in forage production on the ranges there depending upon the amount of precipitation received from September 1 to March 31. This is a piece of excellent research that was done by Forrest Sneva, who worked out of the Northern Great Basin Experiment Station, which is located about thirty-five miles west of Burns, Oregon. Sneva looked at something like eighteen years of data and made the correlation between precipitation and forage production which, as I say, showed a ten-fold variation.[1]

The problem here is that the rancher and the agency managers won't know how much forage is available until, let's say, March 31. Just using 2003 as an example, they wouldn't know how much forage was available until March of

2003. But the rancher must size his herd and have his cows bred by July of 2002—much in advance of knowing how much forage is available. Well, if he estimates, and the agencies estimate that there's an average amount of forage out there, but they get into a drought year, so there's only 15 or 20 percent as much forage as they predicted, then what do they do come April 15–May 1st, when they normally turn out the cattle on the range? If they estimate the average amount of forage, and they put 'em out there and there's only 20 percent of what they estimated, what do they do come July? Do they send the cows home? Or do they leave 'em out there?

Well, if they send 'em home, there's no forage on the home ranch either, because of the drought. And so, if they take 'em to market at that time when the calves are weighing, let's say, only 150 pounds, instead of the usual 450 pounds in October, the rancher takes a beating financially. They often have mortgage payments coming due and that sort of thing.

So if the agency personnel, the forest ranger or the manger for BLM, are trying to protect the land, then they become an obstacle to the rancher leaving his cattle out there. The rancher then complains to his congressman that the district ranger is uncooperative. And so the congressman then goes to the chief of the Forest Service and says this manager is uncooperative, and so on.

The administrators of the Forest Service and the BLM in Washington, DC, are beholden to Congress for their budgets. So the Congress has a tremendous amount of leverage on the agencies to do what a congressman wants. And the congressman wants the folks back home to be pleased with the management there, because that goes a long way toward insuring his re-election. And, in my opinion, most politicians only look as far ahead as the next election. So it corrupts the system. And the land is abused.

One story that I heard from a friend who worked in the Washington Office shows the kind of control Congress has on the agency. There was a congressman from Mississippi who had a disagreement with an assistant secretary of agriculture. And the assistant secretary did not roll over and obey what the congressman wanted him to do. So when the new budget came out, there were no funds for the position of this assistant secretary of agriculture. The congressman was able to just eliminate his position. And that kind of message, of course, becomes widely known within the agency.

So there's no question that agency people realize the power that the rancher has working through his congressman and senators in terms of calling the shots on land management decisions.

CHAPTER 3

The genesis of destructive livestock grazing in Eastern Oregon
The Oregon Trail was at its heaviest use in terms of moving people from the eastern United States into Oregon between 1845 and 1855. In that period, most people came to western Oregon because its agricultural land was more productive than lands in the eastern part of the state.

But then about 1860 or '70, they discovered that there was a lot of grass over on the eastern side of the Cascade Mountains. And it was free for the taking, most of it being unclaimed public land. So, many ranchers moved into that area. And the result was the tragedy of the commons.[2] If one rancher sought to save something for next year—for the future—then his neighbors would move in and consume the forage during the current year. And so there was no incentive to conserve. And that went a long way toward corrupting the management of those rangelands.

CHAPTER 4

Autumn cattle grazing can damage streams
When I first started to work for the Forest Service back in 1969, there was a general policy on most of the forests to remove the cattle from the range by the first of October when deer season opened. That way they'd avoid any conflicts with hunters. But before I retired in 1988, elk hunters were finding cattle in their elk camps on the national forest in November. Perhaps the management plans were changed so that the ranchers could leave their cattle out for a longer period of time. Or a lot of ranchers would leave the cattle out until the snow drove them down into the lowlands, so it would be easier to gather them and take 'em back to the home ranch.

Especially, if you have a dry fall—if you don't get any precipitation to speak of until mid- or late October—then the only green forage for the cattle is right down in the riparian areas. And that's where the cattle stay until they're removed. The riparian areas suffer then with the loss of the woody vegetation and the plant community that holds the soil in place. And we also have the breakdown of the streambanks. The streams take a beating. And I've seen examples of that when I was hunting.

CHAPTER 5

Recovery of the Bruneau River after termination of livestock grazing
I have a buddy in Nevada who has scouted out some of the streams there. And so I've gone over a couple, maybe three times every summer and fished with him for the last five or six years. On one stream over there in northeast Nevada, the Bruneau, there was a buyout of a ranch by the Rocky Mountain Elk Foundation ten years ago or so.[3] According to my friend, and at least in my experience too, the trout fishing has improved greatly since the cattle were removed. And I understand that the Nevada Department of Wildlife, using an electric shocker, found that there was something like a six-fold increase in the number of trout after the cattle were gone, and the willows and the streambanks began to stabilize.[4] A six-fold increase in the number of trout! So that's one personal experience that I've had.

CHAPTER 6

Diminishing efforts to protect grazed streams in Eastern Oregon
Oregon Trout and the Oregon Natural Resources Council—both of those groups, at one time, were active in terms of improving fish habitat through addressing livestock grazing issues on the eastern forests and BLM lands in the state. But over the last several years, and I don't know if there was a cutback in funding or what, but the Oregon Natural Resources Council has dropped their consideration of the grasslands, where the grazing problems are, and instead has concentrated on forest management.

And Oregon Trout, at one time, was a stronger advocate for better range management to protect those lands also. But now they have become less aggressive in pursuing those policies. I'm not sure why.

Perhaps one of the reasons is that most people in this state see little, if any, value in the rangelands of Eastern Oregon. Most people in the state live in the western part, and they never get over to Eastern Oregon. And those that drive through don't see any trees, or maybe only a few juniper. All they see is sagebrush, and they just feel that the land is basically worthless.

The few other Oregonians that do get over there might be hunters during the fall. And they're there for such a short time that they have little personal involvement with the local citizens in terms of land management. Consequently, there's just not a lot of support for changes over there.

Jim Prunty

Having grown up on a ranch, Jim Prunty found it natural to pursue work involving livestock. Then in a major career change past the age of forty, he joined the US Forest Service as a fire management officer. There, for more than twenty years, Prunty worked in fire prevention, fire suppression, and range improvement—tasks from which he learned much about the effects that livestock grazing has on the landscape, before and after a fire.

Since his retirement from the Forest Service in 1987, Prunty has been an independent advocate for the protection of public lands.

Jim Prunty made his remarks on the 24th of August 2003 in Twin Falls, Idaho.

Chapters
1. Prunty's early years in ranching
2. Prunty joins the Forest Service
3. Consequences of burning the Idaho range
4. Suppression of natural fires expands forests
5. Prunty advocates for public lands
6. Don Oman counts cattle on the range
7. Forest Service excludes the public
8. Prunty's work with Red Willow Research
9. The government constructs a water pipeline for ranchers (summer 2001)
10. The unrecognized value of western land

CHAPTER 1

Prunty's early years in ranching

I was raised on an Idaho ranch in a remote area east of Sun Valley. And I went to a little country school twenty miles from any town.

My dad was an old-time cowboy—a dyed in the wool cowboy. And in those days there wasn't anybody that knew what the word "environment" meant. They just raised livestock. And my dad was one of them.

As a youngster, I saw that there was an area of the ranch where a ditch went around a hill and the field above there was not irrigated. Beyond the fence, up above there, was open range. But this particular field only got grazed in the winter when livestock were brought in. And every year, that field would just be covered with flowers and lupines and not much sagebrush. It was so much different from just over the fence. And so I became aware there was such a difference when I was real young.

And I cowboyed. I remember one time we'd gathered a bunch of cattle and held 'em on a big flat that was covered with real heavy sagebrush. One of the fellas there had come in that country to cowboy when my dad did.

He said, "When we first came in here—this big flat—we could hold cattle here and work 'em. Run horses."

And the way it was now, there was so much sagebrush and no grass—just big ol' brush—that you couldn't work cattle there, just trail 'em through.

He told me there'd been a lot of changes just since they'd came in the country.

CHAPTER 2

Prunty joins the Forest Service

Over the years, my dad sold the ranch, and I wasn't prepared to buy it. So I had to drift off and do something else. And I finally got to working for the Forest Service. By happenstance I got into fire management. But I still was not very involved in how to take care of the land. It was just go do my job.

The biggest part of my job was to burn logging slash and put in the water bars[1] and such. There I learned that after burning, some trees didn't come back. There were some areas we were "timber mining." And other places, different trees just came back fine. Even now I can go back and see places where my crews planted trees that are a huge forest—a huge success. And others, there's nothing. So I was getting an education on the ground pretty fast.

Then the stockmen would insist that we burn for range improvement. And since I had learned a great deal about fire—when you could burn; when you couldn't; what would burn and what wouldn't—I was called on to write up plans and actually go do the burning.

And I had lots of things at my command. I had helicopters and Caterpillars—whatever I wanted. There seemed to be no end to burning range.

I burned several thousands of acres of rangeland. And there again, I stuck around long enough to see what came of what I did.

A lot of the other professionals in the Forest Service, rangers and such, move every few years. And so they never have to be actually accountable for what they do. But in my case, I was still there and still doing the burning, and I saw some places where, if it was managed right, the grass came back. The fires did a pretty good job.

Other places, where the range was overgrazed, and then overgrazed after a fire, it would come back to very undesirable species.

Over the years, I got to where I really understood what was happening. But when I'd write up burning plans, I had very little effect on whether it was burned. If they wanted it burned, that was my job—I'd go do it. And if I even made a comment that maybe this was not a good idea here, they'd say, "Well, that's not your decision. Your decision is to burn it. And we will do the rest."

I've seen some drastic consequences just from the effects of burning—wildfires, and, in my case, a lot of controlled fires. So that got to be my specialty. They used to kid me that they put pyromaniacs in jail, but in my case it was job security.

CHAPTER 3

Consequences of burning the Idaho range
I don't know that it's accurate to say that an area responded *right* to burning, because it all burns at one time or another. The result has a lot to do with how it's treated after a burn. Or time of year that it's burned. That sort of thing. But mainly how it responds after a burn is cows.

That's the biggest thing, because all of this western land burned in a sequence of maybe twenty to five hundred years in the trees. And on the range, I think all of it would have burned in a sequence of thirty to fifty years. So it isn't a matter of whether a fire was bad. It was how the area was treated after the fire.

This region had wonderful, wonderful grass before the cows came here. We know it had, because in 1849, and there about, the Oregon Trail came though this country. And there were thousands of animals pulling those wagons, and nobody was hauling hay for them. So they had to live off the land.

There must have been a terrific amount of grass here to supply that westward migration. I can't even imagine what it must have been, because there's nothing here to represent that now. It's gone. And the only reason it's gone is because of cows and our method of grazing.

Cattle grazing after a fire spreads weeds
If you put too many cattle in after a fire, the new plants that are coming up have their roots not exposed exactly, but they're more exposed than they were when they had a plant covering them. And with animals chomping on them, they compact the ground. Then when the little plants come up, they get nipped off before they get a chance to actually re-establish themselves. And if they keep nibbling those off—and it's just the nature of the animal to nibble off the best ones—they leave the worst ones.

The worst ones turned out to be weeds and brush and things that were probably here, but they weren't dominant. But by eliminating all the "ice cream plants," it left the others. And those are what became dominant. While if they'd been left alone, the grasses would have come back thick like they were before. Then these other plants really wouldn't have had a chance to encroach on them.

CHAPTER 4

Suppression of natural fires expands forests
There was sagebrush and there was juniper, but they were remote. And by the time they got started, in thirty, forty years, another grass fire would go through. Even though a grass fire isn't very hot, it did burn off the little new plants. And so they were back to square one as far as the brush was concerned, but the grass was doing fine.

And we see that now. I can point out places up here in the mountains where Douglas-fir is encroaching on thousands of acres of rangeland. And there's absolutely no evidence for the last thousand years of there ever being a tree through there. But now that they've stopped fire from going up on those hillsides, the trees get a chance to grow, and some of them are beginning a forest. Some of them are twenty, thirty, forty feet high. And lots of little ones coming.

So now you see thousands of acres of little fir trees comin' up. It's lack of fire. And I've never seen it mentioned in any planning. I don't think there's anybody who recognizes it. But it's there. I have pictures of it. When you point it out to people, they say, "Oh yeah, I see that. But that's just a change that's going on here in the West—in this particular part of the country."

CHAPTER 5

Prunty advocates for public lands

I started to agitate on one of the forest plans. I could see where we could make some wonderful changes, and I didn't realize that the Forest Service was not ruled by any real intelligence. It was ruled by politics. But I didn't realize that. I should have maybe known it, but I didn't. So I started sticking my neck out before I retired, and was told, "Don't agitate these things."

Then, after I retired, I wrote a letter commenting on the forest plan before this last one. That was twelve, fourteen years ago. For this particular place here in the South Hills, I tried to show that without the cattle it could be a wonderful recreation area. It would bring in literally millions of dollars.

People liked that letter. And so I passed it around.

When I looked where the Forest Service had put down all the letters that were written into the forest plan, there were 135 of my letters in from different people. And they counted them as one. I found out right there that's how the Forest Service works. If they don't like it, they don't count it.

So I started to gripe about management before I got out, but then I didn't do much until I met Don Oman.[2]

CHAPTER 6

Don Oman counts cattle on the range

I worked for Don Oman for just a few months before I retired. I didn't really know the man.

So when Don Oman went with a couple of people to count cattle on the allotment[3] down there, I think he was about as unprepared as anybody could have been for what happened. Because when he got down there to count the cattle, the ranchers came unglued. They accused him of everything in the book—of being terrible, of having airplanes in the air, of having gunmen. And all he did was go down there and want to count the cattle. But that's a no-no. Politics—you don't count the cattle.

The Forest Service has no idea how many cattle are on an allotment. They don't count them. I don't know whether there's anything in the regulations that says the agency's supposed to know how many cattle are on an allotment. But when Don tried to count them, the ranchers went clear to Washington. I have the papers. They had a meeting in Sun Valley. They were going to get rid of Oman for counting those cattle or for trying to get the ranchers to do what they said on their grazing permits. They weren't used to doing that.

Don, I guess, tried to do the right thing. And when I found out how much trouble the ranchers were making, I thought, "Gosh, I can't let that guy go out there and stand alone. I want to get out there and get into this too." I was retired by then. And so then I got into it just because I could see it was the right thing to do.

I don't have the influence that some of the people do, but I have written an awful lot of letters to the newspaper, which, I believe, irritates them—kinda like a coyote nippin' at their heals, while somebody else takes 'em by the horns. And I enjoy it. It gives me a chance to get out and see what's going on and to help the people that are influential.

CHAPTER 7

Forest Service excludes the public

Every spring the Forest Service has meetings with the permit holder. And at those meetings all they do is go through what the rotation will be, and what water troughs and what fences to fix.

So I wrote to the Forest Service and asked that I be allowed to attend those meetings. I didn't want to comment because it was past the comment stage, but I wanted to hear what was going on.

And that went to Congress—enough that congressional aides went to those meetings just to see if I was there. And it caused a real ruckus.

The Forest Service said that we weren't allowed to attend those meetings. I didn't understand, because it had to do with public land. The meetings were held in a public office, but the public was not allowed to hear them.

So I asked one of the congressional aides, "Why am I excluded from that?"

He said, "You intimidate them."

He made my day when he said that. But that's the only thing I've ever heard why I'm not allowed to attend these meetings. Because there was never a meeting—and I went to several—that I ever disturbed or hardly ever said anything. It was too late to say anything. I just wanted to hear what was going on. But that's how the politics of the thing works.

CHAPTER 8

Prunty's work with Red Willow Research
Since my retirement from the Forest Service, I've worked with Miriam Austin[4] when she's required to look at a range project. I travel thousands of miles a year when she does work for National Wildlife, for the Fish and Game, and for the BLM, and Forest Service. Quite often I go along to help with whatever she's doing, and then I'll send in my comments to the appropriate agency.

And I've worked with the Jarbidge Sage-Grouse Local Working Group down here in the Three Creeks country. I've gone to a lot of their meetings, and I'm on the board.

I got thoroughly disgusted with it, because the ranchers and such, all it seemed to me was that they were trying to justify ranching. And there was very little to do with sage-grouse. In fact, they came out and said the biggest problem—they never mentioned *cow* in their whole report—was the ravens and the foxes and crows and hawks and eagles.

And so I wrote back to them and said, "All of those things were here before the cow. And everything was doing fine. We had an abundance of wildlife."

And, of course, they don't like those comments, and that's probably as far as they go.

CHAPTER 9

The government constructs a water pipeline for ranchers (summer 2001)
Don Oman can tell you more about the background of the project.[5] The pipeline is not necessary in the first place. It was just that the ranchers didn't want to take care of pipelines they already had.

This new pipeline was paid for with Drought Relief money. And when I went to the Department of Agriculture with the Freedom of Information Act and asked them how much money was spent in that county on Drought Relief, they said it was none of my business. They wouldn't tell me. I appealed it to Washington, and they ignored it. Eventually, I saw a piece in the newspaper that said the cost was close to $60,000—that's the only reason I can come up with a figure.

But what makes me hot under the collar is that this is a quarantine area for leafy spurge. When I was working with the Forest Service up on the South Fork of the Boise River twenty-five years ago, there was a huge infestation of leafy spurge. I was assigned to work on eradicating it. But it was a losing battle.

When I was up there this summer, leafy spurge was all over the hillsides. And I know the Forest Service spent a million dollars up there over the years.

None of these people here have worked with the leafy spurge program to the extent that I have. When I see them just blatantly pumping the leafy spurge seed seven or eight miles up onto the Forest and nobody seems to care—that's what bothers me. And that's why I am so upset about the pipeline.

What's so bad about leafy spurge?

Leafy spurge is a plant that came in from Europe, I believe—from the Alps. In its native habitat, it has enough predators of one kind or another that it doesn't cause a problem. But in this country, it just thrives everywhere it's found. And some places there's square miles of it.

And it's a toxic plant. It grows from seed and it grows from rhizomes. Nothing here can compete with it that we know of. So given a chance, it takes over. Montana, and a lot of the states, have multi-million dollars worth of damage from leafy spurge.

It's a strange seed. When I was working with it, the Bureau of Reclamation wanted me to get some seed for them, because they were running some experiments. I made the mistake of just getting some seeds and lying them out on a sheet to let them ripen. And when I came back to get my seeds, I found out I didn't have any. They were all over the room.

So I got more plants, and I put 'em on a sheet. And this time I put a sheet over them, because as they ripen they will pop a little beady seed up to maybe fifteen feet in any direction.

They're a little, round seed that can get down into the cracks of the rocks or dirt. And they'll grow in almost any condition from wet meadows to up on the driest, harshest sites you can imagine. And the plants look about the same whether they're up there in harsh condition or down here where it's good. They're a terrific plant as far as plants go.

It's such a green plant that fire has absolutely no effect on it. It's either green or it's gone. When the stem dries out, there's nothing to carry fire. And when it's green, it won't burn either. So fire is of no consequence to it.

There's a sheep experiment station over in eastern Idaho, where they're trying to teach sheep and goats to eat leafy spurge. They're not doing too well because it is toxic. But they say that they're getting some of 'em where they can live on it.

CHAPTER 10

The unrecognized value of western land

The politics of the West is something that's awfully hard to understand, because the worst possible use for this western country is livestock grazing. Absolutely the highest use for planning in the West should be to preserve it as a watershed. Yet it's hardly ever considered. And we're getting less and less water all the time. That land is getting hammered more all the time.

In our case right here—what we call the South Hills was a breadbasket for the pioneers. They could come across that desert and go up there in those hills and get elk or deer or mountain sheep to go on. And now all it's doing is costing us to run cattle there.

There must be five hundred miles of fence in that little area. How many cattle guards are there at $10,000 a piece? It's just costing the taxpayers.

And without the cattle this area would almost be a Yellowstone Park. They're trying to get mountain sheep started again, because they were natural here. The streams had trout in them until they were dried up by cattle. Elk were natural. Lots of mule deer.

If people could stop here now and just drive up there and see those animals, it would be worth probably a thousand times more than to graze subsidized cattle just in this one little area that I'm talking about. So that's one of the things I'm sure that politically will never be done in my lifetime. But maybe someday.

Doug Troutman

As a teenager, Doug Troutman sought solitude from the overcrowding of Southern California by hiking in the nearby San Gabriel and San Bernardino Mountains. Upon graduation from high school in 1962, Troutman began turning his passion for the outdoors into a career by working as a cook at Yellowstone National Park. After serving in Vietnam as a helicopter crew chief, he went on to become a ranger at Yosemite National Park, while obtaining a bachelor's degree in biology from Fresno State. Troutman joined the Bureau of Land Management in 1976, beginning a twenty-three-year career during which he had extensive experience with livestock grazing, first as a wilderness ranger in Arizona and then as a wilderness specialist in Oregon.

Doug Troutman made his remarks on the 14th of August 2004 in Lakeview, Oregon.

Chapters
1. Troutman's life and career
2. BLM's cattle management in Arizona
3. BLM's cattle management on the Lakeview District
4. BLM's cattle management after fire on the Lakeview District
5. Disadvantages of post-fire seeding with crested wheatgrass
6. Cattle exclosures on streams of the Lakeview District
7. Troutman's experiences with BLM range conservationists
8. Politics undercuts sound management at BLM
9. Coyotes made scapegoats for poor livestock management
10. Range improvement funding

CHAPTER 1

Troutman's life and career

I was born in Everett, Washington. Raised in Auburn until I was twelve. Then in '55 my folks moved to Southern California and dragged me down there kicking and screaming. I hated the place from the word "go" because it was too crowded. I was used to remote environment and so spent most of my time up in the San Gabriels and San Bernardino Mountains when I was able to get out.

I graduated from high school in '62. Took off to Yellowstone National Park where I was a cook. Worked there '63–64, then tried to get on as a seasonal with the National Park Service my second year there. I worked for the Yellowstone Park Company (YPC)[1] as a "savage" or concession employee. We were savages; the visitors were the dudes. It was in large part our participation in the sexual revolution. I resolved if I ever had a daughter, she would *not* work for YPC. In those days I was a Republican, so the double standard was easy to accept. During the winter, I was going to school at Fullerton Jr. College.

A couple of people I knew in Park Service had transferred to Yosemite. So in '65 I went there as a cook. I was still trying to get on seasonal with NPS but couldn't arrange it. Those years working for the concessionaire proved to be helpful later as a ranger, because I knew all the party spots where nobody was supposed to be.

In '66 the draft board knocked and I answered, reluctantly, by enlisting. Ended up going to Nam.[2] When I got out in '69, I applied to Yosemite for a ranger position. And I got it! Was hired on as a seasonal in March of 1969.

Met my wife in the park that year, and we were married there in January of 1970. Came back to the park seasonally in WAE status until '76. Spent most of the time after the first year in fee collection—in law enforcement. Did a lot of search and rescue. Worked some with the resource people. My degree is in biology from Fresno State.

We were starting quite a bit of prescribed burning in Yosemite at the time. Took part in some of those early burn projects in the sequoias and then later in the yellow pine. That's how I got interested in fire. But being a WAE, I didn't have any status. So in February of '76, when BLM offered me a permanent job at Aravaipa Canyon, I went with them.[3]

Aravaipa Canyon, about seventy miles north of Tucson, was a primitive area at the time. They hired me on as a wilderness ranger patrolling the canyon, which included some of my early introduction to livestock grazing on BLM lands.

Defenders of Wildlife managed the Wood brother's property, which is the private land outside the west entrance to Aravaipa Canyon.[4] I used to go to the top on horseback once in a while, where I could see that on the San Carlos Apache Indian Reservation that livestock grazing had just completely nuked the territory to the north of the primitive area. To the south was BLM land. That land was also nuked, just awful. Then in the middle was the Wood brother's ranch with lots of native grasses.

The Wood brothers, Fred and Cliff,[5] started rest-rotation grazing in the 1920s. Their habit was that whenever they'd move cattle in and out, they'd always harvest native grass seeds. Then they'd broadcast that seed as they were coming out. So the rangelands that were Wood brothers property were in real good shape even after several years of not being run by them.[6]

Fred and Cliff ran a beef operation. They took fewer cows in and brought out healthier, fatter cows, and more real beef, than their competitors who overstocked the range. This same thing is seen on cow-calf operations that overstock, because the calves are sold off to feedlots to be fattened. You can afford to beat down the range even if the cows are skinny.

The biggest mistake the Wood brothers ever made was putting some perch in one of the stock tanks up on top. It blew out, and some of the fish got down into Aravaipa Canyon—into the creek. Old Fred Wood was most upset when he accidentally introduced a nonnative species into Aravaipa Creek. There's seven native species in the creek that were all on the study list. Aravaipa's one of the few perennial streams in that part of Arizona, and it's the only one in that section of the Galiuros.

It was also a great birding place. Added 176 species to my bird list when I was at Aravaipa Canyon. Used to get a lot of accidentals in also when storms would move through. I saw an aplomado falcon in the canyon, and nobody would believe me until Dr. Minckley of Arizona State University confirmed the sighting. These are fantastic birds.

The vermillion flycatcher was the favorite of my wife and I as the year at Aravaipa ended. My daughter was a toddler running out in the yard to play with scorpions, javelina, and other wildlife, which is why I get upset today when people panic at the sight of any wildlife in the neighborhood, when it is feral dogs that kill kids. Jeffery Dahmer also killed more people than all the lions in the lower forty-eight states in all of history![7]

After a year I went over to Lake Havasu City as an outdoor recreation planner,[8] because staying at Aravaipa wasn't going to get me any promotions. Havasu City is one of the worst places on the planet. I hated it there. (If you die and have the choice of hell or Havasu, pick hell, it's cooler.) But we stayed for a year-and-a-half.

Then when the BLM started their wilderness program, they had a whole bunch of positions open up for wilderness specialists. I applied for one of the positions and came out top on the list. I got a call from the Prineville Office and was told I could have anyplace that I wanted to go. Well, after having the experience of going to Lake Havasu City, having never seen the place first, my wife informed me that I could either live a little longer, or I could take another job in a miserable place we'd never seen before.

We'd come through Lakeview before—we knew the area. So we moved there in 1978. I came in as a wilderness specialist—still classed officially as "outdoor rec planner," but my job description was "wilderness specialist."[9]

The first two years were tied up with the FLPMA-mandated inventory to comply with the Wilderness Act—the first inventory for BLM wilderness. There were four-and-a-half million acres to review here on the Lakeview District. Much of the field work began by flying a lot of the areas. Then we did helicopter work in checking out more specific locations. After that we did it all on the ground—hiking in each of the study areas. When we completed the inventory, we ended up with twelve WSAs and a little over half-a-million acres.

Those were interesting times to be at BLM, because nobody in the local community wanted anything to do with any wilderness. And nobody in the BLM—the range cons and most of the staff—wanted to see any wilderness designations on BLM land either. Management didn't really have an interest in wilderness at all. During the time that I was working on the inventory, and then on the study, I was also pointing out the purpose of wilderness and how it would affect or not affect the management policies. Eventually, a lot of the management and many of the local people turned around their attitudes quite a bit, and were more accepting of wilderness designation for some of the lands.

The Lakeview District is all really remote, and there's a lot of wide-open spaces. One of the difficulties we ran into during the inventory was that some of these areas out there are so exposed and so big and open that you can see anything going on for miles. It was a real problem when we started to ask the question on "solitude"—where did you actually have it? A number of the units lacked outstanding opportunities for solitude just because they were so open that if somebody were out there, you could see them for miles. Same would be true for horses or cattle or anything else that was out there.

Then we went into the study phase and made our final recommendations. From then on we were mostly monitoring WSAs to make sure that nothing was going on that would alter the wilderness suitability of the areas.

One of the things that *was* going on was that we got some new range cons in. And we started having problems with them, because there was always pressure to increase grazing numbers in the WSAs, both from the Burns District and from the Lakeview District.

WSA policy says that you can graze at the level that was occurring at the time you did the inventory. But it doesn't allow for increasing grazing authorization numbers within a WSA. That was one of the big fights we had, especially with one situation, where we have a WSA that's partly in the Burns District and partly in the Lakeview District.

We had the responsibility for doing the monitoring. And when they went to increase from two hundred AUMs of winter use to extremely heavy summer grazing (something around two thousand AUMs, based on observations), we put a stop to it. The Burns District was most pissed off at us for shutting down the increases that they were trying to put on the land.

The area also included one of their own RNAs for *Oryzopsis* and *Stipa* community[10] that they were just beating to death. So because it was in the WSA, we called them back to the numbers that they had at the original time.

On a field trip with an OSU range professor to the RNA, he declared the area was "decadent"[11] until I showed him a picture taken just a couple weeks earlier of grasses with abundant seed heads and vigorous growth before Burns nuked the area with cows. He quickly revised his assessment and said we'd better get the cows out.

This was one of the first times that we actually shut down some grazing based on what the wilderness study policy mandated. We actually took out some stock tanks that were no longer in serviceable condition and restored the land back to the original condition. I can't think of another location where range improvements were ever taken out to benefit wilderness values. We accomplished a few things that way.

After the wilderness program started winding down, I took an interest in disabled access. Being a Nam Vet, a lot of my friends ended up disabled after Nam—multiple amputees, blind, et cetera. So I started working a program nationwide to do a study team to put together a program for designing programs and facilities for people with disabilities. I worked with that a lot the last five, six years that I was with the BLM.

The study team put together a video on orientation to acquaint people with what was needed by the disabled community and the things that they wanted to do. I also created a web program on the BLM intranet to help with design. Its still available on the Oregon BLM internal pages.

In my last two years with the bureau I worked strictly as an access coordinator for the states of Oregon and Washington. Did a lot of traveling around—designing trails and campgrounds, modifying facilities, including putting in toilets that were accessible for people with disabilities. That was the way I ended up my time. I retired from BLM at the end of '99.

CHAPTER 2

BLM's cattle management in Arizona

I first encountered livestock grazing on the public lands at Aravaipa in '76. There was no grazing in the primitive area, but the Salazars had the grazing operation on the east end of the canyon. My first involvement with livestock trespass was in moving the cattle that would sneak into the east end of the canyon. I'd round 'em up and head 'em back out. I had a very interesting day once when my horse, who I called "Dudley Do-Wrong,"[12] because he always got stuck in quicksand, suddenly turned cutting horse on some cows. He took off after them, right into a grove of cat claw acacia. I used up a lot of Bactine[13] after that incident. But the Salazars were pretty good about managing their cattle.

When I moved to Havasu there wasn't a lot of grazing in that area simply because of the poor grazing potential. There's a huge number of acres per AUM on that country along the Colorado River. In some places it's a section of land (640 acres) per cow. There wasn't a lot of grazing management in the Havasu Resource Area.

One of the things that did happen while I was at Havasu was that Gary McVickers, the BLM area manager down south, ran afoul of some ranchers whose cattle were in trespass. And that brought in the Arizona congressman. I can't recall his name right now, but he demanded that McVickers be removed because he was bugging the cattlemen. They did end up moving Gary out of the area. Bob Buffington[14] tried to support him, but there was a lot of pressure brought to bear on the BLM not to interfere with any of the grazing operations.

Bob was a pretty good guy. He had a real strong resource ethic. There were several situations, where he was ready to roll on things involving miners and cattlemen, that people didn't think were politically astute. Yet Bob went ahead and took action against them.

After I left and came up here, Buffington ran into problems with the cattlemen out of the Kingman area—that's where Gary was too. Bob ended up moving to Idaho as state director. Later on he ran afoul of the cattlemen in Idaho too, just by enforcing the regulations. Again the congressional delegation stepped in, took things over, and Bob was moved out.[15] He ended up retiring from BLM and went to work for USAID, because of the pressure brought politically from the cattle association and influence with Congress.

CHAPTER 3

BLM's cattle management on the Lakeview District

When I got to Oregon, I was shocked by how many projects were being done by the BLM for the cattle graziers—water pipelines, seedings, fence construction. That was unheard of in Arizona. There was also a lot more pressure from the livestock industry as to how things would be run—whatever the cattlemen wanted, the range cons pretty much provided 'em.

When A. K.[16] was area manager, his big objective was to have a water hole every mile so the cattle wouldn't have to walk very far for water. He also put in a lot of lakebed pits that caused damage to the ecosystems.[17] Many more lakebed pits were proposed but never completed because they wouldn't hold water or funds weren't available. Other projects were blocked due to rare plant sites.

But there was always pressure to provide more water for livestock. And, of course, the answer was always that this was good for wildlife too. It's true to a degree, but when the cattle overgraze to where there's no forage left for the wildlife, then you're not really doing them that big a favor.

The other big thing here is that there are quite a few wild horses that have been managed for years in two major herds. The livestock operators would always complain about horses being out there. But I never saw impacts from horses that were anywhere near as severe as what the livestock were doing.

The old RMP designated that there was supposed to be forage allocation for livestock, wildlife, and horses. But almost every place I saw, the livestock were taking far more AUMs off than were allocated. And the grazing levels were really intense on all the resource area in the district—really intensive grazing to where native grasses were pretty much extirpated from most of the area.

The worst area, and still is, is out in Beaty Butte country. It's just been really trashed because of the range con that was brought up from New Mexico. I forget what year, but when he got in he really started beating the heck out of the land. Prior to that, the MC Ranch was a corporate operation owned by an insurance company and run by Pete Talbot. He was usually pretty good about keeping the cattle numbers within proportion and keeping the AUMs within the license. There was no fencing out there, so he worked quite a bit with the range con at the time, Willie Street, to use cowboys to move the cattle around. They had a lot of trouble keeping them where they wanted, but at least they were making the effort.

Then when the MC was sold to the MC User's Group, things really started getting bad out in that unit with overgrazing. In drought years, hauling water in allowed the cattle to beat the plants down even harder while they were under stress. The cattle are still foraging and taking the energy out of the plants when they don't have the water to recover. They brought water in pretty routinely. And it led to some pretty beat up condition out there. This stress is hard on the plants, the wildlife, and the wild horses left in the area.

Often the BLM would make the claim that the plants were in better condition than they were. They would say they were getting only 60 percent utilization, which was what they were shooting for. But I would say instead of 60, most of those plants were cropped down 90 percent or above—down to the root system, basically.

Most of the energy of the plant is actually stored below the ground, and as you beat it down, the root ball itself disappears. Then the grasses just aren't able to recover anymore, and you get pedestalling. A lot of that area out there has got heavy soils loss due to erosion because the plants have been just beat down.

CHAPTER 4

BLM's cattle management after fire on the Lakeview District

The biggest problem that I saw with cattle management after fire was that the fire would come in—a lot of them are natural fires—but after they burn through, they either go to crested wheatgrass (I call it "crested cheatgrass") seeding, or they would allow only two growing seasons of rest. Even if it was droughty, they'd still move the cattle back in after two growing seasons, which means, in reality, one year. That just wasn't sufficient time for a lot of the communities to recover from the fire.

I wrote up a plan in the early '80s for wilderness study areas calling for prescribed burning in order to restore the health of some native communities. In that plan we called for five years of rest, minimum, before any livestock would be brought in. Then they'd be brought in at lower numbers than were in there originally for the first few years in order to get good regrowth on the native plant communities. The plan was a big hit at a range conference but was never implemented.

We had a number of fires burn that had pretty good response from the native bunchgrass communities. But then they overgrazed it. One area in particular, Cook Well,[18] came back beautifully after the fire. Had fantastic recovery of the native grasses. But then the range con let the Gale Leaver cattle in up there. And they just beat it down to where there was nothing but cheatgrass[19] and pepperweed three years after the fire.[20] Where it had come back the second year after the fire there was some of the most beautiful native grass you ever wanted to see. But because they let the cattle stay in there and beat the hell out of it, it never recovered after that. Once the cheatgrass and pepperweed and larkspur moved back in, they just took it over. The area still hasn't recovered, and it's been twelve years.

CHAPTER 5

Disadvantages of post-fire seeding with crested wheatgrass

They argued that they needed the crested, first to stabilize the soils. Yet in most of the areas around the Lakeview District where the fires burned, they were pretty flat, nonerosive soils anyway. So there really wasn't the need for them to come in and seed the crested other than that the cattlemen always wanted it, because then they could graze the shit out of it.

Of course, the most huge in scale and scope crested seedings, the Vale Project,[21] were done in the '60s. But crested wheatgrass turns a region into a real low-volume desert, because nothing will grow in it other than some of the weed species—cheatgrass, larkspur, and so on. Even sagebrush takes a long time before it comes back in crested seedings. Once you plant your crested, that's what you'll have until the weeds come in. You won't get good regrowth of any natives, because they can't compete with the crested.

There's thousands of acres of crested wheatgrass seeding in the Lakeview District. When they did the Sharptop Fire in 1983, something like 73,000 acres burned and they seeded over 48,000 acres with crested wheatgrass. They did some study plots with aerial seeding, and another study plot where they did nothing. And right next to the crested seeding nothing much came back except cheatgrass, because the cattle drifted over and hit it hard. Out in the middle, where they'd tried the aerial seeding, it came back to a little degree. But the place that came back best on the Sharptop burn was on the west, where they left it alone, and they didn't do any grazing for over four years. It came back real well with a lot of native grasses and forbs. But even that they eventually grazed the heck out of and beat it back down.

The BLM is finally taking a better approach in trying to restore native grass seedings. There's problems with native grasses, but the biggest one has always been cost. Trying to get a reliable seed source with the bureau's financing situation made it difficult for them to contract with people to grow native seed. They really needed to have a sustained way to produce native grasses so that every year, whenever there was a fire somewhere, they would have the native grasses available to source to at a reasonable cost. They've never answered that problem within the purchasing power of the agency.

Native wildlife diversity declines in crested wheatgrass seeding
Wildlife diversity declines in a crested seeding. You lose your browsers. It provides feed early in the season, but later in the season there's nothing out there with any nutrient value.

Overall diversity, especially in the avian species, really declines in a crested seeding. About all you see are horned larks and a few others such as raptors, which work the seedings because of the sparse comparative cover that makes small rodents and lagomorphs good targets. You've lost the habitat for sage-grouse when you put in a crested seeding.

Economic analysis of the Sharptop Fire
They did a supposed economic analysis for the seeding after the Sharptop Fire. I think they sat down one day and figured out that it would take something like 340 years for the grazing fees to pay back what it cost to seed.

There's no economic benefit to those huge crested seedings. When you increase forage (AUMs) by that much, you are also committed to provide increased water. Every time there's a seeding, pipelines are needed. BLM managers go along with that.

CHAPTER 6

Cattle exclosures on streams of the Lakeview District
There is a lot of riparian area out there on the desert, especially from seeps and springs in the Beaty Butte country. A lot of exclosures had been put in, and the livestock operators were claiming that wild horses were tearing down the fences and allowing cattle into their exclosures.[22]

Actually, one of the more interesting things I ever saw was out at Spaulding Reservoir. The gate was down and cattle were in the exclosure. We herded 'em out on foot and put the gate back up. As we were walking back to the truck, one of the cows actually jumped the fence into the exclosure. I'd never seen a cow jump a barbed wire fence before, but she didn't have much problem getting over it at all. That was an area where it was a constant fight to keep the cattle out. And the blame was always put on horses tearing down the upper end of the fence.

There's just no rationale whatsoever for letting cows into riparian zones out on the desert. They should all be fenced and continually maintained. Ranchers are suppose to do that. They get the money back with 8100. The ridiculously low grazing fees are divided to give 50 percent back to the local ranchers to spend on range improvements. In the BLM 8100 section of the "program and billing process," 25 percent goes to the counties to be spent on range improvements, and 25 percent actually goes to the treasury. Considering the range fee was higher in the 1960s, in real dollars, not adjusted dollars, a simple examination of inflation trend would show the ludicrous nature of the range fee computation. It should be their responsibility for their cheap grazing fees[23] to make sure that those exclosures are maintained. You can pipe the water out for the cattle, but you don't have to destroy every stream and spring.

CHAPTER 7

Troutman's experiences with BLM range conservationists

One range con that came in was "anything the cattlemen wanted." He wanted to do some fencing and some spring stuff to increase the AUMs in the area. And he wanted to license more cattle use.

I said, "No, you're not gonna do that in a WSA, because it's not allowed."

He said, "We do it in New Mexico all the time."

And I said, "You're not in New Mexico anymore, Toto. And you're not gonna do it in one of my WSAs."[24]

I've known a few good range cons in the BLM also. Willie Street was one. And there have been a couple of others—people that had responsible attitude. But by and large, the BLM management was ready to roll over for the rancher at a moment's notice.

150 / Governmental Personnel

CHAPTER 8

Politics undercuts sound management at BLM
No one bucked the system here[25] very much after Art Gerity left.[26] We had a dog in Vietnam that pretty much summed up post-Gerity managers. Her name was No Balls. It just wasn't politically astute to buck the ranchers.

Mineral use is the same way in most of the BLM. And, of course, that's really getting severe now in New Mexico and Colorado with the new policies coming out of the current Bush administration for energy development. Everything is up for grabs.

After I left at the end of 1999, true tragedy has struck the BLM under George W. Bush.[27] All control is now in the hands of the interior secretary[28] and her minions—in all programs. Right after they took over, area managers could no longer sign federal register notices for anything. A common case would be a notice of road closure, or minor action. Norton moved this to an undersecretary in her office, and suddenly there was a six-month backlog in register notices at BLM.

The state directors don't really have control of the BLM lands anymore. The decision-making process has been pulled back to Washington. Now it's undersecretaries that are making decisions out in the field and telling people what's gonna go down. If the managers don't like it, they can leave. It's the most abusive administration I've ever seen, and I go back to the 1960s.

They're stealing everything blind. They're issuing permits for license renewals on hydroelectric projects. They're telling officials, "You will not fight this. You will re-license no matter what they want." Environmental assessments are not being properly conducted. There's all kinds of stuff that's going on that the public doesn't have a clue about.

CHAPTER 9

Coyotes made scapegoats for poor livestock management
There's always been an anti-coyote regimen within most of the ranching community. They claim that they always have a huge problem with predators. But I've never seen it. The biggest problem we have with predators is a lack of them when we get our rabbit booms. They've had plague come in a couple times out in the Warner Valley and up north because of the huge surges in the rabbit population. The predator population wasn't enough to keep them evened out.

It's a boom and bust cycle with lagomorphs. They do that anyway on about a seven-year cycle.

One of the things that always got to me was that because of the winter grazing, you've got cattle out there that are calving in the dead of winter. It's thirty below zero. You've got a calf that dies in delivering, and then there are ranchers that say, "Oh, I've seen coyotes actually rip the calves right out of the cows."

One situation that I know occurred was when a cow had died in calving, and a coyote was chewing on the half-exposed calf. But the coyote was just scavenging. When you have the cattle calving in those conditions, you're going to lose some of your calves. It's a pretty harsh environment out there. And to blame the deaths all on the coyotes is absurd. In fact, most "coyote" predation is actually caused by feral dog packs in most areas near human population.

If a rancher can't sustain predator loss on the open range, he should bring his cows home and stable them, or pasture them at fair market value on private land. Cows are the most dangerous "predator" on public lands if you really look at wildlife losses.

CHAPTER 10

Range improvement funding
Eighty-one hundred money is the money that comes back from the grazing fee for enhancement projects that benefit livestock. I've seen an awful lot of range improvement projects done for livestock, and I've seen none that actually improved the range for total purposes—wildlife and restoration and good management conditions. It's always been whatever could maximize the number of AUMs that they can pull off the land.

The purpose of multiple-use management always seems to be to get more cows out there. In Arizona and Idaho and Nevada, I have yet to see a BLM allotment that hasn't been to some degree overgrazed. And by that I mean if you have a riparian zone that's being beat out within the allotment, you may have some uplands that are not getting hammered, because the riparian's there for them to go to like in the Trout Creeks and the Pueblos.[29] But if any part of the component is being hammered hard—if you're beating up your riparian zones even though your uplands may still have some grass left—you're still overgrazing. You have to manage for the most fragile resource within an allotment.

You can't do this thing of saying, "Well, we're going to get a little improvement in Area B, so we'll be harder on Area A." It's got to be an overview. And livestock grazing levels should be established by the most fragile part of the environment; not the most hardy.

When I came in here, A. K. was trying to put waterholes in everywhere. Part of the reason was that the adjudication done in the 1960s for the BLM in Lakeview District assumed that water was available on all the allotments. This was not the case, not even close. But as a consequence, most of the allotments in the BLM had a lot more AUMs assigned than what the land could actually carry, because there was no water to disperse the cattle use.

We saw that in the RMP, which we did in the late 1970s. Then when the new management plan came on in 2000, they were still using some of the old analysis for how many AUMs were available.

I don't have a problem with grazing on public lands at reasonable levels. It's just that I've not seen acreage on BLM lands that *were* at reasonable levels. They've always been at higher levels than what the land could actually sustain for a healthy environment. But that's the way the ballgame is written by the range program.

Larry Walker

Raised in New Mexico, with much of his early life spent on ranches, Larry Walker seemed destined for a career involving livestock. Following a stint in the US Navy, Walker indeed completed a bachelor's degree in rangeland management at New Mexico State University. And after pursuing graduate coursework at Utah State University, he joined the Oregon BLM in 1971 as a range conservationist. In that capacity, Walker worked on ecological site inventories and rangeland monitoring, along with coordinating the agency's pesticide and noxious weed efforts. After retiring from BLM in 1997, Walker founded the Internet websites RangeBiome and RangeNet, along with some associated Internet discussions groups. These Internet-based tools have surpassed all previous efforts to facilitate communication among conservationists working to improve the condition of America's livestock-grazed public lands.

Larry Walker made his remarks on the 13th of August 2003 in Portland, Oregon.

Chapters
1. Walker's early life, formal education, and career at BLM
2. Inventory and monitoring of lands managed by BLM (1977–82)
3. Political pressure trumps science at BLM
4. Walker establishes the RangeBiome and RangeNet websites (1997)

CHAPTER 1

Walker's early life, formal education, and career at BLM

I was born in northern New Mexico and grew up there mostly on ranches. I went from booties to boots, and I was probably a teenager before I had my first pair of regular shoes.

About the time I turned seventeen, I decided I wanted to see the world. So I joined the Navy and spent a hitch there. When I got out they came along with the Vietnam-era GI Bill. So I was able to attend college and get a degree in rangeland management—a bachelor's degree from New Mexico State University.

I went on to Utah State University in Logan to start graduate school, but I didn't complete that.

154 / *Governmental Personnel*

During summers, while I was going to college, I worked for the Forest Service in northern New Mexico. They called it a "recreation aide." Some summers we were doing range studies. Other summers I was basically patrolling campgrounds. So things were varied, but there was a little range work involved in that.

Then after I left graduate school, I got a job with the Bureau of Land Management on the Medford, Oregon, District as a range conservationist half-time and a jack-of-all-trades the other half-time.

Working on the Medford District as a range conservationist was interesting to say the least, since that is one of the largest timber producing areas in the federal scheme of things. Both BLM and Forest Service in Oregon grow lots of big trees.

After working there for three or four years, I went to the Prineville, Oregon, BLM District in central Oregon working with grazing and planning issues and that type of thing.

Then in the summer of 1976, I went to the Oregon State Office in Portland—where I basically spent the rest of my career until 1997 when I retired—working a number of different areas of the range program including monitoring, inventory, budget, and quite a bit of work on trying to get ADP databases[1] together and interface with geographic information systems, and things like that.

CHAPTER 2

Inventory and monitoring of lands managed by BLM (1977-82)

New regulations are developed just about every time you have a change of presidential administration. The new crowd goes in, and they want to change a few things.

Under the Carter administration,[2] about midway through, we were getting quite a bit of direction out of the Washington Office, headquarters of the BLM. And we were receiving policy statements as such that were trying to shift our emphasis on management more to an ecosystem approach.

The primary way to know what your management is doing is to have a decent inventory and monitoring program in place. So from that aspect we did, under the Carter administration, put in place the Soil Vegetation Inventory Method (SVIM),[3] which is a scientifically sound method. Some people have a few problems with some aspects of it, but overall it was the best that BLM ever had in place. And great emphasis was placed on that from about 1977 until about 1982, which was how long it took the Watt administration[4] to get some priorities shifted and some regulations changed.

So we had a total shift away from that policy. And, in fact, the policy that they tried to—. Well, I won't say it was a policy. But it was said a couple of times that what they would have liked would be if all of the pickup keys in BLM offices were hung on the wall or locked away so they couldn't get out and see what was going on.

And even though the inventories survived the Carter administration, Watt shifted the emphasis to monitoring. Of course, you need monitoring also, but BLM never has enough money to do both things at the same time.

Then we go along several years and when the BLM has something like half of their land, maybe more, inventoried under a common method for the first time in their history, the Clinton administration came in.[5] And, of course, they figured that everything that the previous people were doing must be wrong. So they came along and shifted the emphasis to what they called "rangeland health." But the criteria that got worked up for determining that are pretty soft and pretty subjective. And nowhere near up to the scientific standards of what the Soil Vegetation Inventory Method was, which through several of these phases is now known as the "Ecological Site Inventory."[6]

So that's a couple of examples of how politics affected things from my viewpoint, which was mostly from working in the Oregon State Office.

CHAPTER 3

Political pressure trumps science at BLM

While it's not necessarily the greatest and most accurate approach in the world, one product from an inventory like the Soil Vegetation Inventory Method was the estimation of grazing capacity based upon the kinds and amounts of plants present. And during development of that method there were two different linear program modules developed by Colorado State University for the BLM to make these grazing capacity estimates.

Then, when the shift in policy came,[7] the BLM decided that they didn't like the results that were coming out of these forage allocation models. So they put out a policy instruction for them not to be used. And they subsequently changed the regulations to where they didn't quite say not to use them. They just said they couldn't be used alone.

Of course, this was all about the fact that the allocation models were coming up with grazing reductions. And the political pressure was on. So you had a political determination that the science was not valid. Everybody wants good science. It's just that they want the good science to agree with what they want.

CHAPTER 4

Walker establishes the RangeBiome and RangeNet websites (1997)
When a person starts reaching the end of a career, I think it's natural to look back and kind of assess where you've been, where you're going, where things could be going. What's right. What's wrong. And a few things like that. While I felt that we had made some progress in some areas, it was painfully slow and, I felt, inadequate.

Looking towards retired life and what may, or may not, have been my contribution to the greater good of things, I thought a little bit about what was the single thing, based on my experience, that could be done on BLM lands that would bring about the greatest improvement of the ecosystems, watersheds, wildlife. All of those things. And it was to get rid of the adverse impacts of grazing. Pure and simple.

Essentially all of it is grazed. And grazing is one of the most severe impacts out there. So I planned, when I retired, to get active in the environmental community to work towards that.

Well, when I retired,[8] I found out that getting active in the environmental community is not all that easy, particularly if you come from a questionable background like being a BLM range conservationist.

Anyway, I pulled together a portfolio of a few things that I had done with some color reproduction. Put it into a loose leaf binder. And I made up, I don't know, eight or ten sets of it, which for one thing was getting kind of expensive. Then I started trying to find folks in the environmental community to get a copy to. Some responded; other didn't.

About that time I also started feeling a little bit of withdrawal. When you're working within an agency like that, you don't know until it's gone how much information is being fed to you through your in-box and the various things that are going on.

So I went ahead and did what I had planned not to do, which was to get onto the Internet. About that time the cost of websites had come down to about $20 a month. And with a little amount of exploring, I decided that most of the stuff I'd been manually putting together in a portfolio I could put onto a webpage, save myself a whole lot of money. Save a whole lot of trees. And maybe get some other people to look at it.

So I brought up the RangeBiome website,[9] where I initially just put some of my own stuff. Then I started thinking, "There's some other retired range folks around—maybe we could set up a listserve or something and talk with each other."

Well, generally, if a range con from the BLM or from the Forest Service is of the persuasion that maybe things weren't done so good and there might be some improvement needed, by the time they retire they're burned out. So the last thing that they want to communicate about is range.

But by and by, I met folks through the Internet and through going to this meeting or that meeting. And so I expanded that thought a little bit and said, "Well, if I can't do this with retired folks, how about just doing it with folks who are interested in the management of federal public lands?" At that time there was no group that was really focused specifically and primarily on that.

Primarily then, I started RangeNet[10] to improve and facilitate communication among activists and across organizations through the individuals rather than through the organizations, and to bridge something that happens in every bureaucracy and in every environmental organization that I know of where there is an inner circle at work. Folks who are not part of that inner circle don't know what's going on. So a large part of RangeNet was an attempt to bridge between those inner circles and individuals—individuals who are either affiliated with one of the organizations involved or are unaffiliated.

So I started RangeNet. And started recruiting members using the old National Geographic approach of *nomination, invitation,* and *acceptance.*

From there we grew to where we've got a couple hundred members. A lot of them contribute information, data, opinions, and other things that we post on the RangeNet website, which is getting substantial now. It's something over 250 megabytes of storage. And we get something over 10,000 unique visitors a month to it from the net. That's not fantastic by Internet standards, but it's still substantial. And it's been growing.

Along the way, after RangeNet had been going a couple of years, there were a lot of people on there that I'd never met. And I was sure there were others who had not met others. So I tossed out the idea, "Hey, why don't we get together someplace?" I suggested Reno since I happen to like casinos. And there was enough interest to do it. But along with that interest, I also started reading between the lines assumptions that this was gonna be a more general meeting. So I expanded it, and we had the first RangeNet meeting, RangeNet 2000, in November of 2000 in Reno, Nevada.[11] And as an outgrowth of that we have had annual RangeNet meetings ever since.

Pat Ward

Born in Evansville, Indiana, and raised in Southern California, Pat Ward went on to serve for three years as a US Army Ranger before pursuing a career in natural resources research. At Humboldt State University, Ward received his undergraduate degree in wildlife management and his master's degree in natural resources with a wildlife emphasis. While at Humboldt, he studied habitat use, dispersal, demography, and prey relationships of the northern spotted owl. Subsequent research into survey techniques for detecting Mexican spotted owls in the American Southwest led Ward to the Sacramento Mountains of southern New Mexico. As a doctoral student at Colorado State University, Ward investigated the responses of Mexican spotted owls to several environmental variables. He received his PhD in 2001. At the time of our interview, Ward was employed by the Rocky Mountain Research Station as a wildlife biologist in research. His primary study at the time was to determine, through management experiments, a forest thinning prescription that would reduce fire risk, while benefiting the Mexican spotted owl by improving conditions for the owls' preferred prey.

Pat Ward made his remarks on the 27th of September 2004 in Alamogordo, New Mexico.

Chapters
1. Cattle grazing impacts prey of Mexican spotted owls on the Lincoln NF
2. Faulty monitoring leads to overgrazing
3. Cattle grazing on the Lincoln NF (2004)

CHAPTER 1

Cattle grazing impacts prey of Mexican spotted owls on the Lincoln NF
One of the things that I would like to talk about is the unintuitive connection between the Sacramento Mountains population of Mexican spotted owls[1] and grazing effects. There are some other spotted owl populations for which what I found here may also apply. But generally, spotted owls with forest-dwelling habits don't often co-occur with open meadows to the extent that they get a lot of their prey there. And so those other spotted owls wouldn't

be impacted by excessive grazing of either domestic or native ungulates. But research from my doctoral dissertation sheds light on some of the potential impacts that excessive grazing in the Sacramento Mountains can have on Mexican spotted owls.

Basically I set out to learn if there is any support for the idea that Mexican spotted owls, which as of 1993 have been listed as a threatened species,[2] could be conserved or their ecological situation improved through manipulations of habitat to improve prey populations. My research involved collecting information using an objective and rigorous sampling design to estimate small mammal populations over time, especially the owls' common prey species. I also examined some of the micro-habitat features through time—where owls occurred and where owls were hunting for small mammals.

I found that some smaller prey species, like deer mice,[3] periodically attained very high densities in mixed conifer habitat, which comprised most of the owls' foraging area. I found a similar pattern for two species of voles[4] occurring in meadows adjacent to this type of forest.

All their biomass combined in a given year—two vole species and the mice—was correlated more than anything else with the owls' reproduction. This is very counterintuitive for anyone knowledgeable about spotted owl ecology, especially that of northern spotted owls,[5] whose reproductive success is tied to woodrats and flying squirrels, neither of which occupy meadow areas.

In the Sacramento Mountains there is no endemic flying squirrel. They're not known to range there. There is a woodrat species, the Mexican woodrat,[6] that lives in the Sacramento Mountains, which, it turns out, is the preferred prey of the Mexican spotted owl. The Mexican woodrat is the smallest North American woodrat, averaging about 120 grams, whereas the dusky-footed woodrat,[7] which is the preferred prey of northern spotted owls, can get up to 300 grams. The reason I bring up this point is because it takes only six deer mice in the Sacramento Mountains to equal one woodrat on average. It would take nearly double that number of deer mice in the Pacific Northwest to equal a dusky-footed woodrat. Thus, compared to northern spotted owls, it is easier for Mexican spotted owls to match the energetic value in woodrats by taking smaller alternative prey, particularly in years when these alternative prey are abundant.

One point that followed from my dissertation was that the second-growth forest in the Lincoln has helped create periods of great eruption in the deer mice. Deer mice eat several kinds of foods, but one food that has been shown to be related to eruptions is Douglas-fir seed. There's considerable variation over time in factors associated with seed production and release. This condition helps produce the dynamics I have witnessed in the prey populations, spotted owl food habits, and their reproduction.

Another point that I showed in looking at the prey composition of some remaining old-growth stands, relative to the second-growth stands, is that there is much greater biomass of deer mice per unit of area than there is of Mexican woodrats. Mixed conifer forest on the current landscape available to spotted owls in the Sacramento Mountains (Lincoln National Forest) is now predominately second growth (i.e., dominant trees are 70 to 110 years old). So as we have transitioned from much of the landscape being older forest to primarily being second-growth, we have completely changed the landscape in terms of the distribution of common prey biomass for this population of Mexican spotted owls. The change was from late-seral forests with more woodrats and less deer mice to mid-seral forests with more deer mice and fewer woodrats.

This change in forest condition also likely changed the dynamics of the owls' reproductive success. By having to rely more on smaller rodents, like deer mice and voles, the owls could now only take in enough energy to reproduce when the smaller rodents are abundant. Thus, their reproduction tracks the abundance of smaller rodents more so than abundance of woodrats. And the population dynamics of the smaller rodents are comparatively much greater.[8]

In addition to deer mice, these owls typically consume anywhere from 10 to 25 percent of their biomass in voles. Well, voles, particularly the Mexican vole, occur predominately in open meadows, although the long-tailed vole can occur in the ecotones in addition to the meadows.

I have found that in some years—when the owls' regular prey, deer mice or Mexican woodrats, are low—the owls will take a lot more voles, especially when the voles are high in abundance. During the six years (1991–1996) that I was following these relationships, there were two peaks in Mexican vole abundance, following a trend similar to meadow voles from the Plains and back East. The long-tailed vole population and the Mexican vole population were both high in 1992. Mexican voles again peaked in 1995, but the long-tailed vole just continued on a downward trend. Their abundance peaks aren't always synchronous.

So in 1992 there was quite a bit of biomass of voles available in the meadows. On average, meadows comprise only about 8 percent of what we would call a typical spotted owl home range in the Sacramento Mountains. But if you take that 8 percent and you multiply it by the density of voles, you'll find that their total biomass in that meadow is comparable to all of the biomass of mice and woodrats in the forest in those peak years.

If the owl is an opportunist at all, it is going to take advantage of voles. And, we might say, well, so what? They're just getting their food.

Well, it turns out that, like I said, there is a strong correlation between the amount of small rodent biomass and the owls' reproduction. And if there are years when owls cannot get deer mice, but there are voles around, it will stave off reproductive failure. A difficulty is that we can't say that it's a factor every year.

The bottom line here is that the two vole species can contribute up to a quarter of the owls' food base in certain years and that contribution can affect reproduction or prevent starvation. My research tells me, that for Mexican spotted owls in the Sacramento Mountains, voles matter.

More recent research that I have conducted,[9] where I got involved with range managers and those who make a living off of public grazing, is relevant to assessing if there is enough herbaceous vegetation to provide habitat for voles in sufficient numbers for spotted owls.

The habitat factor that we've found to be the strongest so far is a simple measure of herbaceous plant height. Percent cover has some play as well, but I've found that in comparing this one measure, mean maximum plant height, with vole abundance, that it explains about 77 percent of the variation in the abundance of voles. It's an easy-to-measure metric that shows a linear relationship between the numbers of voles per unit area relative to habitat condition in montane meadows.

As a result of a couple court cases, the Forest Service, locally, was asked to insure by monitoring that they had a certain amount of residual herbaceous plant height. At the time there was no way to determine exactly where the cut-off should be—that is, how much grass/forb height is enough. One local biologist picked 4 inches because in talking with folks from the region, he thought that it might correlate pretty well with their 4 inches of leaf length on forage species that provided good-to-excellent range conditions. As it turned out, my analysis showed that a minimum 4.5 inches of herbaceous plant height (not forage leaf-length) was needed to provide for the maximum level of vole consumption by spotted owls in the Sacramento Mountains.

The correlation of general herbaceous plant height and leaf length of cattle-forage plant species has yet to be studied or established, despite my repeated encouragement to do so.

The court case suggested that the Forest Service was not doing monitoring that it was supposed to for management indicator species or to quantify whether good-to-excellent range conditions were being met on the permitted pastures. Forest Guardians took them to court over the issue saying, "Your monitoring data aren't meeting the conditions that you're supposed to meet under your forest plan." The Paragon Foundation, in conjunction with Otero County, joined in the suit as an intervening party.

The outcome of that case was that the Forest Service needed to do a better job of monitoring. And they were to use the best available science to make decisions about allotment monitoring and management, including the function I'd developed that showed the linear relationship between vole abundance and grass/forb height to assess conditions on the meadows.

CHAPTER 2

Faulty monitoring leads to overgrazing

As a result of the court case, the Forest Service was supposed to give their monitoring data to Forest Guardians every year in the form of a report. What wasn't established, per se, was exactly how they were to do the monitoring. So we had a workshop after the court case in early 2001 at which the range cons were given the information in writing on how to perform the monitoring. And we went over it out in the field.

One of the big points of contention in doing the monitoring, and still is today, is over procedure when you have an intercept point and there's no grass or forb there. According to the information that I provided to the range cons, when that happens, you need to record a zero for plant height.

The range cons don't want to count it that way. The Range Improvement Task Force, a state body working out of New Mexico State University, is a strong proponent of the range cons following conventional range sampling procedures. They advise that when an intercept point has no grass, that they should move off the line to the next plant and measure its height. Well, you can tell right away that if you have no zeros in your sample, you're automatically going to have a much greater average height, unless everything is all the same height, which it never is.

Last year, 2003, I was out evaluating the model and collecting measurements of herbaceous height. We were also sampling vole abundance to see if the relationship held in a very dry period. The range cons were measuring their transects. And the Range Improvement Task Force was measuring their transects. What we discovered was that the methods that both the task force and the Forest Service were using to measure prey cover height were really biased. Their methods may have been fine for getting at forage allocation and leaf length, but their measurements were significantly below their own established criterion of 4 inches of herbaceous plant height for meeting vole-habitat conditions, at which point cattle should have been removed from the pastures.

It really became a problem when the range cons were under the impression that they were meeting the 4-inch plant-height requirement (established in part by consultation with the US Fish and Wildlife Service for maintaining suitable conditions for Mexican spotted owls), because they told Fish and Wildlife Service that they had no problem meeting the 4-inch herbaceous plant-height requirement, when in fact they weren't.

So where the rub is, is that the court has established a guideline based on this herbaceous height that triggers whether or not cows are pulled from the pastures. And last year, cows were left on because it was falsely stated that they were meeting the cover height.

In reviewing the data, what was more stark, and even scary, to me was when I compared my plant measurements from 1992 to '94 with the 2003 data. In the earlier measurements, average horizontal cover of herbaceous plants was 89 percent. Plant heights ranged, depending upon the year, from an average of 4 inches to 12 inches. In 2003 the average height among ten meadows was a little over 2 inches. And the average percent cover was down to 44 percent. Additionally, the cows were being left on.

My emphasis now has been to clarify what the methods are to do the monitoring. Try and insure, the best I can, that people understand how to do them, and why. And I'm finishing up a report on where the models came from, how they were developed, and why it's important that we measure things the way that they're proposed. Otherwise you get faulty readings that result in unguided management decisions.

CHAPTER 3

Cattle grazing on the Lincoln NF (2004)

We had not too much more rain this year than last, but the timing was a little more evenly spread. And I did see some areas, that in 2003 that were down to 30 percent horizontal cover, recover to probably about—I didn't measure it, so I'm guessing here—50, maybe 60 percent. So it wasn't as damning a condition, or as long term a condition, as I thought it might be.

But coinciding with that the district ranger this year said to the permittees, "You will remove the cattle when we tell you to remove them. And you will lower the numbers to what we say."

And even though there's still some fighting going on about that, the cooperation has been better this year than last. Last year was also pretty dismal in terms of a combination of being dry and the timing of the rain. Plants actually grew at the first response of rain, but then withered in the meadows because we didn't get rain until later months than usual. And, of course, there was no change in grazing management. In part that was due to people's perspectives, but also to political pressure.

The county commissioners have been involved in meetings with how to handle grazing without other constituents involved. There's been no invitation, or if there's been invitation, nobody's taken up the invitation from environmental groups to sit at the table to deal with the sort of guidelines that should be used for judging the monitoring results. Or what kind of monitoring should be done or changed. Those discussions have taken place with Forest Service officials and county commissioners. There's been no other stakeholders involved. And the meetings were usually at the request of the county commissioners or from the regional office down to lower officials. That's all I can say first hand.[10]

Bill Worf

In 1929 when Bill Worf was three years old, his parents purchased five homesteads south of Rosebud, Montana, and went into farming and ranching. Over the succeeding years, Worf saw the surrounding open range transformed from short-grass prairie to dense sagebrush cover. Although the prairie's demise was blamed on drought, cheatgrass, sagebrush, grasshoppers, and prairie dogs, Worf came to realize that the actual cause was overgrazing by livestock.

After serving in the US Marines during World War Two, including fighting in the battle of Iwo Jima, Worf attended the University of Montana under the GI Bill and received his bachelor of science degree in forestry/range in 1950. Upon graduation, he began a thirty-one-year career with the US Forest Service, holding positions of district ranger on the Ashley National Forest (Utah), staff officer at the regional office in Ogden, Utah, staff officer on the Fishlake National Forest (Utah), and supervisor of the Bridger National Forest (Wyoming). In 1965 Worf was assigned to the agency's Washington, DC, Office, where he headed the development of policy for implementing the 1964 Wilderness Act. Subsequently, he served as director for wilderness, recreation, and lands at the regional office in Missoula, Montana, until his retirement in 1981.

Finding that no national environmental organization was encouraging the government to better care for existing wilderness, Worf co-founded Wilderness Watch for that purpose in 1989. He then served as the organization's president from its inception until 2003.

Also in 2003, Worf received the Keith Corrigall Wilderness Stewardship Award from the **International Journal of Wilderness** *in recognition of a lifetime of achievement in wilderness protection and stewardship.*

Bill Worf made his remarks on the 25th of August 2004 in Missoula, Montana.

Chapters
1. Worf's early work with the Forest Service (1950)
2. Rancher resistance to livestock reductions on the Uinta NF (1951–55)
3. Worf's experience with ranchers on the Ashley NF (June 1955)
4. The range improvement program on the Ashley NF (1958–60)
5. Worf's tenure as supervisor of the Bridger NF (1962–65)
6. Worf tackles wilderness issues in the Forest Service

7. Some of Worf's activities with Wilderness Watch (1990s)
8. Worf's concern over management of the South Warner Wilderness (1999)
9. Worf's first visit to the South Warner Wilderness (3–5 October 1999)
10. Worf's second visit to the South Warner Wilderness (13–16 July 2000)
11. Worf's third visit to the South Warner Wilderness (31 July–3 August 2001)
12. Worf's fourth visit to the South Warner Wilderness (4–7 September 2003)
13. Aspen meadows of the South Warner Wilderness
14. Livestock trample seeps and streams of the South Warner Wilderness
15. Livestock overgraze Pine Creek Basin, South Warner Wilderness
16. Sheep and cattle trespass on the South Warner Wilderness
17. Diseased domestic sheep thwart bighorn reintroduction
18. Worf's vision for managing livestock in the South Warner Wilderness
19. Federal agencies need to support field personnel

CHAPTER 1

Worf's early work with the Forest Service (1950)

My first job after I graduated from the university in 1950 was an appointment to a range survey crew on the Bridger National Forest in Wyoming, where we finished up the last range survey. That was a process that was put together by foresters, and it was a pretty well-run scientific approach to determining what exactly is on the ground. There was a process in which we'd map out a vegetative type, and then we would go through there and record the plant density, the species composition, and the amount of bare ground. All that sort of stuff. And this was all documented in what we called the "data sheet" for each one of these mapped units. That was then compiled.

The problem wasn't with the quality of the data collection, it was in how it was applied. Of course, the old livestock—the cow and the sheep—don't understand that they're supposed to use everything uniformly. They would use some places heavier and consume some species more than others. As a result, that generally gave a capacity figure that was well above the actual capacity of the land. That's the reason why the system was abandoned. But the data and the picture of the vegetation gave a pretty good representation of the conditions if you had the data sheets along with the map.

I had occasion to use that several times in my career going back to the old range survey where it had been done.

From there I went to the Uinta National Forest as a range con on a district that had originally been two districts. One was a GS 9 district, and the other was a GS 7 district. They put them together and made the first

GS 11 ranger district in Region 4. The purpose of doing that was to put a full-time range conservation specialist on the district. And that was me.

My job was to do basically the first version of Forest Service range allotment analysis—to record, again by types, the condition and trend and proper use. They're supposed to follow "use" through and determine what was proper and what was sustainable. And, in addition to doing it for that particular district, I was kind of the guinea pig to start what later became known as the "allotment analysis process" that carried on through the 1950s and 1960s. Eventually, we started establishing permanent photo transects.

The permanent Parker 3-Step Transect was established to measure trend and changes over time. This had some statistical problems, though. When Parker first set it up, he thought that he had a statistically sound process, where he could go back time after time and get the same answer on the same plot of ground. It turned out that the sample wasn't quite good enough. You could measure it one time and get 60 percent ground cover, and next time you might get 40 percent. But the procedure also involved the taking of photographs, which later became one of the most valuable parts of it, because you had a photo point that you could deal with.

CHAPTER 2

Rancher resistance to livestock reductions on the Uinta NF (1951–55)

When I first went to the Uinta National Forest[1] after the range survey, I went down there as the range analysis person. That district had about 30,000 sheep and 7,200 cattle grazing on it. And in four years there, I completed proper-use studies on the whole district. We started a reduction program taking two-thirds of all the livestock off of the district. The permittees went to the politicians and really put the heat on us. But we made it stick.

CHAPTER 3

Worf's experience with ranchers on the Ashley NF (June 1955)

I made a date to ride with the permittees that had the Farm Creek Allotment. They were all there up at their cabin on the forest. I hauled my horse up in a Forest Service horse trailer. And while I was gettin' my horse out of the trailer and gettin' it saddled ready to go, they all gathered around me. They were all mounted—probably eight or nine of them.

168 / *Governmental Personnel*

They gathered around me in a circle, and the president of the association said, "Well, Bill, we just want you to know that we'll get along fine with you as long as you don't start talking about cuts."

He's kind of layin' down the gauntlet, you know. If I worked with 'em and didn't start talking about cutting their numbers or their season, why, we'd get along fine. So I just kind of laughed, and I said, "Well, we'll talk about cuts if, after I take a look at things, I think it needs it."

CHAPTER 4

The range improvement program on the Ashley NF (1958–60)
One of the big things that we dealt with was sagebrush eradication. Sagebrush generally grew originally in clumps. And I can remember as I was growing up that there'd be a clump of sagebrush here, and fifty yards away there'd be another clump. They'd kind of spread out. That's what the sage-grouse really liked. You'd find them in around these clumps of sagebrush.

But because of two factors—heavy grazing, which reduced the vigor of the grasses and allowed the sagebrush to spread, and lack of fire, which every once in a while would wipe out sagebrush—the density of sagebrush increased. So one of the things that I did a lot of was spraying sagebrush on national forest lands all over Utah and Idaho and western Wyoming.

Other things that we were doing included seeding of crested wheat and stuff like that on disturbed areas.

CHAPTER 5

Worf's tenure as supervisor of the Bridger NF (1962–65)
The Bridger National Forest, at the time, had the third largest grazing load of any national forest in the system. I don't remember what the numbers were, but there was an awful lot of cattle and sheep grazing on the Bridger—in the Bridger Wilderness. And we were working with the permittees on some innovative management programs at the time. That was one of the ways that I used the old range survey that I had been involved with.

There was a "sheep driveway" that ran the full length of the Wyoming Division. And it was creating some real watershed problems. I had some of my people map the old driveway and those areas that were completely denuded by trailing of sheep. We had, if my memory is correct, somewhere around thirty-two or thirty-three thousand acres completely denuded as a result of the driveway.

Well, when I started talking with the permittees they, of course, said it had always been that way, and there was nothing that could be done about it. I told them that I didn't care whether that's the way they thought it always was. I didn't believe 'em, and I went to the old survey work that we'd done there in 1950 and showed 'em that while there was some denuded area already in 1950, by just fifteen years later, it had expanded a lot. We got that from the data sheets off of the range survey. And faced with that, I convinced 'em all they had to take their stock to the allotments by truck instead of driving them the full length of the Wyoming Division.

CHAPTER 6

Worf tackles wilderness issues in the Forest Service

I was supervisor of the Bridger National Forest in Wyoming when the first national wilderness act was under consideration.[2] And I was responsible for management of the Bridger Wilderness, which had been established in 1960. The idea of wilderness interested me, and when the Forest Service was finally in support of the passage of the bill, that gave me the freedom to get out and talk about it and urge support for it.

The idea of wilderness in Wyoming wasn't a real popular one. So anytime I would speak, it usually drew some headlines and some controversy. And that brought attention to the chief's office.[3] When the Wilderness Act passed in 1964, they appointed me to serve on a task force to draft the regulations and policy to implement it. When the work of that task force was completed, they transferred me in on a full-time basis to run the program.

A lot of people think I had something to do with the passage of the Wilderness Act, but I didn't. I knew Howard Zahniser,[4] because he came out and made a trip with me in the Bridger Wilderness in 1963. He died in '64 before the act passed.

CHAPTER 7

Some of Worf's activities with Wilderness Watch (1990s)

Because of my range background, I took on the issues that involved grazing in wilderness situations. That's where I met Mike Sauber[5] on the Diamond

Bar down in New Mexico. He and his partner at that time, Susan Schock, had started an organization they called Gila Watch. And I went down to see what was going on—got involved in that whole issue. Made three or four trips down there.[6] Up to that point, I figured that the Forest Service was truly the leader in professional range management. And I was kind of shocked at the fact that we weren't following through on some of the old range allotment analysis data.

Since then I've been involved in detail in two other cases. One of them is West Fork of Blacks Fork on the Wasatch-Cache National Forest down in Utah. Our youngest son was killed in a plane accident there. He was the chief pilot for American Check Transport, and for some reason he got off course in the dark and crashed into the high Uinta Mountains. I went down to see his crash site, and when I got there I was shocked at the serious overuse by sheep that I saw in the West Fork of the Blacks Fork. So I kind of dug into that, and I've made several trips down there.

One of the things that I've done in all the cases I get involved—the Diamond Bar and on the West Fork of Blacks Fork—I go out and I try to keep it on a strictly professional level. Then I come home, and I write a report that is documented with photographs. I don't have the time to run the transects. But I did make, I think, three trips to the West Fork of Blacks Fork, and my assessment is that at least the Alpine types should not be grazed for a long, long, long time, if ever. They are currently being severely damaged by sheep grazing. Of course, the permittee doesn't agree, and the Forest Service doesn't agree. And the Forest Service doesn't pay any attention to the old range allotment analysis that was done in the early 1960s. They hadn't tried to compare one with the other—what's there now against those old transects. That was really disturbing to me.

CHAPTER 8

Worf's concern over management of the South Warner Wilderness (1999)

Then I got involved in the Modoc—South Warner Wilderness.[7] What got me down there were a couple things. I had seen that country from the seat of a Cessna aircraft in May 1966, when I went down to look at a proposed wilderness. By that time, I was head of the Wilderness Program in the Washington Office of the Forest Service. And we selected the San Rafael as the first area to take to Congress for a new wilderness proposal.

So I went down to look at it, and while I was there I decided to take an airplane tour of all the primitive areas, and proposed wildernesses in the state of

California.[8] I made a circle up over the Warner Mountains and saw this neat little piece of country there.

Then, after Wilderness Watch was formed in 1989, we began to get reports from folks, who went to the South Warner Wilderness for backpacking or for fishing, about what they thought was serious overgrazing. Reports even came from some locals right there in Cedarville and Alturas. But those local folks were intimidated by the livestock people that they live with. That little corner of California is a bit of the old West. The livestock permittees rule the roost, and any local person that spoke up could be ostracized and criticized. I don't know of anybody that was physically harmed, but they didn't feel comfortable taking any kind of a stand.

CHAPTER 9

Worf's first visit to the South Warner Wilderness (3–5 October 1999)

I decided to go down and take a look at the problems. The forest supervisor at that time was a guy by the name of Scott Conroy.[9] Scott had a good solid land ethic—I think he's a pretty good professional. My trip down there was with him, and it was his first look at the South Warner Wilderness on the ground. He was as concerned as I was about some of the things that he saw.

I came back and wrote a report on that. And before my report got down there, Scott was selected by the chief of the Forest Service to run the Roadless Area Review that the Clinton administration wanted to do of all the roadless areas in the United States. That later led to establishment of the famous Clinton Roadless Area Policy.

So, Scott Conroy was not there very long. I think if he had been, maybe things might be different today.

Then he was replaced by a guy who was just there to retire—Dan Chisholm.[10] Pretty nice guy to be out with, but he didn't want to bite the bullet on any problems.

I was really concerned about the watershed problems that I saw and the potential for serious gutting out on Mill Creek in particular. The willows were being gradually eliminated from the bottom, but, more importantly, the uplands had been taken over pretty much by sagebrush. Very little understory and no A-horizon on the soil.[11] Very little litter underneath the sagebrush.

Two things that are kind of accidents of nature have kept that country somewhat together. One of 'em is the soil. The soil is a real loose volcanic ash sort of thing over most of that country. And then there's very few torrential type summer storms. Most of the moisture comes during the winter. As a result the snowmelt is absorbed by this volcanic soil and there's very few summer storms to really build up overflow problems. I think if it was somewhat different, a good summer storm on Mill Creek could rip it apart.

My first trip out there with Scott Conroy, the first or second day, they were in the process of doing a range EIS for the whole Warner Mountains. And, of course, I was interested in only that part covered by the South Warner Wilderness. But the district ranger had been assigned the responsibility to do what they called the "Warner Mountain Rangeland Project." And during that first trip with the district ranger, Edie Asrow, it became pretty obvious to me that she did not have an understanding of watershed and range ecology at all.

I'd done a little research before I went down there and had learned that all of the Warner Mountains had been covered in 1934 by a range survey, which I was real familiar with, having worked on the last range survey in Region 4. In those early days, we had looked at the Modoc as one of the shining examples of leadership in range management.

I asked Edie what kind of changes have taken place since the 1934 range survey. That country was all grazed very, very heavily for over 150 years. A lot of sheep. A lot of cattle. And, in fact, the establishment of the national forests was an attempt to get on top of the problem. I also knew that there'd been a range allotment analysis completed over the same country in the early 1960s. And so I asked her about these two things and whether she'd gone back and checked what's there today against what was there in 1934 and 1960.

And her response was that the Forest Service is thinking from here forward, not going back.

I said, "Well, Edie, when you go back there, you can find out what was on the ground. Has there been any change? Maybe we're way better than we were then." There'd been 70 years of grazing by that time. And now we've had another 70 years. Has it gotten better?

Talking to the permittees, I learned that following the 1934 range survey there'd been an adjustment in numbers. Then in 1950 there was another adjustment in numbers. And then again in the 1960s they'd taken some more livestock off. So there's been quite a reduction in stock. But the district ranger couldn't verify even the old history on numbers. I got it from the permittees.

I've been really pushin' the Forest Service to get their old history together. The district ranger still insists to this day that there's no value in that. That's, I think, almost criminal when a land manager doesn't want to know what the range was like.

I've been told that the old range survey data sheets were in the Alturas warehouse in 1996. So I told Edie that she ought to get those out so we could take a look and see what the changes have been since 1934. She said she had the maps, and they didn't tell her anything except where the different types are.

And I said, "Well, that's right, Edie, the maps are not much good by themselves. It takes the maps and the data sheet. And you ought to get those data sheets and have one of your range cons, or someone, go out and check a few of 'em. See where they are."

Now they tell me they can't find those data sheets. They're no longer in existence. They'd be worth their weight in platinum if we had 'em today.

And the same thing with several old Parker 3-Step Transects that were made in different places. I raised the question whether those areas have been re-photographed. The response I got to that question was, "Well, they're in the wrong place, Bill. And besides that, we can't find them."

They had the transect files, but they said they couldn't locate the transects anymore. And they were in the wrong place. I don't know how you correlate those two answers. But my concern is that the Modoc Forest, and particularly this district ranger that's there now, has absolutely no interest in knowing the history of that country.

I recommended after my first trip to the South Warner that they get a geomorphologist to take a look at that Mill Creek drainage. Up to then, District Ranger Asrow was concentrating all of her efforts at looking at the riparian areas down along the stream. That's important. We need to know what's happening there, but what we need to know more from a watershed standpoint is what the groundcover is on the slopes. That's the watershed.

This riparian area, Mill Creek itself, is the conduit that has to handle the water that runs off of the slopes. But the condition of that water as it approaches the stream is also important. If it's coming in as an overflow with a lot of sediment in it, it carries a lot of power. If it comes out slowly—spread out—streams can handle it.

They had their ecologist and a geomorphologist go in and take a look at it. I don't remember right now the name of the geologist. The ecologist was a lady by the name of Sidney Smith. Sidney and the geomorphologist went out there, and they tended to agree with my concerns. And so I went down for my second trip.

CHAPTER 10

Worf's second visit to the South Warner Wilderness (13–16 July 2000)

We got together in Mill Creek with the permittee, some range scientists, and Forest Service people, and they showed us their concern—how the stream was gradually straightening out.

While I was down there, I was looking for some country that might be on the mend. I found one allotment that had supposedly been in nonuse for four or five years. So I thought, "Well, I oughta go take a look at that and see what no grazing for that amount of time would do." That was the Granger Allotment, and the wilderness part of the Granger Allotment is the headwaters of South Parker Creek. I went in there by myself. I had a friend with me that drove me there and walked through with me, but basically I was on my own. I didn't have any Forest Service people with me. And I went across the headwaters of South Parker Creek up in the wilderness. And what I saw up there really bothered me. The first thing, of course, is that even though it had been under nonuse, it showed signs of very, very heavy grazing. Cowpies were from the year before. In fact we saw a lot of cattle when we walked across there.

But the thing that bothered me more than that was all the bare soil on the slopes and some gullies that were cutting down through the meadows up there on the top.

I had two days that I could spend there. So I was gonna go see another allotment that had been in nonuse for a while. But after looking at the headwaters of South Parker Creek, and seeing what looked so bad to me up there, I decided that instead of going to Emerson/Cottonwood,[12] I'd walk up South Parker Creek.

And what I saw when I walked up South Parker Creek just turned my stomach. It was gutted out. Just tons and tons and tons of material had washed out of there going down into the Pit River, which provides about one-quarter of the water for the Sacramento River drainage. I was surprised, really surprised, because the data and the maps that had been put out with the *Draft Environmental Impact Statement* for the Warner Mountain Rangeland Project had identified those drainages that had problems in the riparian zone.

Then it said all of the others are "properly functioning, satisfactory condition." And Parker Creek was one of those that was identified as properly functioning, satisfactory condition. Nobody, not even the rawest amateur, could look at what I saw in South Parker Creek, both the headwaters and down in the bottom, and say it was properly functioning, satisfactory condition. So I have to assume that this wash-out occurred after the field work for the Warner Mountain Rangeland Project was done.

I wrote that up. Sent a report to Edie and the forest supervisor, Dan Chisholm. Consequently, they took a field trip with Dan Chisholm down through the Granger Allotment. Then they decided to bring in the National Riparian Task Force to look at it.

CHAPTER 11

Worf's third visit to the South Warner Wilderness (31 July–3 August 2001)
My next trip down was when that National Riparian Task Force came. We had a bunch of permittees and a bunch of local RC&D people.[13] A real entourage of people walked down through the South Parker Creek. The National Riparian Task Force, that portion of it that was there, was chaired by a BLM range person by the name of Steve Leonard.

They looked at it, and then they went over the next day and looked at Mill Creek. They found some problems, and they wrote their report.

I can't disagree with anything in it, but it was a politically written report. Steve Leonard was about ready to retire, and he'd already hung out his shingle as a range consultant. He expected that his clientele would be livestock people, who wanted to have somebody working for them on range issues and on their own lands.

He wrote his report in a way that would not reflect badly on how he and livestock people related to each other. But basically his report was sound. He recognized there were some problems, but he was gonna solve them all by better management of livestock.

He didn't think there were too many livestock. He felt that trespass from other allotments was a big part of the problem. The other part of the problem was that livestock were being allowed to stay in after proper use, so they would continue to pound the same areas. All of these remarks were valid.

CHAPTER 12

Worf's fourth visit to the South Warner Wilderness (4–7 September 2003)
I went down again in 2003 with Ed Bloedel,[14] because I wanted somebody else besides me to take a look at it. We went down there and spent a day on

the Granger Allotment and a day with Stan Sylva,[15] Edie Asrow, and the permittee.

The permittee had made a major effort to reduce the impact on the Granger Allotment and had done some good things. One thing he'd done was reduce the grazing below his permitted AUMs just about 50 percent. And that had really helped. In fact, I think if we could continue to manage them like that on a yearly basis, that maybe we could start bringing that country back.

Stan Sylva, as we were walking across there, stated that a lot of what we were seeing is natural geologic erosion. When we got down in the drainage, where the big gully is in the bottom—six, eight feet deep and twenty feet across—he didn't argue that was geologic. But the district ranger, Edie Asrow, said, "Well, this is just the way it is."

Incidentally, while we were riding through that day I noticed her picking plastic ribbon off of the trees as she was going by—sticking it in her pockets. And so I asked her what that ribbon was. And she said, "Well, I had the wilderness ranger come through and mark this so we wouldn't get in trouble."

She is not familiar enough with the country. She'd been there thirteen years and had to have somebody mark the way so she could lead a party through there.

My major concern with the area is the uplands, because they're the ones that have gotta deliver the water to the riparian areas in a way that they can handle. Of course, the riparian areas have to be in a condition to handle them. But it was the uplands, the condition of the uplands, that cut out the bottom of South Parker. It wasn't the condition of the riparian.

Edie's answer to that was that the uplands now carry all of the vegetation that they can handle. It isn't the result of grazing. It's just the way it is.

I've been telling 'em that what needs to happen is that they need to graze the uplands lightly enough that there'll be a buildup of fine material underneath the sagebrush that will carry fire. And that some good natural fire in there would eventually convert that back from solid sage to a grass/forb combination.

I had a conversation with Edie on the telephone just last week in which she told me, "Bill, when we had the big Blue Fire[16] in 2001, it got to the edge of the sagebrush and went out." She said, "It won't burn through there."

I told her that I understood that, and that the reason it wouldn't burn is because there isn't anything under the sagebrush. That's when she insisted to me that it wouldn't make any difference if it wasn't grazed for forty years. There still wouldn't be anything under that sagebrush.

I am very, very concerned that the Forest Service is ignoring what I think is good sound range science in the way they're managing the grazing on national forest land today. Same thing that I found on the West Fork of Blacks Fork down on the Wasatch-Cache National Forest. For some reason the Forest Service does not want to deal with the old studies. As Edie tells me, they want to start with where they are today and look forward.

I told her the other day that that's kind of like a mountain climber that's stuck half way up a mountain. If he doesn't know where he came from—whether he came from the bottom or fell off the top—he doesn't know how to get out. And this is the problem that we're in now.

Henderson Meadow Allotment, at one time, had 2,000 pairs grazing for a full long season. Today it has 650 pairs grazing. If we put 2,000 back there today, they would starve to death. There's no way that they could survive, which tells me there's been a terrible difference in the amount of forage that's produced there. And that's something they don't want to deal with.

CHAPTER 13

Aspen meadows of the South Warner Wilderness
One of the things that the rancher points to is that at one time we didn't have any aspen reproduction in here. And now look at all those young aspen that are coming.

My reaction to that is that aspen is not a key grazing species. It shouldn't be much of a forage producer at all. And the fact that we're getting aspen reproduction indicates that we now have a lighter amount of grazing than we had at one time. No question about that. But it still isn't light enough to start an upward trend on the good palatable forage grasses. I think in one of my letters I used the comparison of a wife beater—a woman who's abused by her husband. If he used to beat her every night and he only beats her on Saturday nights now, that's still beating. And that's kind of where we are on some of that. There are some places where they're no longer hitting the aspen quite as heavy.

CHAPTER 14

Livestock trample seeps and streams of the South Warner Wilderness
The seeps come out mid-slope and they don't have a lot of water. But there's kind of a little riparian area around each one of them. Some of 'em actually produce water that runs into the stream. Most of these seeps are very badly

trampled, and the black mud comes up, which, in turn, absorbs a lot of sunlight and warms the water. There's hundreds, if not thousands, of these seeps. And all of them are producing warm water, because of the black soil that's not covered up with grass. I think that has a significant bearing on the temperature of water in Mill Creek and therefore a bearing on its fish-producing capacity. Trout don't like warm water.

And then, of course, Mill Creek itself is wider and shallower than it would normally be. Under good conditions the stream channel would be narrower, and more winding, and a lot deeper. Its current structure helps also to warm the water, or to keep it warm, as it comes down through there. So there's tremendous warming effect due to grazing. Also, some of the willows that used to shade the stream are no longer there.

CHAPTER 15

Livestock overgraze Pine Creek Basin, South Warner Wilderness

Pine Creek Basin is one of the more popular recreation areas. It's really popular for fishing and day trips. An awful lot of the grazing comments from the public come out of people who have visited that area.

The basin itself is pretty small. And I think the permittee puts eighty head in for supposedly about two weeks. So something like forty animal unit months are produced from there. In the process, even if you graze the basin properly, the route in and out of the basin is right up the trail that all the recreationists hike that go in there for fishing. And it just looks terrible. That route up the drainage is just denuded. It can't help but reflect badly on people's impression of the impact of grazing in there.

My suggestion, and Ed Bloedel agreed, was that the area ought to be closed to grazing unless they can find another way of getting those cattle in and out of the basin. It's probably producing less than a $100 worth of forage as far as return to the taxpayer is concerned. And it can't be justified from an economic standpoint.

Another thing about Pine Creek Basin that we found was that there'd been water and sediment running off of the uplands into a big floodplain out on the riparian area in the bottom of the basin. And that was coming from heavy sheep use on the uplands. It's another example of how overuse in the uplands creates watershed problems.

The recreationists have basically been told that there isn't anything that can be done about grazing—that the grazing guidelines in connection with the Colorado Wilderness Bill don't allow the Forest Service to adjust grazing.[17] This is not true. But that's the common theme that's been given to the recreationists that complain.

The Forest Service, specifically Edie Asrow, allowed the permittee that runs into Pine Creek Basin to post a sheet in the Forest Service bulletin board—glass-covered bulletin board—at the trailhead saying what great things cows are—how important they are to the forest there, which is an indication, I guess, of the importance that Edie puts on grazing.[18]

One of the concerns I have is whether or not Edie can really stand up to the pressure, or the resistance, of those strong permittees. She lives right there in that little town of Cedarville, fifteen hundred people maybe, something like that. Her daughter goes to school there. I think she's pretty near out of high school now. And Edie, if she takes on these grazing issues at all, is gonna anger some of her neighbors and her daughter's friends. There's tremendous temptation to tread lightly.[19]

CHAPTER 16

Sheep and cattle trespass on the South Warner Wilderness

There's an allotment line between the Henderson Meadow Allotment and the Wilderness Sheep Allotment. It's just an imaginary line across the slope about half way up. The sheep come down supposedly to that line, and the cattle go up to the line from below. Well, of course, the cattle can't read signs, and there's no fence or anything between the sheep and the cattle. So the cattle go on up onto the sheep allotment. And, in fact, the first trip down there,[20] we found cattle sign a mile inside of the sheep allotment.

Then the second summer I was there,[21] the cattle permittee told me he'd seen sheep clear down to the bottom of Mill Creek. What that does, of course, is put both classes of livestock grazing the same country and creating common use—kind of like having common use of a checking account. Each person that's there has to see that they get their fair share. The sheepmen go a little farther into the cattle allotment than they should. And the cattle just naturally go up onto the sheep allotment. All of that creates some real stress on the range. And the sheep and cattle are owned by different people.

CHAPTER 17

Diseased domestic sheep thwart bighorn reintroduction
The bighorn sheep get disease from domestic sheep. And it's been fatal to bighorns in a lot of places. A few years ago there was one part of the South Warner that was closed to sheep grazing in order to allow some bighorns to be established there. It wasn't very successful because there were domestic sheep—I don't think the sheep ever completely stayed out of the area that was supposed to have been sheep free. And then there were sheep around the edge of the forest that also carried the diseases. So I don't think there's been any successful re-introduction of bighorns into the South Warner yet. That's another case where maybe we ought to give up the sheep grazing in South Warner in order to get bighorns in.

I've proposed that. And, of course, the district ranger's reaction was that that was something she didn't have the authority to do.

CHAPTER 18

Worf's vision for managing livestock in the South Warner Wilderness
I would reduce the amount of grazing in there considerably. I don't think it would put any of the permittees out of business, but it would reduce their income at least in the short term.

I think the sheep oughta all come off of the South Warner. That would be a real impact on the sheep permittee. But I don't see how we can justify grazing those uplands as we do with sheep. It's just one of those things, I think, we're gonna have to do.

Now, the fellow that has the Granger Allotment, it looked to me that if he had half the cattle that he was permitted, and took very good care of how they were managed, I think that country would probably come back.

CHAPTER 19

Federal agencies need to support field personnel
I think the way that the agency can help is to just be sure that the local field people are given full support from the top. When I was on the range job on the Heber District of the Uinta National Forest,[22] where I did the studies that eventually led to taking two-thirds of the stock off of that district, the first

thing that happened when we started the study program was I came home from the office one day and there hanging in my garage was a freshly dressed lamb. No sign on it to say where it came from or anything. So I took it over and gave it to the Relief Society at the Mormon Church.

And then when we got down the road a little bit, figuratively speaking, and I began telling the permittees what the proper use information was showing, things got really tense. Our children weren't in school at the time. But it got pretty tough on my wife. She couldn't go to the grocery store in town to get groceries, because nobody would wait on her. Everybody in town was either a permittee or closely related to one. And we ended up driving to Provo, Utah, to do our shopping, because she was ostracized in Heber City.[23] We had some staunch friends that stood by us, but it got pretty rough there. And we were there four years. If we'd had older children that were in school, particularly in junior high and high school, that would have been terrible on them. We just had to suck it up and live with it.

The first winter that I was there, livestock people stopped by the office all the time, all winter long. In fact, it was a problem a lot of the time. People would just stop in and want to chew the fat for a while.

By the second year, they were beginning to see where we were headed with the range studies, and we had a winter without any interruptions in the office. But I belonged to the Lions Club there, and District Ranger Andy McConkie also belonged. The president of the National Wool Growers Association, Don Clyde, and the president of the National Cattlemen's Association, Levi Montgomery, were also in the same Lions Club. You know, it was a little town. We couldn't avoid contact with people. And things got pretty tough.

Part II

Nongovernmental Conservationists

If the environmental movement doesn't pay more attention to the public lands issue in the West, and really gets serious about it, and starts taking some political risks, grazing will end when the land is ungrazable, the cheatgrass has taken over, and the land's productive capacity has been destroyed in forty, maybe fifty years.
—Patrick Diehl, co-founder, Escalante Wilderness Project

What did George Bush do when he gained power? What did Ronald Reagan do when he gained power? They bought what they called "ranches." That just shows the hold that the ranching mystique has on the United States even though there are very few traditional ranchers left.
—Ralph Maughan, professor of political science,
Idaho State University

What needs to be understood about all of these resource managers is that while they may have gone to any number of very excellent schools, they're embedded with the local community. They are obligated to get along with the local people. If they have children, this is especially a barrier to their objectivity. Because if their kids are going to school with the cowboy's kids or the "timber beast" kids, those kids are vulnerable. And they're often singled out and attacked physically if their daddy, now, I suppose, their mommy too, is advocating a policy that the local populace disagrees with. Very powerful, very powerful factor.
—Steven G. Herman, faculty, The Evergreen State College

Joy Belsky (1945–2001)

Joy Belsky spent much of her career as a grassland ecologist conducting research in Africa. But in the early 1990s, she and her husband moved to Portland, Oregon, where she devoted the rest of her life to enlisting science in the cause of protecting public lands from bad management—particularly management that involved livestock.

I first met Joy at her Portland office in the summer of 1997. It was my first information-gathering excursion throughout the West, and I'd heard that Joy was someone I should meet. We spoke for a couple hours, during which time she let me inspect her extensive folders of research articles about grasslands and grazing. Many of those articles I took to the local copy shop to provide me with future reading material.

Before Joy died in December 2001, we had met twice more, spoken by phone dozens of times, and exchanged a couple hundred e-mail messages. I would have loved to have interviewed her for this book. Unfortunately, I did not begin the research for it until more than eighteen months after her death. So, in her absence, I sought out a few of the people who knew her best in her role as a scientist: her husband, her last scientific collaborator, and her last boss, the executive director of the Oregon Natural Desert Association. From them we learn about her education, her work, her passion for science, and some of her frustrations and successes in dealing with the scientific and land-management communities.

Robert Amundson

Robert Amundson, husband of Joy Belsky, received his BA in biology from Whitman College, and his MS and PhD degrees from the University of Washington. Subsequently, he held positions in research at the Boyce Thompson Institute at Cornell University before relocating to Portland, Oregon, in 1992. Since that time he has served as an environmental consultant and as a lecturer at Portland State University. Amundson's research studies have included carbon sequestration as mitigation for carbon dioxide emission, economic impacts of ambient ozone concentrations on crops, and impacts of ultra-violet B radiation on crop growth and yields. He is the author of more than fifty peer-reviewed articles in scientific journals.

Robert Amundson made his remarks on the 12th of August 2003 in Portland, Oregon.

Chapters
1. Belsky receives her master's degree from Yale University
2. Belsky's doctoral research at the University of Washington (1970s)
3. Belsky's interest in livestock impacts on grasslands
4. Belsky's discrimination complaint and lawsuit against the BLM (1992–93)
5. Belsky opposes killing coyotes at Hart Mountain NAR (mid-1990s)
6. Belsky co-authors article about livestock grazing's degradation of forest structure and dynamics (mid-1990s)
7. Cowboy politics in academia (late 1990s)

CHAPTER 1

Belsky receives her master's degree from Yale University
Joy did a project on the Branford Ponds near New Haven, looking at siltation. But there wasn't a lot of manipulation or research. It was just going out and doing survey. And it didn't flow into her research at the University of Washington.

Joy ended up at the University of Washington for her PhD program[1] because she took a summer job with the Student Conservation Corps in the summer of 1971 at the newly opened North Cascades National Park.

She and a few other female graduate students at Yale had applied for a research assistantship on Isle Royale, but she was told that the male students needed to be taken care of first. In addition, the faculty member was concerned about sending a female assistant out in the field with a married male student. That was just the way it was back then.

So, instead of getting an RA, she ended up at North Cascades National Park and became the naturalist for the Park Service there.[2] Loved the North Cascades. Loved the Northwest. And did quite a bit of hiking and camping in the new park. She fell in love with alpine. And so when she came back to graduate school in Seattle, I think, that was probably one of the motivations for her to pick a project that dealt with alpine ecology.

CHAPTER 2

Belsky's doctoral research at the University of Washington (1970s)

Joy first started research on grasses when she was doing her doctoral dissertation at the University of Washington.[3] She chose a topic contrasting a grass, *Festuca ovina*,[4] that is found all around the Northern Hemispheric high latitudes. And she contrasted that with a local grass, *Festuca idahoensis*, which is found here in the Northwest. She looked at how they would compete under similar situations. So that's what got her started studying grasses.

I don't think she had a great passion for the topic that she studied for her dissertation. And in fact, I think, she got only one publication out of it. It was a very difficult, complicated question to ask, and the manipulations that she did were incredibly difficult, especially since she was working with an alpine plant, and a lot of them were in a greenhouse at sea level in Seattle.

We decided to move to Upstate New York in 1978 because my employer, Boyce Thompson Institute for Plant Research, was relocating to the Cornell Campus. Joy contacted Sam McNaughton at Syracuse University, who was doing quite an extensive research project in the Serengeti in Tanzania. Her background in looking at populations of grasses, and why they're distributed the way they are, fit with some of the work that Sam was doing in the Serengeti. She went up to Syracuse, talked to him, said she was moving to Upstate New York, and asked if she could write an NSF grant with his approval to fund some research that she was interested in doing.

It seems a bit off the wall, but, in fact, Joy had spent two years in Kenya in the Peace Corps.[5] A very high percentage of Peace Corps volunteers that go to some third world country usually end up doing something in that country later in their life. The Peace Corps had been a wonderful experience for her, and she felt this was an opportunity to get back to Kenya, while using her background in grass ecology. And at that time also, Sam's work was very cutting edge in terms of looking at ecosystem function of grasslands.

Joy wrote the NSF grant and returned to Seattle to defend her dissertation. On the day she was defending, she got the word from NSF that she'd been funded to do the research in the Serengeti. So that's where she got her start looking at grasslands and ecosystem function in grasslands.

She worked with McNaughton for, I believe, four or five years in the Serengeti. Then when the research and just the day to day logistics of getting food, petrol, and other supplies to work in the Serengeti became too rough,[6] we wrote a joint proposal to NSF to do work on the influence of individual trees on savannah grassland productivity. And we shifted the research in 1985 to Tsavo National Park in Kenya.

Most of the researchers, either in Serengeti or Tsavo, were identified by their research subjects. In Swahili, "grass" is called *majani*. And so Joy was called "Mama Majani"—the Grass Woman.

CHAPTER 3

Belsky's interest in livestock impacts on grasslands

We should not confuse Joy's passion for grasslands and grasses with her passion for science and honesty. If Joy was anything, she was very passionate about science. She identified herself as a scientist first and foremost. And her study of grasses was just part of her study of science. She was very passionate about plants. Every place we would go, Joy would be looking down at the flowers, at the plants. Not just grasses. The whole field of plant biology was her passion.

I think the reason that the passion came across in terms of grazing issues was the fact that there was so much dishonesty in the field of grazing management throughout the West. Everyone in the rangeland community would agree that cattle do cause problems, but most range scientists start from the premise that all we have to do is more research to find that perfect scheme for how we can put cattle out on the land to do the least amount of damage.

Joy saw the issue the other way around. That is, that the western grasslands had had very little natural herbivory, very little in the way of large grazing herds, as opposed to what she'd seen in Africa, where the grasses and the grazers were co-evolved.

188 / *Nongovernmental Conservationists*

Here in the Northwest there are environmental groups that were being managed by people, who were very passionate about getting cattle off these grasslands that were very obviously not able to support cattle. And what passion Joy had for that effort was motivated by the lack of honesty from the scientists, who were making careers out of trying to keep cattle on these lands that shouldn't have had cattle in the first place.

Her passion there was to look at the science that was available—and there's quite a bit of science—and to then write reviews showing where cattle are so damaging to these very dry, non-grazing-tolerant ecosystems.[7]

CHAPTER 4

Belsky's discrimination complaint and lawsuit against the BLM (1992–93)
Joy applied for a position at Oregon State that would have been with the Bureau of Land Management to manage a program to discover ways to mitigate for the spread of weeds throughout the dry West.[8] She was rated number one by the panel that screened the original applicants. And she was one of only three people who were interviewed by the Department of Range Science, and BLM.

When she didn't get the position, she started looking into the qualifications of the applicant who got it. She felt that she was still a superior candidate. She looked into the number of publications, the type of research that had been done, and she decided that she had a good enough case to sue the federal government.

But she lost the lawsuit. The Supreme Court had just ruled in a case with a black man in which the justices basically said that unless he had it in writing, or someone was willing to testify, that he was not hired because he was black, that it didn't meet the criteria for discrimination. And even though Joy's lawyer showed that the people hiring her had underplayed her qualifications, and overplayed the qualifications of the person who was hired, that wasn't enough for her to win the case.

So Joy wrote an article about her pursuit of information about this position, about the people who did the hiring, about the department that she would have been in.[9] And in the process, I think, she found that there was only one woman in all of the range science departments across the West, that the unit that was hiring her had no minorities and no women at the senior level, and that the Department of Range Science at Oregon State University had no women.

Then she started looking at the number of citations of her work versus the work of the applicant who got the position. And she found that her work was cited much more than the applicant that got it.[10] On top of that, she then looked at the citations of the work done by the people in the Department of Range Science at Oregon State University, and she found that she actually had more citations than the entire department.[11]

For her it was very cathartic in putting this together. She felt there was no better person for the position, because she knew she was highly qualified. She knew she was more qualified than the individual that got the position, and she had all the numbers to back her up on that.

So even though it might not have been sex discrimination, this is not a field that is amenable to women working in it. There are very few women.

When Joy went to work for ONRC,[12] she and Sally Cross and Diane Valantine wrote an article about sex discrimination in the environmental movement.[13] But that was, I believe, prompted by the fact that Joy had written this other article about her sex discrimination case against BLM.

CHAPTER 5

Belsky opposes killing coyotes at Hart Mountain NAR (mid-1990s)

There'd been, I believe, a series of droughts, or very dry years. And the Hart Mountain Refuge was so hammered by the cattle that the refuge manager and his chief biologist[14] had decided to remove the cattle, I believe, for a period of fifteen years[15] to let it rest and recover. And about three or four years into that management plan, after the removal of cattle, the managers were concerned about the series of years when the survivability of the fawns was low.

As it happened, Joy and I were down at Hart Mountain in the spring of one of those years when there was very low, if not zero, survivability. We couldn't camp at Hart Mountain in early June that year because there was six to eight inches of snow. It was clear that weather was a big factor in the survivability of the fawns that year. However, what took place at that point was that the manager[16] wanted to start shooting coyotes, because coyotes eat young fawns. So they were proposing to cull the coyote population to achieve higher survivability.

Joy, as an ecologist, was saying, "Well, wait a minute, you've got this wonderful ecosystem study going on in one of the few places in the West where cattle have been removed from the ecosystem. Don't go in and do another manipulation before you've let this system come into some sort of equilibrium. If not equilibrium, just watch this experiment go on, mainly because there's nothing in the scientific literature saying that shooting coyotes would increase the herd."

ONDA stopped the coyote shoot for three consecutive years,[17] either by lawsuit or through negotiations with the managers of the refuge. And over this time period the survivability started to come back. And, in fact, this year the number of pronghorn on the refuge are now at an all-time high.[18] The number of sage-grouse have gone up appreciably.[19] I believe, bighorn sheep are coming back in good numbers.

And now, the managers are saying that a lot of the reason for this improvement is because they've been able to do controlled burns and improve the habitat. If they had been allowed to do the coyote shoot, you can bet that this manager would have been saying they'd been able to manipulate the coyote population to improve the pronghorn numbers. And so, by preventing that from happening, we have a lot more information about dynamics of these animal populations on the refuge than we would have had with a bogus manipulation.

As to why the local managers were so adamant about doing coyote control, I think it was just the culture of the locals and of the Fish and Wildlife Department itself. They see themselves as "wildlife managers." And with the reduction in the numbers of pronghorn that they were counting in their annual aerial surveys, they felt they had to do something. And it is clear that coyotes eat pronghorn fawns. So, on the surface, it makes sense that you would then go out and manage the coyote population. But Joy found in the literature the idea that if you kill the alpha female of a coyote pack that, in fact, the sisters then start to breed and become breeding populations. So you can, in fact, kill the alpha female and end up with more coyotes rather than fewer.

It is a testament to Joy's scientific credibility that she was able to make a good enough case to stop that coyote shoot even though the refuge manager, Fish and Wildlife, the hunting community, and the local community wanted, in the worst way, to go in and do those shoots.

CHAPTER 6

Belsky co-authors article about livestock grazing's degradation of forest structure and dynamics (mid-1990s)[20]

Once Joy had one or two papers that looked at areas that had never been grazed because of the topography, versus grazed areas of similar elevation, slope, and aspect that showed very different demographics of the plant communities, she realized that she was onto something really big. And so she sought out papers where people had looked at either exclosures in the forest or looked at areas where cattle could not get into because of the topography.

And what was interesting is that there were several papers about these plant surveys that show how cattle really did change the composition of forests, and dramatically so. It's so logical you would think that the forest managers would have been very concerned about the change from these open pine forests, with low-intensity fires going through them, to dense thickets of firs that are more apt to be chewed upon by insects and diseases that make them more prone to these stand-replacing fires.

But I think that we've got a history of grazing in the forest now for so long that it's very similar to people looking at a lot of the streams in the high desert where they don't realize that this isn't the natural state, but it's the altered state. And that's all they've ever known.

CHAPTER 7

Cowboy politics in academia (late 1990s)
The private land ranchers seem to find that they are in exactly the same tent with the public land ranchers. To me it doesn't make economic sense for the private land rancher to be promoting production of cattle on public lands. One, it's costing them their tax dollars. And two, it's overproducing cattle.

The ranchers don't look at it that way. Joy was invited to speak at Texas Tech University in Lubbock, Texas,[21] by a faculty member there. These are all private lands in Texas, but when a benefactor of the Range Science Department at Texas Tech found out that Joy was coming to speak about the damaging impact of cattle on public lands, he insisted that they "uninvite" her to give her seminar, or he would pull his funding from the department. I've never heard of anyone being invited to give a speech at an academic institution, and then be uninvited. It is the antithesis of what one would consider to be the academic life. And it just happens that these cattleman will not hear anything bad about cattle and their impact on the land.

Jonathan Gelbard

After graduating from Cornell University in 1995 with a BS in natural resources, Jon Gelbard pursued several field research positions in the western United States. One of those positions, beginning in late 1996, was as a research assistant to Joy Belsky at the Oregon Natural Desert Association (ONDA) in Portland, Oregon. There Gelbard authored a petition to list Great Basin redband trout as threatened or endangered under the Endangered Species Act. And he co-authored, with Joy Belsky, a review article about livestock grazing's promotion of exotic plant invasions throughout the Intermountain West. As such, Gelbard became Belsky's last scientific collaborator, a position that provided him with a unique perspective on her methods as a scientist and writer.

Since completing his work at ONDA in August 1997, Gelbard went on to obtain a master's degree in environmental management from Duke University, and a PhD in ecology from the University of California–Davis. He is currently the executive director of Conservation Value, and consults in the areas of conservation biology, management and communication, and in the production of "green" events.

Jon Gelbard made his remarks on the 9th of August 2003 in Berkeley, California.

Chapters
1. Gelbard's undergraduate study and field research positions
2. Gelbard's tenure as a research assistant to Belsky (late 1996–August 1997)
3. Belsky as mentor
4. Belsky's interest in grasslands
5. Resistance to publishing an article about livestock spreading weeds

CHAPTER 1

Gelbard's undergraduate study and field research positions
I did my undergraduate work at Cornell. I actually started off as a communications major there. My minor was environmental studies. What happened was that when I started taking my environmental studies classes, I really liked them a lot. My work almost became play, which was something I hadn't experienced before. With my communications work, I'd be sitting there reading

my homework and it was like, "Oh, how much longer is it gonna be before I'm done with this? How many more pages do I have?" With my work in environmental studies, it was like, "Wow, this is really great stuff. This is fascinating." And I think that relates back to my interest in the environment. It's been with me ever since I was a little kid.

So about halfway through Cornell in 1993, I took a semester abroad at the University of Wollongong in Australia. And while I was traveling around there, I got this moment of inspiration when I decided I was gonna switch my majors. I said, "Well, I know how to write. Now, I want to learn a lot more about what I like to write about."

And so I came back to Cornell, and I changed my major to natural resources. I graduated a couple of years later with a bachelor of science degree in natural resources, and minors in communications, creative writing, and American Indian studies. The American Indian studies approach is something that I did because I wanted to expand my understanding of how different cultures, and even how people, come at seeing issues, or even just the world around us, from different standpoints—really whole different ways of valuing the world. Whether it's a difference between a Mohawk and a white person, or whether it is a difference even between a rancher and an environmentalist, there's different ways of seeing and valuing the world.

I think that's something that always struck me as being valuable information, especially because to solve environmental problems you really have to get past those biases, past those blockades of "No, I see it this way, and you should too."

I think it is much more fruitful to look at things like, "This is what we're looking at." Especially, as a scientist, "This is what the data indicates," rather than saying, "This data indicates *this* biological importance."

To a rancher you might have an economic importance—a quality of life issue. For different people the need to solve environmental problems really will be for different reasons. So the American Indian studies program kind of exercised my little brain muscles that get that way of thinking across. I find that very valuable.

After Cornell I had a job on the Gray Ranch.[1] Actually, I got hired first by the Animas Foundation to do some mapping on the ranch itself. It's a 320,000-acre ranch down by the Mexican border.

Then I got transferred to a Nature Conservancy/New Mexico Natural Heritage Program crew that was ground-truthing satellite imagery of not only the Gray Ranch, but also this whole surrounding Malpai Borderlands Group, which is a group of ranchers and environmentalists. The Nature Conservancy, state and federal agencies, and private foundations were getting together in one of the more ambitious private-lands conservation efforts that I've seen in terms of trying to get different people focused on improving the health of grasslands and deserts and oak woodlands and savannahs and pinyon-juniper woodlands.

Ground-truthing satellite imagery is basically what we were doing. There's all these different colors on the satellite imagery, and "ground truthing" means going out to areas of land that show up as all those different colors and finding out what kind of vegetation—what plant-community type—each color reflects.

For example, grasslands might show up as red; Chihuahua pine forests as a shade of dark green. Once you find out what kind of plant community each color on the satellite image indicates, then you can get another image taken ten years or so down the road, and take a look how, at this large, regional scale there's been an increase in grassland, and a decrease in woody invaders like mesquite and creosote bush. Both have really been expanding into grasslands over the last 100 to 120 years in large part because of grazing. They used to be beat back onto the sides of the hills by fire. But if a lot of the grass and the biomass is being consumed, there's really not enough fuel to carry a fire.

And being out there every single day surveying vegetation for months on end, I really become pretty intimate with the land, and got to know what a more heavily grazed, or more lightly grazed landscape looks like. We surveyed lands on some ungrazed ranches, which were really quite a whole different world in a lot of places. We also saw ranches where fires had burned through recently.

For example, Drum Hadley, down at his Guadalupe Canyon Ranch, had been running controlled burns, and the grasses were tall and vibrant. And he'd taken cows off of his ranch too. It was really a whole different world than some of these other ranches where the grazing practices varied—some were better; some were worse.

So anyway, being down there really gave me a good feel for a lot of the grazing issues and this new trend of different interest groups coming together: ranchers, environmentalists, and agency people trying to work towards common solutions. They're trying to figure out where they have common understanding to help solve environmental problems like desertification, which is essentially what that whole process is called of grasslands being replaced by mesquite and creosote.

From there I moved on up to Steamboat, Colorado, and tried to work with another group of people there on an air pollution issue—air pollution impacting biodiversity.[2] I was gonna do some work basically trying to determine what the biological impacts of that air pollution were. But the money never came through on that. So I just ended up learning a lot about it, because I wrote the proposals and did a whole lot of research.

Then, for a couple of months, I traveled around Colorado and visited some friends in the mountains there. I also looked at landscapes. Looked at the land management to see if indeed the logging practices were what people had recommended in our forest management classes; if the grazing practices were really in line with what the professors in the university courses had recommended as part of their coursework.

And sure enough, probably not too surprisingly, I found that we really have quite a ways to go in terms of getting the management out there on the land in line with what's actually taught in a lot of the courses. We're still seeing logging on steep slopes, and really bad overgrazing in quite a few places all around the West.

Then I moved on to Portland. Actually went down to Eugene to survey noxious weeds in the McKenzie Resource Area, which is Bureau of Land Management forestland.[3] They were paying me only to survey along the roads, but I would get out and take a look, just out of my own curiosity, to see how far in the weeds would get away from the roads. And for the most part, the weeds were along the logging roads, in the clearcuts, and in power-line corridors—any kind of opening in the forest cover. A lot of the gravel piles that they would use to fill in holes in the roads had St. John's wort, and all sorts of weeds coming out of them. It gave me a good background on what the weed problem was like on national forest lands.

And at the time, I was also very interested in landscape ecology. It kind of came naturally being out there in September in these BLM forests. I'd drive through a clearcut and I'd be brutally hot and I'd start sweating. And I'd be like, "God, if it's doing this to me, what is this fragmentation doing to the nutrient cycles like decomposition, and nutrient release?" Both are really very temperature controlled. "What is it doing to the vegetation? To plant species composition?" And so I became increasingly interested in biological effects of fragmentation.

CHAPTER 2

Gelbard's tenure as a research assistant to Belsky (late 1996–August 1997)
That was a short contract with BLM, so I was still looking for work. And I ended up being recommended to Joy Belsky.[4] I had a meeting with Joy one day. I guess it was the end of 1996. We talked about my background. And she was talking about what she was looking for out of her interns, and about writing a paper on the effects of livestock grazing on the spread of noxious weeds.

She also talked about the other work that she was doing—kind of being a watchdog for the public interest by looking at the science and the management that the federal and state agencies are doing based on the science. She was looking at it from a scientific standpoint and asking, "Does the management that they're proposing actually reflect what the science says? Is that a scientifically appropriate way to manage the land."

Joy said to me, "You're really not that interested."

At the time I wasn't that interested in grasslands. I was much more of a forest person. I had done my undergraduate work at Cornell. I had just done that work in these beautiful Pacific Northwest forests. I was very much intrigued by forests. I love the grasslands and the deserts, but my experience with them was for just a few months.

But Joy said, "By the time you're done working with me, you're gonna love grasslands and deserts. And you're just gonna get sucked in by this whole spread-of-invasive-noxious-weeds type of issue. And you're gonna end up doing your PhD on it."

And I said, "Yeah, right."

But, sure enough, she was right.

CHAPTER 3

Belsky as mentor
I came in to working with Joy having an idea of how to write. And I remember the first draft that I wrote of something for Joy. She looked at it and said, "This stinks." She had no bones about saying anything like that. She'd be like, "No, that's not what you're supposed to do. You need more examples. You need more points."

She'd be very specific about it—"This is what you need to do to make this better." And because of that, she made me a much, much better writer. She taught me to this day more than anybody else, who I've worked with, as far as writing, science, anything. To this day, Joy is the best mentor I've ever had.

And when I'm sitting here working on my publications, when I'm up at four in the morning pounding away on my PhD thesis, burning the late-night oil, listening to Bob Dylan low on the radio, I'll still hear Joy's voice over my shoulder, "Oh, a run-on sentence. Oh, you need more examples here." Those lessons still stick with me. Different things to look for.

Having since then worked with other mentors—and had my papers reviewed by other people—none have provided me with anywhere close to the feedback that Joy did—they really don't say too much.

I would get back a paper from Joy, and there'd be more of her handwriting then there was my typing. And it wasn't just scientific comments. It was grammatical stuff. I'd type in those corrections and it taught me the whole process of reviewing scientific papers. To this day I haven't found anybody who I've been that comfortable with—who I could just go back and forth and debate issues with, and never have it be awkward in any way.

Whatever she said. Whatever I said. We just took it back and forth. We were friends. We were colleagues. She was my mentor. I had a ton of respect for her.

CHAPTER 4

Belsky's interest in grasslands

Joy did a lot of work in Africa, in the Serengeti, a system that's actually very adapted to heavy grazing by large herbivores. But they're not restricted within fences as they are here. And she looked at the effect of herbivory on the grasses there.

She expressed to me that after twenty years in academia, she felt that it was time to really give something back to the public—that she didn't just want to be in the ivory tower. She wanted to take that knowledge, the connections within academia, out of the ivory tower, and really help make the land healthier—help cure the land of a lot of the environmental problems that are afflicting grasslands and other semi-arid systems: the grasslands, the shrublands, the woodlands, the forests all over the West.

She expressed a deep frustration at reading some of these management plans that were coming out of the agencies. She'd just be rubbing her forehead and say, "They're doing *what*? We've known for twenty years better ways to manage the land, and they're still doing *that*?"

And I think that's what really inspired Joy—a deep respect for science and the scientific process. But then she saw that in the management and policy arenas, the science is essentially ignored a lot of the time. And she wanted to do something about that. That's what she expressed to me—a real strong desire to fight for the science. And fight to make sure that the management was more in line with the science.

Some of the interesting conversations we had were why she thought some of the science wasn't getting in there. She talked about the nature of the scientist—the cautious scientist that wants to maintain objectivity. Just the very nature of our training will make people want to stay away from management controversies. If we become involved in a policy issue we've lost our objectivity. But she said the land doesn't care about objectivity. And if it's getting hammered out there because they're ignoring the science, somebody has to stand up.

Another thing she said, was that either the history and the English majors, who are working and lobbying for a lot of environmental groups, can be the ones who go to hearings and testify before Congress, or we can, the true experts, the ecologists, the biologists, who have the best training in the world about how these ecosystems work. We, as scientists, should do more to get involved in this decision-making process, just in the same way that you'd actually want medical PhDs or MDs testifying about what an appropriate medical treatment is. You wouldn't want activists being the ones who give all the testimony to Congress on behalf of a medication. You'd want the scientists, the real experts, in there. And that's what Joy was doing.

And we're already starting to see that this desire of hers was, I think, foreshadowing the trend of increasing scientific involvement in the policy and management processes. Just in the last couple of years the Society for Conservation Biology has opened an office in Washington, DC, to address this issue. The Ecological Society of America has been putting out white papers on a lot of important scientific issues: everything from biological invasions to disruption of the global nitrogen cycle. And they're also helping to communicate the point that what's environmentally important is also important for our quality of life, and the good of our economy.

Frankly, the scientists who don't have communication skills maybe shouldn't get involved. But for scientists who do—absolutely. For the sake of the land. For the health of the land. And really for the integrity of the science.

Can we in good conscience stand by and watch, as all around us the management practices that we see on a lot of lands today leave a lot to be desired?

And Joy got involved in that. She was trying to help by saying, "No, this is what the science actually says."

And that's really the way to think and look at things critically. I really learned a lot from that. It was quite illuminating to see just how big was this gap between what the science says about how we should manage lands, and what was actually being done on the landscape.

CHAPTER 5

Resistance to publishing an article about livestock spreading weeds[5]

Land management plans will pay attention to the need to avoid introducing nonnative seeds into an area. They'll say, for example, we need to wash off our cars before we go into uncontaminated areas.

Yeah, who's gonna do that?

"Hikers, wash off your shoes!"

Are most people going to take off their shoes and sit there with a little brush and make sure all the weed seeds are off their shoes? That's just not realistic.

With cows, livestock, you can quarantine them for ten, fifteen days. And those strategies are important for keeping weed seeds out of an area.

But there was nothing really being done in terms of reducing the vulnerability of the actual environments—the disturbances to native plants and to soils that increase their susceptibility to invasion. Especially in the Intermountain West, people weren't paying attention to biological soil crusts, for example.

In that particular paper, Joy and I had differences in terms of strength of language. She was really gung-ho about the weed problem and just getting the word out there. It really wasn't Joy's nature to tiptoe around anything. But the grazing issue is very polarizing, and I wanted to soften up the language so that the skeptics wouldn't be so turned off by hard language that they couldn't see the science that we were presenting.

To me, the point was to get the information across to people who weren't yet convinced, and say, "Hey look, the science here is so overwhelming that when devising livestock management practices, when determining how to manage grasslands, rangelands, shrublands, you really have to pay attention to the livestock issue." It's the predominant land use all throughout the West—hundreds of millions of acres of land in grazing. And really very little of that management out there is in line with the science. A lot of the management is very poor. But by incorporating the science into the management, there are a lot of things that we can do to make the situation better.

I wanted to present the science and address some of these polarizing questions: can ungrazed habitats fend off weed invasions? Oftentimes yes. Sometimes no. Then say under what conditions yes. Under what conditions no.

Can invaded habitats, in general, recover after you remove cows?

These are questions where the answer is really "it depends." In a lot of cases yes, if there's enough moisture; if there's a native seed source.

Sometimes no, if it's a really dry, south-facing slope. If there's no native seed source the weeds aren't going anywhere, at least not in the immediate time-scale. It might take twenty, thirty years. Fifty years. In the case of recovery of microbiotic soil crusts, longer.

Unfortunately, with the language that was in those papers, it was turning a lot of the reviewers off. They didn't want to put it into *Ecological Applications* or *Conservation Biology*. The same reviews said that the language is too strong; it's too opinionated. So that's what I'm working on.

I think it's very important to get that paper into the scientific literature. And now, at this point, I think we can get that in there preserving Joy's ideas and her spirit, while at the same time taking out some of that extra fire of Joy's language that was turning off some of the reviewers, and causing them to reject the manuscript. But overall, as far as the science that's in there, we were very careful to look for the best science around.

But in some cases there really wasn't good science available. Like, for example, some of the older papers: not good replication; not good statistics. They didn't necessarily stratify sites to ensure that ungrazed and grazed sites, for example, were the same soil type, topography, and so forth.

But on the other side, that type of information helps me and my colleagues today understand how to better approach these types of questions with good scientific and statistical methods. We need to ask how we can do a better job managing these lands—in this case, to avoid grazing practices that favor weed invasions. And that's what I've been doing on my master's and PhD work.

Bill Marlett

Bill Marlett, born and raised in Wisconsin, graduated from the University of Wisconsin–Madison, School of Agriculture with a major in soil science. Subsequently, he worked for the Wisconsin Department of Natural Resources in floodplain and shoreland management. A co-founding member of the Oregon Natural Desert Association (ONDA) in 1987, Marlett became its executive director in 1993. He speaks here from the perspective of having known Joy Belsky since the early 1990s, and from being her boss at ONDA from her arrival there in January 1997 until her death in December 2001.

Bill Marlett made his remarks on the 16th of August 2003 in Bend, Oregon.

Chapters
1. How Belsky became staff ecologist at ONDA
2. Choosing topics for Belsky's articles
3. The controversy over coyote control at Hart Mountain NAR

CHAPTER 1

How Belsky became staff ecologist at ONDA

She came to us, collectively, when she and Bob[1] moved to Portland. Then sometime after settling in, she wanted to get involved in the grazing issue. My first recollection of her was at one of our grazing roundtables up in Portland. It was around '92, I believe. And she was there, this new shining star, so to speak, with a lot of background and education and history and experience, able and willing to help us on the grazing front.

So she sought us out. I don't know who she originally talked with to get into that loop. My guess is it would have been whoever the ONRC director was. It might have been Larry Tuttle.[2] There were quite a few transitions at that point, but I do recall Larry being in the room along with Alice Elshoff.[3]

Shortly thereafter, Joy was trying to work out an arrangement with ONRC where she could be their staff ecologist. She was not interested in the money issues as much as having a position within which she could be a grazing advocate.

So, I recall, Alice and I sent Andy Kerr[4] a letter of recommendation on Joy's behalf, urging ONRC to hire Joy, so to speak. Though, of course, they did not have such a position. ONRC, historically, never really did grazing. To the extent that they did, it was work done by their Eastside field reps Don Tryon (way back), and then, subsequently, Tim Lillebo (a little bit), and Wendell Wood. But it was secondary to their primary forest work.

I don't know the specific arrangement between Joy and ONRC, but she was brought on as their staff ecologist.[5] And did actually work on some forest related issues. And published several papers on forest issues that were germane to ONRC's agenda. But at the same time, she was working on grazing issues. That lasted three years 'till late '96.

ONDA brought Joy on, officially, on January 1st '97. It was one of those situations where the financial commitment to her was that if we could pay, we would. But, if we couldn't, she still wanted to fulfill the responsibilities of the position that we had agreed on. That worked out well for both of us, though the question of obtaining money for a staff ecologist was problematic in that most foundations were not keen on nonprofit conservation groups being in the business of publishing papers. That was something foreign to most foundations that we were familiar with.

So fundraising around Joy's position was always problematic. But it became easier to the extent that we moved towards issues that were more in line with the activism and advocacy role that we are normally associated with—whether it's filing a petition or filing a lawsuit. Kind of the normal stock in trade of the conservation community.

To the extent that we could couch her position and her responsibilities in that light, then funding became easier. But definitely, the paper issue—the publication of papers—was an ongoing problem that we never were really able to overcome in terms of the funding community.

As I said in the beginning, we were blessed to have Joy come to us. Neither ONRC, nor ONDA or anyone in Oregon, but for the Nature Conservancy, has had the luxury of affording a staff ecologist. There are a lot of advocates in the community with strong ecological backgrounds, in terms of their training, but they're doing advocacy work.

Joy was in the education and the research communities, and was moving into a different realm that wasn't her background, but one where her background was sorely needed. There was definitely, and still remains, a vacuum in the conservation community with respect to our need for people of that caliber with that background.

CHAPTER 2

Choosing topics for Belsky's articles

Joy was well aware of our agenda and what was priority. And given that, it was more or less her ideas coming forward. But there was always dialog in terms of "Is this a good idea? Do we really need this? Who's the audience? What purpose will it serve?" So I would say that the ideas for the papers were largely hers.[6]

Holistic resource management—that paper still unpublished[7]—was more of a personal issue for Joy because of the pseudoscience that Allan Savory advocates. And, as a scientist, that certainly rubbed Joy the wrong way. The other papers are more of the straight-on scientific nature, whereas there's definitely a mission quality to the Savory paper. She was more or less targeting the Savory system to expose a fallacy that was being perpetrated onto the public at large, but more specifically upon the ranching community and the range managers. That one was very personal to her.

CHAPTER 3

The controversy over coyote control at Hart Mountain NAR

Subsequent to removal of livestock from the refuge[8] there remained questions about the pronghorn population on the refuge and the fact that it was in decline.

So the refuge managers decided, in their infinite wisdom, based on a lot of past management practices, to use coyote control to affect the coyote predation that was occurring on the pronghorn fawns in the early spring. The rationale, of course, was that with fewer coyotes there is less predation, and more fawns will make it through the summer and onto adulthood, thereby bumping up the population on the refuge.

Well, there's enough information out there in the scientific literature that suggested that that may not actually occur. Not that it would have necessarily unintended consequences, but it certainly wasn't the panacea that the refuge had portrayed. And Joy took that particular issue on, which really was not in her primary area of expertise. But again, it was one of those issues of using pseudoscience—"Well, you know, superficially it looks good. Let's shoot the coyotes." This harkens back to some of the things that Aldo Leopold had experience with in the Southwest where "fewer wolves meant more deer." It's the exact same mentality sixty years later.

So the refuge prepared a draft environmental assessment and through a number of legal maneuvers on our part and their part, it went through a number of iterations within the Fish and Wildlife Service.

Through Joy's work, we were able to raise the profile of the issue and their proposed actions to a point where persons within the Fish and Wildlife Service, going up to the regional director, were leery of getting into a situation that would appear unpopular and rile the great masses of folks for whatever reason.

Obviously, there's the animal-rights faction not wanting to see coyotes shot for the sake of shooting them. And then there's the more conservation-based faction, taking the notion that this is not the way you deal with population dynamics—applying a superficial analysis and tabletop study of what's really going on in the refuge, especially since the refuge was now cow free.

Having removed a huge variable, ecologically speaking, from the refuge, nobody had any data about what was really going on from the ground up. It was all just looking at pronghorns. And it was a very superficial assessment by the Fish and Wildlife Service, and specifically by Refuge Manager Mike Nunn, of what was going on.

He was schooled in the old tradition of game management. And he was applying the best available science that he was familiar with, irrespective of the fact that it was dated and certainly not in keeping with current literature.

So here we are in 2003 and the refuge has yet to produce their environmental assessment amending the management plan to allow for any sort of coyote control. I would say that so far it's been a huge success in that even though we haven't had to go to court to ultimately resolve this issue, through Joy's work we were able to raise the profile of what the refuge wanted to do to a level where internally, through peer review at the Fish and Wildlife Service, there were enough questions raised about this action that it's kept this at bay. That's not to say that it won't come out next week. It very well could. But my guess is that even if it does come out, it will be so tempered in its action that I'm not too concerned that any sort of coyote control on the refuge is going to be forthcoming any time soon.

And, if they say they're gonna do it, my guess is they will only if the pronghorn population heads toward extinction. I think it's gotten to the point where that's probably the only viable legal option that they could produce, while still saving face. And, frankly, saving face is probably one of their motivating factors.

Patrick Diehl

As a graduate of Harvard, Oxford, and the University of California at Berkeley, with a background in medieval literature, Patrick Diehl, would seem an unlikely person to emerge as a significant anti-grazing advocate in the desert of southern Utah. Yet he and his partner, Tori Woodard, drawing upon many years of experience as anti-nuclear activists, brought an unorthodox style to their environmental activism that generated significant media attention. It also brought them the wrath not only of their rural neighbors, but even of some environmentalists. The latter experiences, especially, reveal what may be a significant factor in stifling efforts to build a stronger grassroots movement to end public lands ranching.

Patrick Diehl made his remarks on the 16th of September 2004 in Escalante, Utah.

Chapters
1. Diehl's early years and formal education
2. Diehl's academic career
3. Diehl becomes an anti-nuclear activist
4. Diehl joins the Ward Valley Campaign
5. Diehl and Woodard settle in Escalante, Utah
6. Diehl and Woodard's property is vandalized (April and July 1999)
7. Diehl and Woodard are shunned by environmentalists
8. Diehl comments on the Southern Utah Wilderness Alliance
9. Tabling at the Escalante BLM Visitor Center (27–28 May 2000)
10. Drought prompts BLM to curtail cattle grazing at GSENM (fall 2000)
11. Relict aspen on the Rock Creek/Mudholes Allotment
12. Trespass cattle on the Steep Creek Allotment (February 2002)
13. BLM ignores public comment on their grazing management
14. Establishment and management of GSENM
15. The future of livestock grazing on federal public lands

CHAPTER 1

Diehl's early years and formal education

I was born in Texas and have a ranching background on my mother's side. Her parents actually operated a ranch well into my own lifetime west of San

Antonioa near the Mexican border between Uvalde and Eagle Pass. I moved with my parents to eastern Tennessee when I was just a little under three years old. We traveled back to Texas for Christmas vacations, so I definitely had a strong connection to my Texas relatives all through my childhood.

I went to a prep school in Tennessee and graduated as valedictorian in 1964. Then I went to Harvard. Graduated summa cum laude in three years in "classics and allied fields," which actually meant Latin and French. In '67–68 I received a scholarship to study in Paris that was funded partially by Harvard and partially by the French government. I was there for the excitement in April–May of '68. Part of it, at least. Actually I sailed out of there just before everything shut down in May.[1] I'd bought a ticket on an ocean liner two months earlier, so I had to leave.

By then I'd already gotten a Marshall Scholarship[2] to Oxford, where I studied for the next two years. Took a second BA in English. And I got a First Class in that.

I was admitted to the University of California at Berkeley graduate program in comparative literature in 1970 on a Ford Foundation Career Fellowship. I was one of the lucky fellows in the old days when you could still get solid funding for a graduate education. Very, very difficult since then at least in the humanities.

I was hired in classics and in comparative literature at UCLA in '74. Then completed the dissertation and received my doctorate in 1975 in comparative literature. Latin was my main language from its origins up to the Renaissance. My other languages were Middle English and Byzantine Greek.

CHAPTER 2

Diehl's academic career

I taught at UCLA for a year, and then I got hired back at Berkeley in comparative literature and English. Taught there for seven years. Didn't get tenure.

I published a translation of Dante's lyric poetry[3] except for the *Vita Nuova* poems. And I also wrote a book, which appeared after I left academia, based on my doctoral dissertation about medieval religious lyric poetry.[4] After leaving academia I was hired to do a translation of a modern Italian novel by a guy named Siciliano.[5]

CHAPTER 3

Diehl becomes an anti-nuclear activist

As a graduate student I'd become aware of nuclear weapons and the threat of nuclear war. Peter Watkins' *The War Game*,[6] which is an amazing film, had a powerful impact on me. It's a pseudo-documentary filmed as if it were a documentary during and after a nuclear war. People who've seen it will certainly never forget it.

Anyway, *The War Game* really brought home to me what a nuclear war would be like. And other things that I read and saw during the early '70s, not to mention the Vietnam War, convinced me that I was against war in general. So I became a pacifist in my graduate school years.

While I was teaching at the University of California at Berkeley the contract with the weapons labs came up for renewal. The University of California was then the nominal overseer of the laboratories that develop nuclear weapons for this country (Livermore, Los Alamos, and also Sandia). I had not previously realized that there was this connection between the place where I was teaching and the nuclear weapons system because I'd been very wrapped up in academic pursuits.

Once I found out that that was the case, I immediately joined up with the people who were trying to convince the faculty to vote for terminating this relationship. I think I was the only non-tenured professor to sign the ballot argument. That was in 1979 or so.

Then when I did not get tenure and my teaching contract ran out in the summer of '82, I joined a very rapidly growing, powerful civil disobedience group, the Livermore Action Group, which was focusing on exactly the same issue that I'd been working on as a faculty member. So I basically stepped right out of the classroom into full-time organizing with the Livermore Action Group.

We had one of the largest civil disobedience actions in US history in June of '82, right after the big march on June 14th. A million people, roughly, turned out in New York City. We then had something like ten thousand people out at the lab and approximately fifteen hundred arrests. A very spectacular mass action.

Then for the next three years, I was totally involved with the anti-nuclear movement. I went to Europe in '83 as one of the two organizers for the Livermore Action Group working with Mobilization for Survival. And we organized coordinated actions all over the world. Actually, in the Pacific (Japan, New Zealand), as well as in Europe and, of course, North America on the summer solstice of '83, there were over a hundred actions.

But the anti-nuclear effort in the US declined partly because the movement in Europe fell apart after the Cruise and Pershing missiles were forced in by the European governments over the objections of their people. That was very demoralizing to the Europeans and to the worldwide anti-nuclear movement.

Another factor in the movement's decline was that Reagan[7] began to back off from his more frightening pronouncements about Armageddon and how the world might be ending soon.

Also, 1984 was a presidential election year. And, as is usual, election activism sucks energy away from citizen movements, and generally neutralizes most of it for a good long period thereafter. It's a pretty predictable cycle.

Anyway, all those factors combined to cause the decline of the group I was involved with. So in '85, around August, we disbanded and transformed into a much smaller project. Eventually, even that just bled away over the years.

I'd separated from my wife a few months before that. Had a four-year-old child at that point. So we co-parented him. I moved to West Berkeley into a collective household that was run by consensus process, and which provided very low rent and high quality of life for its members. We owned the house also. Several of us put in money and bought it together. I lived there for thirteen years.

I became involved in war-tax resistance[8] in '89 for tax year '88, because I was just really disgusted with the Central American policy of the US government, and also with the subsidizing of Israel to stomp all over the Palestinians. That was the time of the First Intifada,[9] which was a very impressive nonviolent movement. I was very excited about that. Of course, I'd been involved in nonviolent disobedience for a long time, so this was inspiring. This was quite different from the current Intifada,[10] which rapidly turned into violence and then, because of Israeli provocation, I think, was pushed over the edge into suicide bombing. My view anyway.

So the money I didn't give the government, I put into a kind of "citizen's trust fund," the income of which was used to finance good causes, like groups working with the homeless or literacy projects. Finally, I started just taking the money and giving it away. One year I gave $3,000 to the *Nuclear Resister*,[11] which was about to go under. It's a vital publication for the anti-nuclear movement in this country. And they used the money to fuel a fundraising drive to get more money.

That's the main, big thing that came out of my war-tax resistance that I feel very proud of at this point. And I gave lots of money to other groups. I continued to give to the *Nuclear Resister* for years as long as I had an income.

I had formed a relationship in '85 with my present partner, Tori, whom I met in the Livermore Action Group. In the meantime, she had developed multiple chemical sensitivity,[12] so I'd been commuting between the Bay Area and places with better air where she lived, such as down in the Mojave Desert or up north of the Bay in Sonoma County. I'd spend half my time in my collective household. Then my son would go join his mother, and I'd go join Tori. It was stressful, including on my house. Not ideal, but that's what I did for several years.

CHAPTER 4

Diehl joins the Ward Valley Campaign

Once my son came of age and moved out, I was free to go live full time with Tori someplace. We decided to move to Utah. But during the years before we did that, the Ward Valley issue became a hot one. There'd been a proposal for some time to build a low-level nuclear waste dump in Ward Valley, which is just west of Needles, California, close to the Colorado River.

There'd been considerable energy put into the campaign a few years earlier. The whole town had been against it—the Native Americans who lived there (the Mojave Tribe), as well as Hispanics and Anglos. Everybody was down on the dump.

Well, then the disposal company, U.S. Ecology, used its economic clout to buy people off in the city government. They made a big donation to the school. The usual stuff. Needles is a very poor town. It's actually a dumping ground for welfare recipients that people push out of the Greater LA Area.[13]

So I gave a few hundred dollars to a couple to go back down there and try to pick up the pieces and get things going again. That indeed happened because the Indians were still really against the dump. They had a really good tribal government, which is almost unique. There was no split between the traditionalists and the tribal government in that tribe.

That fall a bunch of us organized a gathering in Ward Valley. And hundreds of people turned up. A regional coalition started forming including Greenpeace, and groups in Nevada and Arizona. All the tribes up and down the lower Colorado River were soon involved in it. There was a big contingent from northern California, of course. And there were groups from LA.

It built and built once we started holding these gatherings and doing outreach. There were some very effective organizers on the payroll of Greenpeace who were very important at that point.

In any case, the Ward Valley Campaign became quite powerful. We were getting resolutions passed against the dump by the Los Angeles City Council because they were afraid it might affect their water from the Colorado River. And finally, starting in February of '98 the tribe and other activists began a large occupation of the site.

Actually, nonnative activists had started a peace camp there during the first gathering in the fall of '95. A peace camp is kind of movement jargon for an occupation camp which is nonviolent, but you're there to resist what you consider to be wrongful use of the land you're sitting on. It was on BLM land. So the BLM eventually gave us a permit to stay there permanently, because it was politically infeasible for them to kick us off—there was too much local support for what we were doing.

That peace camp went on for years. And I spent about half of each month the first year living there and working in the office. My partner, Tori, was the staff member there for quite a while.

Then in the winter of '98, as I said, the occupation became primarily a Native American one, and there were over a hundred people up there instead of four or five as it had been during the previous years. There was huge media. *New York Times, LA Times* were all over it. NPR was out there. It just really grabbed people's attention.

Eventually, the state of California decided to withdraw its support for the dump. Gray Davis was elected the state's governor. And, although I haven't got much good to say about Gray Davis, he responded to political pressure on this issue and basically shot the project down.

One result of my being involved with Ward Valley and living in the desert so much was that I became more interested in general environmental issues, particularly for the desert. There were cows grazing in Ward Valley, unbelievably to me. This is really severe, creosote bush, bursage desert.[14] Water is extremely scarce there. Yet there were cows. God only knows what they were eating out in the creosote and the bursage and the yuccas near the peace camp.

And, while I was there, I sometimes had some time to read in between running down to the office and trying to earn a living by summarizing depositions on the computer there. One of the books in the peace camp library, covered with dust in a plastic crate, was Lynn Jacobs' *Waste of the West*. So I picked that up. Started reading it. It looked like a really solid, serious book.

My grandparents on my mother's side had been in ranching in Texas. And I already knew what ranching had done to the ecosystem there: converted it basically from grasslands to thorny scrub. Massive change. So I wasn't totally ignorant, but *Waste of the West* really filled in all the details for me. It's a remarkable book. I read that in '95-96.

We went to Australia in the winter of '96, and I saw more grazing damage there. Half the country has been devastated by cows and sheep. It's the worst continent on the planet for mammalian species extinctions. Very, very badly hit.

We returned to the US. And, as I said earlier, my son came of age during this period. So Tori and I decided to move to Utah, because I loved the landscape here and Tori likes it too. She would have been all right in the Mojave also. But we wanted to find a town of some size. So we ended up in Escalante.

CHAPTER 5

Diehl and Woodard settle in Escalante, Utah

We moved here around Labor Day of '98, and shortly thereafter set up our own nonprofit, Escalante House, for the purpose of creating housing for people with multiple chemical sensitivity like Tori. And we also started working on the grazing issue. We set up an environmental group, the Escalante Wilderness Project, under the umbrella of Escalante House.

We began to express our views in the local county paper in February or March of '99. And immediately doors that had been open to us closed. I'd been very involved in music and theater that was sponsored by the Mormon Church here in town. I sang in the church choir. In fact, I was the male soloist in December of '98 for the Christmas cantata. I'm a trained singer, and I've also done a fair amount of acting.

Then a melodrama was thrown together for the Valentine's Day festivities at the church. And I was the male lead in that.

They had me read a Civil War era letter from an officer who was killed in the war to his wife for the Relief Society. It's an annual thing they have in early March.

So around then we started writing to the county paper about some things, and the people started realizing that we were environmentalists.

CHAPTER 6

Diehl and Woodard's property is vandalized (April and July 1999)
In late fall of '98 some people who were not born here in Escalante, business people basically, came to us about a reservoir project. One of the two irrigation companies here wanted to build a new reservoir to irrigate several alfalfa fields that provide cattle feed.[15] The alfalfa hay would also be sold to the horse operations down in southern Nevada. There's a valley there where a lot of race horses spend the winter. And then, of course, there are the mule operations in the Grand Canyon. Some of the hay goes down there.

So they wanted to build a new reservoir, which was going to drown a whole six to seven hundred acres including prehistoric Indian sites.

And we also made the connection: Oh, the cows around here that are destroying the landscape—what do they live on when there isn't anything to eat? (Which is true a lot of the time.)

Oh, alfalfa.

Where does that come from?

Oh, the fields.

And where does the alfalfa get its water from?

Oh, that reservoir which already exists (a small one), which is silting up, because it's an instream reservoir in a valley coming off the Kaiparowits through some extremely erodible formations of rocks. They're actually mudstone and dirt. You've got twenty-five-foot-deep gullies up there behind that reservoir. It's really quite a sight.

So these other people came and told us about this reservoir project. And we said, "Well, let's have a look at it. It doesn't sound too good." So we read the EA on it and decided this was really bad.

We got in touch with the Wilderness Society and with SUWA and other groups. They didn't like the reservoir proposal either. So we basically served as the "trip wire" for the organizations with money and lawyers.

Then, in April, I got a call from somebody at the Wilderness Society in Denver saying that a stringer writing a story for the *Salt Lake Trib* wanted to talk to a local person about why they opposed the reservoir. They asked whether we would be willing to speak to him. I thought about it for about ninety seconds because I knew this was gonna be it. It's definitely the spark in the powder magazine. And I said, "I'll talk with him."

In the article he wrote on the issue, what basically appeared, as far as my contribution, was just my concern that a new reservoir might fuel explosive development in Escalante, which at that time seemed a realistic prospect.

The monument[16] had been set up in '96 over intense local opposition including hanging Clinton and Babbitt[17] in effigy for a week outside the local grocery store. And there were big plans for the monument to be a kind of showcase for the "new BLM."

There was a lot of excitement about a big project called the "Escalante Center," which was going to incorporate a scientific center with a kind of museum—a curatorial-level lab. So there'd be a lab here to deal with specimens brought back from field trips in the monument. University students could come down here and take courses working with professors from Brigham Young University, which actually has a pretty decent environmental program. Also, Southern Utah University, in Cedar City, would have been involved.

There was this whole idea of using the monument as a basis for research. And having the facilities for the research at least partly right here in Escalante.

On top of that, there was also an arts and humanities group which had formed, and which I immediately joined, which was lobbying for an arts and humanities center. The proposal was for a "Little-Theater" space, art classrooms, and a video studio. Our understanding was that we were supposed to get something like ten million to twelve million bucks from Congress through Senator Bennett[18] for the project.

And the BLM was gonna have its new visitor center as part of the Escalante Center. The visitor center is presently being constructed in 2004. Finally.

So there was gonna be a sort of campus out near the high school. Big money, big future for the town because, of course, there'd be a lot more people coming here. There'd be more of an economic base for services. There'd be a better base for tourism with the visitor center and all this other stuff going on.

But the Escalante Center never happened. And the person who tried to make it happen was totally defeated by the local political establishment, basically, because if this project had happened it would have undermined, if not destroyed, their political control over this eight- or nine-hundred person town.

It was "too environmentalist" for them. Scientists are a threat in themselves, because, of course, they deal usually somewhat objectively with the physical reality of the earth. And the ranchers can't afford that. So scientists that aren't being paid by the livestock industry are definitely a threat to them politically.

And I don't know whether this was genuine or not, but there was a huge outcry from parents of daughters in the town against the idea of having male university students living here, because they would seduce their daughters and we would end up with lots of pregnancies.

So every possible argument was used to defeat the project. Of course, with the loud redneck part of the town opposed to the project, you can imagine what happened to the prospects of a congressional bill. Bennett developed cold feet. The amount of money that was being discussed dropped to six million dollars. And then four. And then the whole thing went away.

The BLM visitor center, instead of being built in 2000, which is when it was scheduled for, is being built four years later. And the town has stagnated pretty much economically as a result of the rancher-dominated political establishment. It's been very frustrating to the business people here who are quite bitter about how the people who run the town have behaved.

So that's a long digression, but I was just explaining why it seemed possible that Escalante might suddenly burgeon as of 1999. With all this in prospect, it seemed reasonable to think that lots of people might want to move here. And, if they did, they might be able to get a doctor here, and better support services. And it could become a really thriving retirement community.

We were also working on our multiple-chemical-sensitivity housing project, which we'd envisioned as involving something like a hundred people. We'd found a site down south of town, and we did an architectural charette as they're called. We hired an architect from Berkeley who's into the safe building movement—"green buildings." Developed a really great site plan. Had a web site. We were getting questionnaires filled out by people from all over the country.

So from our personal point of view, we were involved in a promising, really large project for the town.

Anyway, with all of this either happening or being considered, it seemed like building a great big reservoir with lots of water in it could create a basis for condos and subdivisions, and out-of-control growth that has happened so often in western small towns that become destinations.

One thing I forgot to mention was that the cash center of the Escalante Center was supposed to be a destination resort with rooms like kivas in it, and this whole kind of fake Southwestern thing. And a golf course was being proposed. This would be a huge thing for a town of eight or nine hundred people. So actually as we found out more about the Escalante Center project, Tori and I became less and less supportive of it.

The Escalante Center, and our own housing project, encouraged us to think that maybe a lot of people would be moving here. So we wanted to be sure that the growth was planned and controlled. And not just that the pigs line up at the trough the way it usually is in this country. And the town goes to hell.

So I came out publicly against the reservoir in the *Salt Lake Trib* article, even though what they quoted me as saying was basically that I wanted to protect the quality of life in this town, and keep it from turning into a long strip overrun by cappuccino swillers.

Tori has a share of water in the irrigation company that was proposing to build the reservoir. We had irrigation pipes with water in them on our property. And somebody came in, probably about three in the morning, in April after the article appeared, opened up the run of pipe from the riser, moved the pipes just enough so that the water would run into an excavation we'd had dug for a building we were hoping to construct, and turned on the riser full blast. Consequently, the excavation was totally filled up with water—approximately forty thousand gallons. Water was still running out at eight in the morning when the water master happened to come by, or knew to come by—we don't know—and informed us that there was an open pipe. I'd say, two hundred gallons a minute coming out of it. So of course, we immediately closed it.

And soon thereafter the irrigation company sued Tori for having an open pipe. She was basically being sued for being vandalized. I just went into orbit when that happened.

We fought it in court for the next year-and-a-half. First the irrigation company went to small claims court improperly. They were trying to collect a fine, and small claims courts do not consider fines by a private organization against its members. But the local judge found for the irrigation company. So we had to appeal to the district court, where the case should have been filed in the first place. And the district court judge told the irrigation company that it had to come before him if it wanted to try to collect the fine.

By that time the irrigation company had spent over eight thousand dollars trying to collect a fine of eleven hundred dollars. And we'd spent a little over two thousand. It was well worth it to us because they discovered that they couldn't crunch us. They couldn't beat us. They couldn't drive us out of town, which is what they'd done over the decades to people who'd come here and tried to stand up to them.

Anyway, the situation became one of profound hostility between the town and us. The following July our house was attacked and the front door was kicked in. Windows were smashed. They cut the phone lines. Tore up part of our garden. That was on Pioneer Day—the big Mormon celebration here in Utah.[19]

The vandalism continued off and on up 'til last summer. Smashed windows on our cars. One slashed tire. Threats. I had a death threat at a meeting in spring of '99 during all this hullabaloo.

It's kind of tapered off. They kept doing it, but it really wasn't having any effect. I think they began to realize that they weren't gonna drive us out. It was just costing us some money, which we absorbed. Well, $4,500 over the years. Not insignificant, but not enough to make us leave.

CHAPTER 7

Diehl and Woodard are shunned by environmentalists
We had noticed already that the environmentalists in town with whom we'd gone hiking in the winter and early spring of '99 had all made themselves very scarce after our property was vandalized.

I remember talking to one of them and saying, "Well, you could come out and support us and our free speech rights. You're an environmentalist too." And the guy basically said, oh, no, we're not gonna do that.

So essentially we were shunned by all the people in town who have environmental views.

This was a real learning experience for me about the human race. It definitely darkened my view of human beings considerably—that people who agree with you, in theory, on an issue, but who are not ready to stand up to social pressure, will run from you. They will attack you behind your back, because, of course, they're justifying their own cowardice, basically.

The right thing for them to do, of course, would be to stand up beside you and say, "It's wrong for Patrick and Tori to be vandalized because they're environmentalists. I'm an environmentalist too. I think this is a violation of their civil rights and mine. And it should stop."

But instead, these people not only ran away from us, they also, in order to justify their own moral failings, had to explain to themselves and the world what awful people we were. Consequently, what happened was, all over Utah in the environmental network, we were being badmouthed. And we could tell—real fast.

So what were we gonna do? Well, in 2000 we decided to join an insurgent Sierra Club group, the Glen Canyon Group, because they were taking on the environmental establishment in Utah.

CHAPTER 8

Diehl comments on the Southern Utah Wilderness Alliance
We got involved with the Sierra Club and its Glen Canyon Group because of the Glen Canyon Dam issue. Glen Canyon—the idea of bringing it back was quite thrilling to us. And we also got involved because there really wasn't anything else down here to get involved with that we weren't generating ourselves. This was a year-and-a-half after we'd moved here, and we were beginning to realize that we were not going to get the support of the environmental organizations, specifically SUWA, in Utah.

We got the impression that they did not like the fact that we were openly confronting the ranchers in the press. Anybody, of course, who supports a big wilderness bill in the American West cannot openly support people who attack public lands grazing. The wilderness bill guarantees the continuation of grazing. And if the organization, which is seen as the prime sponsor of a statewide wilderness bill like the Red Rock Wilderness Act (now 9.1 million acres here in Utah), comes out openly against public lands ranching in general, then they are basically telling everybody that they want more than just wilderness. They want cow-free wilderness, which would actually be real wilderness. And the Wilderness Act, of course, does not give you that.

So, politically, an organization that makes wilderness designation the centerpiece of its activities becomes, in practice, an organization that must oppose open opposition to public lands grazing.

Now SUWA, of course, engages in lots of harassment of the ranchers through its legal staff and its field representatives who comment on grazing decisions and sue over them. That's good. But the policy of the organization does not include opposition, in general, to public lands grazing.

Here we were down in Escalante, and ranchers, or ranching sympathizers, were vandalizing us. The people at SUWA, I believe, saw this as creating heightened antagonisms in southern Utah, which would make it more difficult politically for them to get their bill through—that the ranchers would perceive that the bill was a threat because other environmentalists were causing them trouble. They don't make distinctions in rural America between one environmental organization and another. They really lump everyone together.

SUWA, I think, saw us as rocking the boat. Making waves and causing trouble. So they rapidly became quite hostile to us. We could feel the temperature drop down to Arctic levels within a few months of when we were vandalized.

CHAPTER 9

Tabling at the Escalante BLM Visitor Center (27–28 May 2000)

When we decided to table about the public lands grazing issue at the BLM, they tried to stick us off in a "free speech area," which has become the way that institutions in this country try to handle citizen groups doing things they don't like. They put you off where nobody will see you. In this case they wanted to put us off by a shed at the far end of a very large parking lot. Of course, all the visitors parked near the door of the building and went straight into it. Meanwhile we'd be about a 120 feet up the hill. So nobody would ever come near us.

Anyway, we had already set up our table right next to the front steps in a graveled area. No plants. Not damaging anything. Anybody walking into that building would see us. And a lot of people were stopping and talking to us.

Within a couple hours, BLM employees were out there telling us to move. And we said no. And they said, "If you don't move were going to arrest you." And we said, "Okay, arrest us." So finally, that afternoon they did arrest me. The BLM law enforcement hauled me off 150 miles to the nearest state facility that had a contract with the feds to accept federal prisoners. And then another member of our group, Juniper Allison, was arrested the next morning and transported the same 150 miles.

Meanwhile I'd started a hunger strike. Tori and Daniel Patterson, who was in town at the time and is on the staff at CBD, and is on the board of our nonprofit, were e-mailing press releases and talking to the press. God, it was incredible. I was in jail, so I didn't directly experience it, but I did see a TV broadcast in jail about our action down here. That really brought home to me how inexperienced this state is with any kind of nonviolent direct action.

This was just a little free speech action where you try to table, and they try to move you. And you refuse, and they arrest you. Happens all the time in California. But out here it was enormous. It was all over the papers in Salt Lake City. And the radio stations were going on about it and interviewing us. They interviewed me after I finally got out of jail.

But I started a hunger strike. And this was very exciting. Apparently no one had ever done a hunger strike in Utah. Or, at least, not lately. They'd forgotten about them.

Anyway, we basically terrified them. They let me out of jail after about thirty-six hours, and they said, "Please come meet with us up in Salt Lake City. We really want to meet with you."

We went to Salt Lake City the following Friday. There, in a big conference room, was the BLM state director along with their top lawyer and all the top officials under the state director. About twenty people. And we, by that time, had a lawyer from the ACLU. The ACLU in Utah is quite good. They're basically defending people's rights against the very oppressive theocratic situation here. So they're pretty feisty.

The BLM was getting creamed in the press. A lot of attention was being drawn to the grazing issue, because that's what we'd gotten arrested over from the viewpoint of the public.

Of course, the BLM was trying to frame the matter as, "They're supposed to be in the area where we want them, which is better for the public." But I don't think people bought that.

We also did a press conference next to the federal building in Salt Lake that was very well attended. The Salt Lake City TV stations—I think all of them came. And radio. And newspapers were shoving their microphones at us. So again, we were able to talk about the grazing message.

That was a very successful action. And, to me, was a great demonstration of how a fairly low-cost, nonviolent direct action can be very powerful. It certainly didn't cost us much money. We did have to drive to Salt Lake City, but that was great because we got incredible media for the action and for our issue.

I spent a day-and-a-half in jail. Big deal. I've done probably forty civil disobedience actions since the early '80s and this one was relatively less painful than some of them. I've sat in jail for a couple weeks on at least one occasion. I've been in federal facilities with bank robbers and murderers and all those scary people. Although the really scary people were the federal marshals—they've got the guns and keys to your shackles. Really nasty characters.

Anyway, that action was probably the high point of our activism here in Escalante. But we didn't know it at the time.

Tourism in the region declined. I tried tabling the following year in 2001. But there were very few people coming to the visitor center. I'd sit there for hours in the blazing sun and nobody'd show up. The few people who did go in would generally not come over and talk to me, even though I was sitting right next to the steps. I did that around Memorial Day—on the anniversary of our action in 2000. Then I tried Labor Day. Finally, I said, this is ridiculous. I'm just sitting out here melting into a puddle and accomplishing nothing.

So that was really sad. After we'd won the right to table on that little piece of gravel next to the steps, it didn't really turn out to be very useful. And that's kind of how it's been down here ever since.

CHAPTER 10

Drought prompts BLM to curtail cattle grazing at GSENM (fall 2000)

Already by midsummer of 2000, BLM had started telling their permittees around Utah to get their cows off and into feedlots or somewhere, because the land was in such terrible shape from the drought. They were talking about 90 percent utilization, which to you or me would mean 100 percent plus. Cattle were literally pulling the grass out by the roots at this point.

The Griffins[20] (one of the politically and economically dominant local families) are the permittees on one of the allotments way down at the end of the Kaiparowits Plateau, which runs for, must be, sixty miles all the way down to above Glen Canyon. The end of the plateau is protected by cliffs. It's definitely a "sky island" up over seven thousand feet above sea level. There are four steep trails on the east side up the escarpment. I've heard stories about horses falling and people getting hurt. And about a horse breaking its leg. And cattle falling off the trail and dying. The escarpment's around eight hundred to one thousand feet. It's pretty impressive and those trails are pretty nasty. Quite steep in places. Bare rock that an animal can slip on. And also lots of loose, broken rock and dirt.

The other permittee is named Mary Bulloch,[21] who lives in a trailer down in Kanab, and as far as I know is without any actual base property. She, to get her cows in and out, has to go around to Big Water on the west side, drive in forty miles and go up a canyon about ten miles before she can finally get to the top of the plateau. This is not easy.

The Griffins come up from the east side. So the drive to the foot of the trail for them is relatively easy. You can be there even with a cattle truck in probably three hours from Escalante. You just go up the escarpment and you're on top of the plateau.

Both the Griffins and Mary Bulloch refused to remove their cows, I think, partly because it would have been so difficult for them to round them up. They didn't have an ongoing system for moving the cows off the plateau down to the lower parts of their allotments in the winter when the plateau gets snow and cold. They just kind of let them run wild up there. They weren't really set up to get them off, which is wrong. They were breaking the law and the BLM had been looking the other way for probably decades, I'd guess. I think the situation goes back way before Mary Bulloch appeared. And the Griffins have been there for about a century grazing sheep, and then cattle after World War Two. So there's a long-standing custom and practice here, which is not only a violation of the BLM regulations, but also ecologically very damaging because you have year-round grazing.

So BLM got a refusal from both of them to get their cows off. And there were letters back and forth. And trespass notices were issued, and it started getting into the press.

And, of course, here we were sitting in Escalante. The Griffins live in our town. So we jump in as Escalante Wilderness Project and condemn their behavior, because Tori and I had been up on the Griffin's allotment in June of that year. And it's like, "My God, this place is fried to a crisp already. And look at the cheatgrass and wild lupines all over the place. And look at all the pedestalling of the plants."

It's intrinsically a gorgeous place. You're three thousand feet above the Escalante desert at that point, which drops off as you approach the river. But the plateau doesn't drop off at all. Some of its highest places are right down at the end. So you're looking down across the whole Escalante drainage below you. At sunset, when we were up there at the solstice, we could look from where we were camping at the top of Willow Tank Slide Road and see Stevens Arch, opposite the mouth of Coyote Gulch, down in the Escalante Canyon. We were at least ten miles away. But it was still very impressive.

Incredible views from up there. And no roads. North of this section of the Kaiparowits are a series of deep canyons that have eroded all the way back to the Straight Cliffs rim on the west. So to get a road in you'd have to drop down into a series of canyons and climb out again, over and over. There's no town there—there's only been wild country and then sheep and cattle. So it's not only an unroaded place; it's a place that feels like it will never have roads.

To me it was very exciting because there are really very few places left in the West that you can day-hike into that don't have roads in them.

So it could have been a wonderful natural area, but instead what we found when we went up there in June of 2000 was this cow-beaten, thoroughly trashed place. Arroyo bottoms had no riparian vegetation. Only a trickle of water. Cow trails going up and down. The springs were beaten into a mire.

So we didn't know what was gonna happen. But we had definitely noticed that this place had been really abused. It must have been a wonderful place a hundred years ago and now—kind of the "bones" are still there. You're still three thousand feet in the air, and there are canyons in the top of the plateau. And all these wonderful views. And there are places where there is some vegetation like oak brush. And it feels really wild. But if you look at the detail, you can see the terrible impact of public lands grazing on these allotments.

So then it happens to be those allotments where the permittees refused to move their cows off. And we know, first hand, what's going on up there. So we could speak with authority, and that's what we did to the press. Of course, this didn't endear us to the environmental establishment either. But by this point our view was that they could go take a hike, because we knew what we had seen. We live down here. We knew what the cows were doing, and we weren't gonna shut up about it no matter how much they didn't like it: the ranchers or the environmentalists up in Salt Lake City. We'd had it with them even this early on.

The Griffins finally did make an effort to get most of their cattle off. In fact, at one point, Quinn Griffin reportedly rented a helicopter and had the cows shot from the air in order to not be fined for trespass for those particular head.

Mary Bulloch, on the other hand, did nothing. But the monument manager, Kate Cannon, didn't back down. She's got some guts unlike a lot of people in these agencies. And, of course, in the end she lost her job because she stood up to the ranchers. With Bush[22] coming in, her days were numbered. She was out by late 2001.

So Mary Bulloch didn't take her cows off, and there were still some cows up on the Griffin allotment.

In any case, a lot of the cows were rounded up by the BLM and brought down the trails before the snows came. And Bulloch's cows were taken up to a stockyard in west central Utah, in a town called Aurora, which I've gone through many times on my route from here to Salt Lake City.

The cattle were supposed to be sold there in order to pay the costs of removing them. They'd been impounded. The ranchers, of course, were raising holy hell about this. And a bunch of them went up to that stockyard when the cows were gonna be sold and took them out of the stockyard across state lines to Arizona and Nevada. And the local sheriff, who was there, did nothing to stop them. After this rustling operation, BLM threatened to file charges against them. But did not. Again showing the strength of the "cowboy caucus" out here in the West.

You want to try rustling a cow openly with the sheriff there and take it to another state? You'll probably be dead if you try to do that. These people, nothing happened to them. And, of course, Mary Bulloch was the beneficiary of this.

Snows came and along about January it became so difficult to remove the cows by taking them down these ice-encrusted, snow-covered trails that BLM actually went out with helicopters and airlifted the cows out, which costs a bit of money.

We were up there on a field trip for two days the following summer with the monument staff.[23] Tori and I were on foot and the monument staff were being ferried around by helicopter. I think they estimated over thirty cows were left on Bulloch's allotment, even after all these efforts. I think the Griffin's allotment was down to just half a dozen by then as far as we could make out.

So we got to hear from them how many cows they were spotting and where they were. We got to see them make faces—the biologists and the wildlife people—at the state of the moist meadows. Yeah, there were fewer cows up there, but the drought had continued. So the moist meadows were really getting hammered even by a reduced number of cows, illegally present, of course, by this point. Also, there were horses that Mary Bulloch was running illegally, not just on her allotment, but also spotted on the Griffin's allotment. What a circus.

So the BLM was airlifting cows with a rented helicopter off the top of this place. I think the final bill was like $60,000 or $70,000 for all of the fines and the expenses of having to remove the cows at public expense.[24] Of course, Bulloch, at least, did not end up eventually paying very much of all these costs, as we now know in 2004, when the cases were settled finally.[25]

Anyway, Bulloch ended up being fined a few thousand dollars. And the good thing is that she may not run any cows up there on her allotment until 2010.[26] So at least that area will get a reprieve for another six years if this is actually carried out and not overturned by the politicians, who are basically controlled by the ranchers out here.

CHAPTER 11

Relict aspen on the Rock Creek/Mudholes Allotment
Bulloch's allotment has relict aspen groves on it, which survive from the ice ages. They are way below their normal elevation. You start seeing aspen around eighty-five hundred or nine thousand feet here in moist ravines on the side of the Aquarius Plateau. Yet the highest elevation on the Kaiparowits, at least the part of it I'm talking about, is around seventy-five hundred feet.

There's a particularly magnificent relict aspen grove in front of Mary Bulloch's line cabin on public land at Mudholes Spring. She used that as a horse pasture. So the aspen were left alone. I don't think horses attack them, whereas cows certainly do. So the grove had lots of young aspen in it and generally was very healthy. There are lots of seeps there, so there's adequate water. And it's down in the head of a canyon. Quite shallow at that point, but still sheltered somewhat from the winds, so it doesn't dry out as much.

And there are other relict aspen groves up there. Most of them dying now, basically because of pressure from the cows plus droughts, which have come and gone. But now we have a really serious drought, which may be around for decades. And I think very few of those aspen are gonna survive that. Whether they would have without the sheep and the cows, I don't know. But with the sheep and the cows, it's like curtains. And it's really sad.

There's a place called "Indian Gardens" on the Griffin allotment, on the western side of the Kaiparowits with this phenomenal view of the Rock Creek Basin three thousand feet below. The plateau is only a couple miles wide at that point with huge drop-offs on either side. So you're looking out at the Rock Creek Basin. And Navajo Mountain is over there with its incredibly rugged flanks. Canyons after canyons including the one where Rainbow Bridge is. Unfortunately, the reservoir, "Lake Foul," as we like to call it (Lake Powell) is down there too, so you can see the houseboats running around on this weird blueness. But it's really a sensational place.

Indian Gardens was actually homesteaded around 1900—incredibly remote in those days. There was an Anglo man and an Indian woman who lived up there for a while. Water comes out of the springs and goes over the cliffs dropping hundreds of feet. There are lots of wildflowers there. The first time I went up there I didn't see the aspen, they were in such terrible shape.

The second time I went up, I knew to look really hard. And there were dying aspens with a few green leaves on them all along the sides of a very shallow canyon. The bottom of the canyon is a moist meadow that looks a lot like a golf fairway. The reason the meadow looks like that is not just that it's grassy. It's because cows mow the grass down. And even after the cattle removals this continued to be the case. In 2001, some of the springs had stopped running, but this one was still running, so the cows were coming there big time—the trespass cattle on the Griffin allotment. They were hammering the spring there. You could see the aspen dying along the north-facing edges of the moist meadow, which is the last place you'll find aspens as you go down in elevation.

And there are other places up there where you can see the same thing happening. There are no young aspen. The cows have been in there and trashed the spring near the aspen. And they're not gonna reproduce themselves with the cows up there. With the cows off Bulloch's allotment for a few years, maybe they'll start producing some new shoots from the roots, because that's how aspen basically propagate.

CHAPTER 12

Trespass cattle on the Steep Creek Allotment (February 2002)
The Grand Canyon Trust bought out an allotment, called the Steep Creek Allotment, just east of Boulder, Utah. Not far from Escalante as the crow flies, though it's a fair drive around by road.

I think they finally acquired the permit in about '99, if I remember correctly. Then, in early 2002, our activist friend in Boulder, Julian Hatch, while hiking into Steep Creek, found that the fence at the mouth of the canyon was down. Cows were circulating freely in and out of Steep Creek Canyon. And there was massive cow damage inside. Cowpies went some distance up the canyon.

So Julian submitted a letter of complaint to the BLM. And shortly thereafter, we had a meeting with the new monument manager, David Hunsaker, who was the replacement for Kate Cannon. She's, I think, deputy superintendent at the Grand Canyon National Park now, where I hope she's a lot happier. She deserves to be. She did try to do the right thing a lot of the time here.

Anyway, we went and met with Hunsaker, Julian Hatch, Tori Woodard my partner, and I.[27] So Hunsaker pulled out a letter from his range cons saying that there's no sign of cow damage, or the presence of cows in the allotment.[28] And he handed the letter to Julian. Julian threw it back at him, and said this is a lie. Hunsaker, I noticed, grabbed the thing and kept it after Julian said that. And we basically sat there and watched Julian and Hunsaker have an argument.

Unfortunately, Julian should have kept the statement rather than throwing it back at Hunsaker, because it truly was a lie, and we could prove it. We went out the next day after this meeting, in which Hunsaker basically told us that he couldn't work with either of us or our organizations, because we might use information we gathered against BLM, I quote.

And, of course, the information we collected would have been used against the permittees. This was a typical and complete identification between the agency, and the people the agency is supposed to be regulating. So I was really angry at the way Julian had been treated by this bureaucrat.

We all trooped out to the Steep Creek Allotment the next day.[29] And it was just the way Julian had described it, of course. The fence was down. There were areas that looked like they were corrals, there were so many cowpies. We went up Steep Creek. We found live cows in there and a dead cow. I think two, actually. From the state of the corpses, you could see they'd been in there for quite a while. They hadn't died yesterday.

We took photographs.

We came out of there just seething. The BLM had had the gall either to falsely claim that they'd sent someone there to check it out. Or what the person, who did go there, said was a lie. You'd have to be blind not to see that cows were in there. And the vegetation all gone. It was the worst conditions that I've seen anywhere in the Escalante area. And this was the allotment that the Grand Canyon Trust had bought the grazing permit on.

Well, we also learned that the attitude of Hunsaker's range staff was that because Grand Canyon Trust had bought it out, the BLM didn't have to pay any attention to it. But that's not the law. The BLM is still administering the allotment. There was no monitor assigned to it by Grand Canyon Trust. Nor were they required to. That was BLM's job.

But BLM's response was, we don't have enough staff to go around. It's been bought out, so why do we have to look at it?

Well, the reason the BLM needs to look at it is because it's being destroyed by trespass cattle, possibly put in there deliberately by a permittee who wanted free forage. All that was necessary was to beat up on the fence a little bit at the mouth of the canyon. And in went the cows.

So anyway, we came out of there seething. Tori and I were both officers of the Glen Canyon Group of the Sierra Club. And Julian Hatch is also a member. We went home and wrote a press release in the name of the Sierra Club in which we said that the BLM lied. And we described the situation on the Steep Creek Allotment.

Julian contacted the Grand Canyon Trust, and told their representative over in Moab, Bill Hedden, what was happening to this area they thought that they were protecting—that because the grazing permits had been bought out, the area was being treated worse by the BLM than areas that were officially being grazed.

So we wrote the press release. Tori and I have written many a press release over twenty years of activism. And we sent it out. We also sent a copy to the Sierra Club's Utah Chapter.

The release got some coverage locally. We also sent it out to national outlets like the *LA Times*, because they sometimes pay attention to grazing. And the *Denver Post*. You never know.

So what do we hear? We don't hear a commendation from the Sierra Club chapter, "Thanks for your really excellent on-the-ground investigation and for catching the BLM lying about an allotment that our ally the Grand Canyon Trust has paid good money to buy out the permit on so it can be protected." No. It's like, "Oh, my God, you didn't run this press release by us first! And you can't call the BLM liars."

Oh well, the government never lies. We know that. Agency officials can be trusted implicitly.

So there was a big uproar with the Sierra Club chapter. Went on for months. The chapter drafted a rule requiring that future press releases be run by them before being issued. And that a release can't go outside the group's area. So it's basically a gag rule. And we essentially told them, "You know where you can put your gag rule."

BLM's response to the actions of Glen Canyon Group members

The BLM finally did an EA and put a fence up on the west side of the canyon. And I think they did something to fix the fence at the mouth of the canyon. It dragged on for over a year.[30]

CHAPTER 13

BLM ignores public comment on their grazing management

At this point, BLM seems to be ignoring everything short of PhD ecologists going out there and walking transects and counting the species.

Basically, you're dealing with a moving target—the more interested the public gets and the more competent they become, the higher BLM raises the bar.

So on that point, I think we have to stop believing that commenting on EAs and EISs by the BLM or the Forest Service is gonna do much good. And we have to talk about serious, fundamental change, which means going after the agencies.

As a candidate for the Green Party for Congress in 2002 and again this year, I've been pushing for a Department of Conservation, which means eliminating the present land management agencies, and creating an institution at the federal level that actually focuses primarily on conservation, not on purveying resources that the public owns to private, extractive interests.

And, of course, the environmental groups generally are nowhere near talking about such a thing. They still want to deal with the existing agencies, tweak things a little and win what they call "little victories," because that helps the foundation money to continue flowing.

As an employee of an environmental organization you can put on your résumé that you accomplished this, that, and the other thing, even though it's tiny. But if you described what was really happening out there, if the foundation cared (sometimes I wonder), they wouldn't give you a cent, because basically while you're winning little skirmishes, you're losing the battles. And you're definitely losing the war, as are we all, unfortunately.

CHAPTER 14

Establishment and management of GSENM

The Forest Service's management here in the Escalante area has been worse, if anything, than the BLM's. I haven't said much about them because we focused mainly on the monument. At least there was a higher standard there, we thought, because of the monument plan. Whereas on BLM land, in general, like the Kanab Resource Area, it's really difficult getting management to improve. Well, it turns out that it's very difficult to get anything to happen in the Grand Staircase-Escalante National Monument, as well. It's the same BLM running it.

They originally brought in a lot of National Park Service people to do the inventory work and draw up the plan. So things looked not great, but at least halfway decent. But then the people implementing the plan are the same old cowboy sycophants who run the BLM. And it's very quickly devolving back into just another large BLM mismanaged resource area, alias being Grand Staircase-Escalante National Monument.

I partly blame Clinton and Babbitt for setting up kind of "Potemkin village"[31] national monuments [32] during their regime for political effect without any real substance—without adequate protections.

They managed, perhaps, to stop the Andalex Coal Mine on the Kaiparowits. Maybe. If that was ever really a genuine threat, given the economic non-viability of trucking coal out of there. But they certainly didn't do anything for the monument in any other respect that I can see. And the environmentalists had succeeded in stopping coal mines in this area a couple times before that without a monument being declared.

So it would have been nice, when the monument was established, to have had some teeth in the proclamation that would actually have protected the lands to a higher standard than the existing one which is, of course, disastrously low.

If the ideal was that the BLM would somehow pull itself up to the level of the National Park Service, as far as managing lands because it was given the opportunity to showcase its new self, it's totally failed here. And from what I've heard around the country, I don't think the BLM-managed monuments are showcases anywhere.

At this point, I see the monument designation as just more Clinton flimflam. And I don't like people being fooled. I might hang him in effigy myself now, though not for the same reasons as people did here in 1996.[33]

CHAPTER 15

The future of livestock grazing on federal public lands

Tori, my partner, was working to get some grazing restrictions on allotments down in the BLM's Kanab Resource Area, very close to the eastern boundary of Zion National Park. There were some tiny allotments there that had been declared unsuitable for grazing in a BLM land use plan twenty-five years ago. And which, apparently, were still being grazed, though they were pretty inaccessible. These were like islands—scraps of an allotment. Although they are called allotments, they're only sixty or eighty acres surrounded by private lands. So it's very hard to get in there to see what's going on.

There was a very badly damaged allotment on the Orderville Gulch—tributary to the North Fork of the Virgin River. The riparian zone was very seriously trashed by the cows. And the uplands were in bad shape.

There was a hearing on it before a regional adjudicatory judge from the Interior Board of Land Appeals. And he found against us on all points. So now we're talking about maybe appealing it, or at least appealing some aspects of it.

The judge basically said that unless the plaintiffs can demonstrate that there was no rational basis—none, zip—for BLM's decision, the decision shall not be overturned. This is a fairly extreme position even under the Bush administration.

At this point, our feeling is that just commenting on NEPA documents, that either the BLM or the Forest Service issues, is futile unless you can sue them. And you're gonna have to go into regular court. The internal process is interminable, and just produced a totally negative result for us even with an allotment, which has been declared "unsuitable" in an agency document. You cannot get the BLM to say that they will not allow it to be grazed, because you are not an expert. Or because you haven't been on it. But, of course, you haven't been on it because you can't legally get on it. Maybe you could rent a helicopter or parachute in from a plane. But you have to trespass across private land to get to it. It's really quite Kafkaesque dealing with the BLM at this point.

I would be more interested in a national campaign to change opinion in the East about grazing, than spending more time trying to get the agencies to do a better job. If they do a better job, it will only be marginally better. And that's not good enough. The deterioration of the lands out here will continue, and it's already pretty far advanced.

If the environmental movement doesn't pay more attention to the public lands issue in the West, and really gets serious about it, and starts taking some political risks, grazing will end when the land is ungrazable, the cheatgrass has taken over, and the land's productive capacity has been destroyed in forty, maybe fifty years.

Global climate change may take its toll as well.

So I think we'll get a convergence of harmful processes out here in the West that will basically make the land valueless. At that point the ranchers will finally let go of it, because they'll be unable to squeeze any more blood out of it. Then the management agencies will probably change. A little late then. That's where we're going as I see it.

Julian Hatch

Utah native Julian Hatch traces his ancestry to Mormon sheep ranchers who settled the region in the nineteenth century with Brigham Young. After graduating from the College of the Atlantic with a degree in human ecology, Hatch returned to Utah—settling in Boulder to pursue a back-to-the-land lifestyle. Repeated trespassing by cattle on his private property, along with a concern for the health of the land, soon brought Hatch into conflict with local ranchers and public lands managers. In 1983 Hatch responded by forming the Boulder Regional Group to facilitate the monitoring of environmental conditions on the public lands of southern Utah.

Julian Hatch made his remarks on the 19th of September 2004 in Boulder, Utah.

Chapters
1. A brief history of ranching in southern Utah
2. Hatch's early experiences with cattle grazing
3. Hatch relocates to Boulder, Utah
4. A grazing EIS for Grand Staircase-Escalante National Monument
5. Winter grazing at Grand Staircase-Escalante National Monument
6. Cattle grazing damages streams and vegetation on Boulder Mountain
7. Grazing of school trust lands generates little funding for education
8. Hunting revenue drives wildlife management on federal public lands
9. Predator control in Utah
10. Inadequate monitoring of livestock
11. Counting cows
12. Using FOIA to obtain information about livestock management
13. Dishonesty of the Forest Service
14. Why livestock should be removed from public lands
15. Persecution of environmental advocates in the rural West

CHAPTER 1

A brief history of ranching in southern Utah

My family was among the pioneers that came with Brigham Young to settle these areas. I come from Mormon cattle ranching roots. My father and my grandfather were big sheep ranchers. They were the founders of the Deseret Livestock Ranch up in northern Utah. There were several other ranchers that were with them, but they were part of the original four or five who put that operation together. Now the Mormon Church owns the ranch—all the ranchers have sold out their interest in it.

I live in the town of Boulder, a hundred and fifty people—the last town in the continental United States to have received its mail by mules. That would have been after the Second World War. The first paved road into this area was built in 1972. And that was a road that dead-ended here.

I wasn't raised in this area, but I've been living here for more than twenty years. I came here because of the public lands, the beautiful canyons, the pristine air—all the good qualities, ecologically. There have not been many jobs here, although there are a few more in the last ten years because of tourism and recreation. People are discovering the area. It is very beautiful slickrock country.

This county, Garfield, is larger than the state of Connecticut. And inside of our county there are about forty-seven hundred people. So it's one of the least densely populated areas in the continental United States. One of the reasons why I came here is because about 96 percent of the county is federal public lands. Another 1 percent is state land, and another 2 percent is private land.

Boulder was not one of the earliest settlements that the Mormon pioneers had in the state of Utah; it was one of the last settlements—about 1898 or something like that. It was hard to reach because of the deep canyon country and the high mountains. They mainly were bringing in Mormon Church herds of cattle and sheep. During the early 1900s—1910 through 1920—there were approximately forty thousand head of cattle here. There were probably sixty thousand to eighty thousand head of sheep brought into this area on Boulder Mountain. They came through the area, and they just hammered it. They also built big herds of cows that they spread throughout the area.

The major streams that come off the mountain run all year. The Native American Indians, who lived here seven hundred years ago, made ditches that diverted the streams right through the village they built. If those diversions were shut down, there'd be no water here where I'm living. And although it looks very green, it's artificial because all the streams would normally be down another hundred and fifty feet below us—down in the deep canyons. But they diverted it off through this little valley here that sits up a little higher. And they were able to grow maize and make a living for a couple hundred years.

When the Mormon pioneer settlers in our valley came, they brought small numbers of cattle. They were looking for homestead lands, an opportunity to lay claim to some property. They were destitute—poor people.

When they dug here, they found the old Indian settlement. They cleaned out the old Indian ditches and ran water right in the same place. They expanded those ditches over the years, and then fought over the water. Certain families got the water, and other families were driven out.

I can understand them having herds of cattle. Each family probably had twenty head of cows if they were lucky. And they built those herds up. They went from five, ten, fifteen, and they would sell three or four cows every year.

They all spread out. There's 1.7 million acres of land in the Grand Staircase National Monument. Probably half of that is here in the Escalante drainage. And so you're talking about fifty families, each picking up a part of the canyon system. On the Aquarius Plateau or Boulder Mountain area there were thousands of animals from Mormon-Church-operated herds, and other private herds from over the other side of the mountain, and Richfield City area primarily. They moved them in during the summers for several years until they ate up all the feed and severe erosion began. These operations came into, and moved onto, new areas all across the state. They would take their sheep and goats out there in the lower canyon country. Eventually, they also had cows.

The turn of the century, around 1905—that's when Teddy Roosevelt designated this national forest on the mountain. But that still left what became the Bureau of Land Management land—the "lower country" we call it—where it was pretty much "do what you wanted to do."

The Taylor Grazing Act came into being because there was so much erosion. The herds of sheep had denuded the land, not only in the Boulder area, but throughout Utah. North of Salt Lake City there were fantastic floods. And it was because they were taking thousands of sheep from their Deseret Livestock Ranch, and other herds, into the higher mountains in northern Utah. They would drive them in the fall through Salt Lake City and around the Great Salt Lake out into what was called Skull Valley. That's where they would winter their animals. And as they came through, they chewed everything down. You can go there today and see where they built rock walls to stop the major flooding.

During the mid-1900s most ranchers got into beef production and out of sheep, although there are still some large sheep herds. But the same grazing process has always been followed here in the Escalante River drainage as elsewhere—fatten the animals on summer range in the mountains, sell off animals and cull the herds in the fall, keep the best bulls and impregnate the females, then put them in the desert country on federal public lands for the winter months, so you don't have to pasture them on your personal ranch property.

In the first part of the twentieth century there was a lot of trouble from grazing and overuse. So the government put some controls on these people and their animals. They reduced their livestock numbers a bit. And apparently it helped. We don't have the problems that they had back then, but the remnants of those problems are still with us. We still see some of the erosion. And some of it is continuing to get worse. It's not been stabilized.

"The major damage was done in the past and it is our legacy today"—the government officials and ranchers use this as an excuse to continue public lands ranching, saying that things can't get much worse. And with the reduced livestock numbers, they claim things are getting better.

Really, it's that things are not getting worse as fast. If all the animals were eliminated, then we might see some real improvement and recovery. But the ranchers want to keep their herds going, because it is difficult to get back in the business once they have no more animals. The herds have been built up from a few cows to between forty and two hundred animals.

The legacy today is that some of these families are still here. The ones remaining have "grazing rights" or their privileges—permits to still put animals out on the public lands. Those numbers are way down from what they once were, but they still put the animals out there. They impregnate the cows, which then have calves during the winter in the lower country. It's a little bit warmer there, but it's still pretty cold. Then in the summer months, they put them on the mountain, into the aspen forests, where there's a lot of lush green vegetation. There they chew down the aspen shoots, and they eat the grass.

They say that Boulder Mountain is one of the largest aspen forests in the western United States. I don't know if that's true, but it's beautiful and it's big. They claim that it's one clone that is all rooted together.

The US Forest Service, and the Dixie National Forest—the Escalante Ranger District here in this area—are claiming that the forests are dying, that the aspen are being reduced. So now they want programs where they go in and clear areas—forty-acre areas. They burn them sometimes, hoping that it will regenerate this aspen clone.

We have applied pressure to them during their environmental assessments by saying, "Look, if you're not going to take the cattle off, then you're wasting your time." And so we have forced them to fence those forty-acre pieces of ground.

You can go in and see where it's been five years or so now since they've been doing this. And you can see inside the fenced enclosures that there's a lot of aspen shoots. Outside the enclosure, there's nothing.

They would say that's because they cut it, they thinned it, they burned it, and all that. But what it is, is keeping the cows out. They're not keeping the elk out. The elk are jumping those fences—it's no problem for them. The deer are getting in there too. So it's not the wildlife. It's the cattle grazing that's the problem.

If they took all of the cattle grazing off Boulder Mountain, five years later we would have no problem with the aspen forest. It's that simple. It doesn't cost any money. Well, I guess, the six or seven ranching families would be hurt by that. Each one of those families probably has an average of less than a hundred cows.

So figure out the money on it. It's not very much. If that's what we need to do, I propose that the government pay these people to not put their animals on the ground. Ranchers pay $1.43 per month for an AUM.[1] That's one cow and a calf. We're not getting enough money back to the government to even pay for monitoring the cattle that are on there. It's a bad program. Taxpayers are losing money.[2] The environment is being destroyed. And what for? There's no reason for it.

Another thing about the cattle ranchers here in our area is that for none of them is this their exclusive job. They're doing this on the side. I'm sure they make $20,000 or something per year off of their cattle business, but that's not very much money today.

One of the biggest ranchers here in our valley is the county commissioner. He gets $30,000 for that. And he also gets to manipulate all of the county funding to improve roads to their cattle allotments and to help the cattle ranchers.

We have three county commissioners, and every one of them are, or have been, cow operators. That takes care of pretty much the Grand Staircase-Escalante National Monument and the mountains here of the Dixie National Forest. That's our county.

Can you imagine three county commissioners operating Connecticut? Or a state that size? Well, that's the kind of power that these people wield.

CHAPTER 2

Hatch's early experiences with cattle grazing
I grew up in Utah, and I spent a lot of time hiking in southern Utah when I was a kid. I obviously saw cattle. Didn't realize the serious effects of them that much, I guess. But they were there. Most of the cattle operations, I think, back

then were more a subsistence type of situation. It's become more of an agribusiness kind of operation now. Even if there's a family that's still operating those grazing permits on a ranch, it's usually one son that ended up having all of it. The rest of the children went away and got jobs and lived in cities and never came back other than to visit.

So the ranchers tended to consolidate into certain individual's hands. And most of those, even in this area, have now, pretty much, sold to someone else—a bigger fish, you might say—a rancher with more money. Most of the big ranchers—there's only a few of them here—came out of California. There are only a couple of large operators in the Escalante, and they are businessmen from out of state who have bought out many of the small local families. Seventy percent of operators are still old local families. But these ten to twenty families only constitute about 45 percent of the total animals on the public lands.

CHAPTER 3

Hatch relocates to Boulder, Utah

I didn't think when I moved here in the early '80s that cattle grazing was gonna be a big focus. I came here to live a subsistence lifestyle myself—raising gardens and having my own home. And then go out hiking a lot.

But within six months of getting here, I realized that the cattle were a major problem. I also realized that the people that ran the cattle were sort of a problem. Although, I will say this, they used to have more of a "live and let live attitude." The ones that are operating now do not have the live and let live attitude.

Either you go along with what they want and keep your mouth shut, or you're gonna get it. And they're gonna take you out and they're gonna destroy you in whatever ways they can.

On these properties I own here in the middle of Boulder town, I have problems every year. They push cows into these places, and they run them into my property. I've taken them out of here twice, three times this year.

Here in the West, the law is you have to keep *their* cows off *your* property. Believe me, it's hard, because even when you have fences, and you maintain them, cattle still get in. And so it puts the onus on you to keep their cows out.

I have had experiences over the years where someone actually cut the fences and pushed the cattle into my gardens and tried to wipe me out. It was quite a few years ago. The cows had been there for a week. I'd been gone. I immediately pushed the cows out of the property and tried to repair the fences. Then when I went to the Bureau of Land Management and I complained to them about this, they laughed in my face.

They said, "They've got cows out there. You've got to keep them out."

And I said, "Well, they cut the fences."

And they said, "We don't believe that. They wouldn't cut your fences. Why would they want to do that?"

I said, "They wanted to get in where the streams are and where it's greener. And they had been used to grazing illegally there anyway. But when people came onto that private land and started developing it, like myself, that made them angry, because we fenced the cows out."

Eventually, I looked up at the bulletin board behind these government officials, and I said, "When are those cows supposed to be down on these areas around this property?"

So we looked up on there and saw that there were not supposed to be any cows on the public land around the private land. Then the government official said, "Oh, yeah, well, we'll get somebody to check that out."

They came out. They checked the fence. They said they saw that the fence had been cut. But they accused us, as environmentalists, of cutting our own fences. They then said, yes, there weren't supposed to be any cows out there on the public lands. And they did see three cows. But only three out of the forty-five or fifty that I had seen, because they had taken two or three days to get there. They identified who the brand belonged to, informed the rancher, and told him who had reported it.

The next week, when I came to town, the son of that rancher shot and killed my dog right in front of me. Shot it six times. Said, "This is what you get, environmentalist."

That's the way these people are. And that's where the problem is with the government. The government will not do its job, and it continues to this day. They are in cahoots with these folks. Most of the grazing staff here in the Grand Staircase are ex-cowmen, or they're children of cowmen, or of ranching families. But the bottom line is that it's a bad situation when you have the foxes guarding the henhouse. That's exactly what we've got.

Any BLM official I've ever seen that has a botany degree or a grazing degree, and has come from outside this area, doesn't last more than a year. They leave because they can't take it. They start saying, "Wow, look at how denuded this is. It's overgrazed. We need to do something about it." And pretty quick they're gone. They find some other place to get a job in BLM.

At the Grand Staircase Monument there was a person in charge there for a few years—a woman.[3] That was a problem right off. They don't like women. And she had some other staff that were trying to do the right thing. They were then accused of being environmentalists. The ranchers eventually applied pressure to their congressmen, and the Republican Party here in Utah, and in the western United States. She was removed from her job.

They replaced her with their hand-picked man.[4] It wasn't a couple of years later they were fighting with him too. They want total control like they've always had. If anybody is gonna work those government positions, then they want them to be their patsy. And that's what you have mostly are patsies.

CHAPTER 4

A grazing EIS for Grand Staircase-Escalante National Monument

It has now been eight years since the monument was established,[5] and they have now started working on a grazing EIS.[6] The draft of that is due in the next year, by the spring of 2005.[7] I think that some of the staff understands there are problems. And we'll see, politically, what happens. I'm sure it will drag on. It's already dragged on for a couple years. It was supposed to have been done by now.

What the government basically does is understaff or underfund these kind of positions. I'm not sure of the exact number, but it's something like ninety positions they're supposed to have here at the Grand Staircase. I think they have only forty or fifty. They don't have the funding for the rest. They, of course, fund the grazing. But they don't fund other things. And that's all political. That's what it comes down to—the Republican Party and the ranchers that control it. They want to have it just like they've always had it, which is to do what they want to do, until, of course, there's some government money. Then they want the government money. But they want it without strings.

I understand that. I do think, though, that these government people have a responsibility and a duty to the public, to the taxpayers, to the environment, to the landscape, to try to do the right thing. I think that some of them try to do the right thing. But it's overwhelming, the political problems they have.

The government employees don't stay very long. "I got to move up, get a better job, a better job, and a better job. And to do that I need to now move to another state, pick up this job, then go to another state, another area, get a little higher level, and keep moving around." They move people all the time. Well, that's better than having somebody here all the time who becomes corrupt.

But it would be better if they had somebody who would stay here, who could actually do the job and do it right, and not be swayed by political persuasion. Not environmental persuasion. Not cattle ranching. Somebody who would do the job and have a stake in the area. And in a scientific way analyze and decide whether this land is suitable for cattle grazing. What are the number of cows that could be on there? Can't we change some fence lines? Can't we change some allotments? Can't we get an allotment management plan?

There are allotments on the Grand Staircase-Escalante National Monument, which was once BLM land, that have never had an allotment management plan. Never! The rest of the allotments in our area have management plans that were last done in 1983 and 1981.

If you look at the files, you will find that very few trend studies have been done on the plants or anything else. Any studies from the past are either static or in a downward trend.

The way they think about it is that there's no way but up. When the land is in poor quality, hey, it can only get better. So things are getting better.

But it's my opinion that it is deteriorating—that when you look at the plants on the Grand Staircase, it *is* getting worse.

I said earlier that I thought things generally were getting better here—I'm talking about the number of cattle. What's happening is, they've destroyed the plant life so much that they cannot put the cows out there. There's nothing for them to eat. They're now chewing the sagebrush down. Animals really don't want to eat the sagebrush, but when they're starving to death they'll eat it.

Now even the sagebrush is going, and all that's left are some invasive weed species—Russian thistle (tumbleweed). Nobody wants to eat tumbleweed. All the native plants—fourwing saltbush, and things like that are gone. What willow growth comes out of some stumps is then chewed off the next year. And then the plants are dead. The endless trampling around the plants has caused erosion and pedestalling, so the roots are exposed and the grass clumps die.

Now they want to supposedly save the sagebrush, which is dying here throughout the Colorado Plateau. And so Congress has allocated a bunch of money—millions of dollars apparently.[8] It's hard to get the facts out of them. I've tried, and they've refused to give it to me. You can FOIA them, and you'll still not get the facts. But they evidently have the money.

What they want to do is "chain"—take two bulldozers, put a chain between them, drag it through the sagebrush, and then kill 50 percent of it. What's left will sprout up. Usually sagebrush will come back fairly well if there's enough water or precipitation. But mainly they're doing this to seed other plants in there, even though they don't have enough seed to operate with. So they need to stockpile it. They're working on that apparently, but they're working very slowly.

We've got them to at least put a lot of native plant seed in there—Indian rice grass. Cows love to eat Indian rice grass. So they're happy to do that. Problem is, if cows eat the Indian rice grass, pretty soon it's dead.

What the government really likes is crested wheat. But with the drought—we've had a six-year drought here—crested wheat's not gonna do well. And even when you do crested wheat, it will only last twenty to thirty years. Then it will die off by itself.

So there's not any really good solution to the problem. Once you've destroyed the native plant community, I don't think you're ever going to bring it back. You might be able to bring it back in one-acre, five-acre pieces that look like little pastures or something. But generally, throughout the thousands of acres, you've lost all of those native plants.

CHAPTER 5

Winter grazing at Grand Staircase-Escalante National Monument

Cows, what they do all winter is wander. And they try to get to water. So they move out into the wilds chewing down anything that's edible that they like, which are the native plants. Then they walk back to get water. They do that back and forth every day. And they cause lots of trails.

In our area of the Grand Staircase National Monument and most of southern Utah, we have cryptobiotic soils, which are very much sandy soil. There's a life form that grows—kind of a lichen cover—over all the sandstone and biotic soils. Those cryptobiotic crusts are really thin and brittle. Every time cows walk through, it just destroys them. Then it takes years to re-develop. In some places, where they trample it enough, it never comes back. It just starts blowing sand.[9]

And so what took millions of years to develop here has been destroyed. Most of it was destroyed by the beginning of the twentieth century. Most of the damage was done a hundred years ago, a hundred-and-fifty years ago even, when the first animals were brought in. That's the excuse that the government gives to me all the time. "All the damage has been done. We're not hurting anything anymore. We might as well let them continue to denude the area."

I think the government's wrong, and I'm talking about what's in government studies—their environmental assessments—not about just what they've told me. I think that there are some remnant species out there, and relict communities, that still can be salvaged. We don't want them fenced off. These are wilderness study areas. To get out there and put a fence in, you have to destroy the area some more. And it costs taxpayers money. The real solution is to remove the cattle from the land. Pay the permittee to remove, or just say it's unsuitable, and kick their asses off the landscape. They don't need to be there. If they want to go on welfare, they should go on welfare like everybody else. Because that's all it really is, is a welfare program. We're giving them the opportunity to put a cow out there for a dollar a month, and trample and trounce the land, and kill off all the native species. I understand that that's where the real problems are. And that's where all the political power is.

Last, but not least, why would these people continue to put their animals out on the Grand Staircase-Escalante National Monument in the winter if it wasn't making them money? The answer is, it doesn't make them money. It's a place to babysit the cows while they starve, and they have the baby cows. Then in the spring they put them up in the aspen forests on the mountain for six months. That's where they put the meat on the animals. And then they sell the animals at the end of fall.

The Grand Staircase Monument is just a place to babysit the cows so the ranchers don't have to keep them on their private land in town. They could do that, but they would have to feed them hay or something. Or they would just eat everything down on their private land. So what they would prefer to do is eat the public land down to nothing. And that's what they've been doing.

I have to mention the latest thing that we've seen the last couple years—they look like half of fifty-gallon drums. They're yellow plastic, and they've got supplemental feed in them. It's like molasses mixed with other nutrients—like vitamins for cows. We're seeing the winter allotments having these all over the place. And, of course, the cows stand around and lick them until they're empty. Then you see those plastic barrels blowing around out in the wilderness.

The point of this is that the cows lick these nutrients out of these barrels, and then they can just eat sagebrush. They have the nutrients, and they have something in their stomach to keep them going through the winter. It's like they could eat cardboard, and lick these nutrients and do fine. That's how bad it is. These animals are actually starving to death out on the landscape.

CHAPTER 6

Cattle grazing damages streams and vegetation on Boulder Mountain
There's beautiful little lakes, and beautiful streams that are coming off of Boulder Mountain, which could be very pristine. You go up there, and there's cows. They're defecating in every stream. They're chewing down every edge of the stream, because that's where there's something to eat. And then we're getting severe erosion in the streams.

The aspen is probably the one thing that the government's pretty much worried about.

We do have a candidate threatened or endangered species—the Indian paintbrush.[10] There are concerns that the cows are chewing that down. Of course, the government says that they don't believe it is. But no one's really done the studies, in my opinion, to prove one way or the other.

We have some endangered fish here that are still in the area. Most of the Colorado cutthroat trout[11] have already been damaged by people from the state of Utah transplanting nonnative rainbow trout[12] and German brown trout[13] that have out-produced the native cutthroat. But there are still some relict communities out there. And they're trying to enhance that now. They're going in and rotenoning the streams and killing off all the nice beautiful rainbow trout to reintroduce the cutthroat. So there are attempts to try to save these species. My guess is that they're doing it because they're getting federal funding. It's jobs and some money for the state of Utah and its programs on wildlife. Therefore, they're taking advantage of it.

CHAPTER 7

Grazing of school trust lands generates little funding for education
One of the problems that we have in the West is that when each of these states was created, they took 640-acre sections of land every so many sections, and made it a "school section." Certain states like Montana, I think, it was a little less; Utah a little more.[14] But every state has these school sections, which are owned by the School and Institutional Trust Lands Administration—a nongovernmental unit. If they want to develop that isolated section, they pressure the federal government to build a road in there. And the government, pretty much, has to give them access to their property so they can develop it.

Most of those school sections in Utah were used for cattle grazing. They would rent them out to some rancher, who'd also have a grazing permit for the federal public land around it. And so the rancher would pick up a permit, and he'd pay $25 or something to have this six hundred acres of state school land. In any case, there wasn't a lot of money being generated by that over the years.

Utah is one of the worst-funded school systems in the United States. I think we're 49th, maybe 50th, in per capita expenditure for students.[15] You would think that the Mormon people would want to have more money to educate their children, but there are large school class sizes because of the large families.

When you look at the money they generate from school sections in the state, it's probably 1 percent or less of the school budget as it is.[16] So they're not getting a lot of money for it. They never did.

Many of those sections they have decided to drill for oil on, or things like that. You can go out throughout the wilderness here, and you can find roads going to drill holes that have been capped because they failed to find oil. For most of the drill holes here there was nothing.

When they made the Grand Staircase National Monument, they got the School Trust Administration to trade out all of those school sections from inside of the monument. That was a very good thing.

These are mostly all wilderness study areas that have now been enhanced by not having these school trust lands in them. That's just the Grand Staircase-Escalante National Monument. That's 1.7 million acres. But there's millions of more acres out there that have these problems that need to be addressed. It's possible they can graze these lands, and drill oil on them at the same time. That's a big threat and a real problem for us all to address.

CHAPTER 8

Hunting revenue drives wildlife management on federal public lands
Hunting's a big thing that they've really pushed. They want to have elk. There were elk here when the pioneers came. We can tell from petroglyphs and pictographs that the Native Americans, seven hundred to eight hundred years ago, were using some elk. But it seems to be pretty artificial now. They're just pumping up as many elk as they can put on here. And there's a lot of effects from the grazing of elk.

They're also very upset about the deer herds. They're not what they were twenty, thirty years ago. But when you look back seventy-five years, a hundred years ago, from the old-timers that I knew, who have now passed away and are dead, they would tell me there were not that many deer. There were deer herds, but there was not the amount of deer that there was thirty years ago.

They're artificially pushing this for hunting. And they're trying to produce as many animals to kill as possible. That's where the state gets the money to operate their wildlife—mostly all from hunting licenses. I don't know why they can't just appropriate some money for wildlife in the state of Utah. It's a big problem. And apparently all of the United States is under the same policy.

It may be federal public land, but the wildlife on those lands, millions of acres, is all completely controlled by the state. The managers of that federal public land have really very little or no say whatsoever about the policies that the state of Utah is driving on the wildlife on the public land. That also extends to the predator control. And grazing has a lot to do with the predator control.

Let me just step back and say that the sheep and cattle ranchers do not really like the elk. They think the deer and the elk are competing with livestock. In the winter the elk herds are going to the lower elevations below six thousand feet. And they are trashing the area. Sometimes it's worse than running cow herds out there because of where they can get to, such as high up on the rocks. The land in the desert is not only getting beat by the cow grazing, but also by artificial elk herds. Love to see the elk, but it's all about hunting.

CHAPTER 9

Predator control in Utah

We have a predator control program operated by the state of Utah. And we do have the Wildlife Services, which is the federal government entity, as a part of that. Basically, the state of Utah controls what the Wildlife Services does in Utah. That includes aerial gunning and trapping.

It's not just a matter of somebody, who has their livestock on public land, calling up the Wildlife Services and saying, "Wow, I think I've got a mountain lion here that's been killing my sheep." Instead, they're doing this as a concerted program where they go throughout the state aerial gunning. They're just killing every animal they can find, especially the coyote. They hate the coyotes. There's a bounty in Utah right now for coyote ears. Twenty dollars per pair. You kill the coyote. You cut its ears off. You take them to any one of the county offices, and you get twenty dollars.

The state of Utah has a committee that operates the predator control.[17] That committee is comprised of the local representative of the Wildlife Services—the federal agent. Then you have the Department of Agriculture for the state of Utah. You have the US Forest Service. The BLM representative. Then you have one sheep, and one cattle rancher that sit on it. Then you have also the Cattlemen's Association representative. The Wool Growers Association representative. The State Wildlife Resources will have a representative. And no environmental representation whatsoever. In fact, it's hard to get the minutes of the meetings or to even know when they take place.

They have a memorandum of understanding that they operate under, which just coordinates all the agencies under this committee. No one that I've seen in the minutes ever dares to say nay to anything.

They were not real happy about the $20 coyote bounty, though. It really doesn't help the deer herds. Besides, people kill coyotes anyway. They just kill 'em for fun. So why pay 'em money? They'd rather see the money go to some wildlife biologist for the state of Utah, or something like that.

But the legislature is what funds all this stuff. And they do get federal funding and matching grants. The committee is operated by those kinds of people. There's no other oversight. They're all appointed by the governor of Utah. Theoretically, if we had a Democratic governor, we might get some better people on there. But mainly these agency heads are all what have the power. They wouldn't be those agency heads if they weren't under the thumb of the ranching community. And the ranching community runs the legislature. That's how they're politically elected.

We have what's called the "cowboy caucus" in the state of Utah. Yes, there's the Salt Lake City area. There's a bunch of Democrats. There's even a couple women and some minorities that are representing our state on the legislature. But most of the areas, and the majority, are in the cowboy caucus.

So who's running things? It's the cowmen. And I don't see that changing any time soon.

The predator control is operated by that committee, and they basically just allocate the funds. They meet once a year, and they talk about what troubles have come up.

One of the troubles that came up was when I brought to them what I discovered when I was out hiking here in the winter. I saw the helicopters coming over shooting. I got upset about that because it's a recreational area. It's the Grand Staircase National Monument. So what did I do? I contacted the Grand Staircase personnel. And they said they have no control over it—that I would have to talk to the state.

They sent me to the wildlife resource folks at the state of Utah. I talked to them. They said they didn't have any control either, and that this committee, and essentially the Wildlife Service's agent, is who determines it.[18]

What they do is, wherever there are cattle grazing areas in the Grand Staircase, they just fly over with those helicopters and they shoot coyotes. It costs money to run those helicopters, and they wreck 'em sometimes. And sometimes people die.[19] But it seems to be a fun program they like to do. Obviously, killing five hundred coyotes and spending $500,000 is not very smart. And it doesn't really work very well either. From what studies that we've all looked at, when you kill the dominant coyotes in the area, more coyotes will come in, and they'll reproduce even faster. So it's really not a solution to the problem.

What we could use here, if they really don't like coyotes, is wolves. The wolves take care of the coyotes pretty well. So that's one solution to their problem. But, of course, they don't want the wolves, probably because of their cattle grazing on public lands.

CHAPTER 10

Inadequate monitoring of livestock

Ranchers usually don't put the animals on until the date that they say. So that's not a problem. But one of the problems is they don't take them off on time. There's nobody going out there herding them back down to their private property. And when I've complained to the Forest Service about it every year, they just basically say, "Well, we'll get somebody up there. We'll let 'em know." Then they'll send a letter to the rancher that says, "Your cows are still on the mountain. You've got two weeks to get them off of there."

I've seen it where they've given 'em two weeks. And another two weeks. And another two weeks. And by the time the six weeks are up, the cows came home anyway because it was too cold. And so they basically have no enforcement. They don't have any real monitoring.

The Forest Service here, the Dixie National Forest, Escalante Ranger District, has one person. And one of his many jobs is to deal with the grazing on the Forest Service land. We're talking thousands and thousands of acres here on the side of the mountain. And that person does not monitor the grazing. They have a memorandum of understanding to operate the grazing on the mountain with the BLM. So the Grand Staircase-Escalante National Monument staff is actually the people who are supposed to be monitoring, and dealing with the grazing on the Dixie National Forest.

But every time I've talked to them I hear, "We're understaffed. We can't do it."

I would like to see some more monitoring. We're basically doing it on our own, and if we had money we could do more. But we're going to do whatever monitoring we can. And if that means just going out and hiking in the canyon, and seeing what's going on, we'll look around and keep our eyes open.

What we'd like to do, and what we're working on at this point, is actually setting up a grazing monitoring program under Western Watersheds Project, the Escalante Wilderness Project, and the Boulder Regional Group in the Escalante River drainage system. We are going to see when the cows go on those allotments. We will go out and check them once during the season just to count them to see what's going on. If we see utilization's already surpassed the limit, we'll try to get the BLM to come in and do something about that. We'll try to take a lot of photographs. We'll try to meet with them. We will then see if they remove the cows on time, and what it looks like after they're finished.

We'll do our own monitoring program. That's how bad it is. The government won't do any monitoring. Bigger environmental groups that have lots of money—Sierra Club, Grand Canyon Trust—don't seem to be doing any monitoring. And so it's left to a handful of local people who are going to take care of this on their own.

When the Grand Staircase comes out with their grazing EIS and they're working through it, we'll come out with our own photographs, and we'll try to prove where they're wrong about this or that. My big fear is that the grazing EIS will be about determining whether or not these lands are suitable for cattle grazing, and that almost every one of them *will* be found suitable for cattle grazing.

They'll say, "If it's winter grazing, how's it hurting? The plants have grown. They've sprouted all this stuff. They're dormant in the winter. We put cows on there. We chew 'em way back. We only utilize 'em 60 percent. Everything's fine. The next year they'll sprout out again. We'll just keep chewing 60 percent off every year."

I think that generally doing that year after year after year is going to destroy the plants anyway, but the government will say that's fine, until twenty years from now when they decide that it didn't work out fine. What I've seen here is that twenty years ago they said *this* was a good project, whatever it was. Today it's not.[20]

It's time to stop making mistakes. And the solution to the problem is so simple. It's the removal of the animals that don't belong there. And it's also not allowing the state to drive the mechanism of wildlife—federal land managers should have the right to decide how much wildlife will be utilizing those areas.

If the state game department pumps up an elk herd, and they wipe out the plant life, there's nothing that the federal land manager can do about it. That's a real problem there, and we hope to see a lawsuit about that. We're right now, and have been for several years, on appeal about some of those issues. But it takes so many years—it's taken three or four years just to work it through the BLM's decision process. And if they decide against us, then we'll have to sue them. There's another five years. So it goes on and on.[21]

CHAPTER 11

Counting cows

Who's going out there and counting the cows? They're spread all over the hillside.

The government says, "We saw that 'Rancher Bob' brought his cows up. I saw eighty head go in there."

Well, who's to say that Rancher Bob didn't bring twenty-four more cows the next week and didn't tell the government? How do they know? They don't know. And how do I know? I don't know either.

I go and I count them. In a valley or on a hillside maybe I can see a bunch of cows. Sometimes I've been on the Grand Staircase-Escalante National Monument where it's bigger vistas, and I can see farther. It's hard in the deep forest to see every cow. But in certain places on the forest, and down there on the monument, I have sat there, and I've counted more animals than they are permitted to have. And that's not counting all the animals that are out there.

But I've been able to see that there are a certain number of cows, especially on the Grand Staircase, where they hang by those water troughs and those feed-bucket stations. They will stay there all during the day. And there'll be hundreds of cows just right there. So you can count quite a few.

Is the government counting? Well, if you look back—and I've gone through their records here—the only trespasses that have ever been found in the last five years have been ones that I've turned in. So the government's not doing anything.

My goal is just to get the cows out as soon as I find a trespass. That's the solution. I don't know what good it does to "slap their hand" and fine a rancher $150 or whatever. But that's up to the government.

This last year I was told quite frankly by the BLM of the Grand Staircase-Escalante National Monument that unless I could prove who owned the cows by identifying the brands and bringing them a digital photo, they wouldn't get in a truck to even go look. They were gonna do nothing about trespass.

And my belief is that they know that I'm right. They don't want to go and find that trespass. They'll do anything they can to not find it.

As an advocate who's been active in the area, what should I do about that? I could call up the superintendent of the national monument. I'd say, "Superintendent, you've got staff members that are saying that they won't even get in a truck and go find it." And the next thing I'll get is, the superintendent will turn around and say I'm a liar—that his staff never said that.

They'll put it in a letter. They'll write me and they'll say I'm lying. They've been out there. They've looked at it. There's no cows there. There's no footprints. There's no defecation. There's nothing.

And I will go right back out there the next day after I get the letter or phone call, and I'll see the cows are still there. And I'll see cow shit all over the ground, and I'll see that they've trampled the stream. They're right by the fences. They're trying to get out if nothing else. They want to go home. There's nothing to eat down there.

I'll take photos. And then I'll go talk to the superintendent again. I'll call him back, and I'll say, "I saw them again today."

And he'll say I'm a liar. He'll put it in writing that I'm lying.

I'll say, "Bring yourself out there Mr. Superintendent. I will prove it to you."

After a while maybe they'll eventually say, "Oh, yes, there were a couple of cows. What's the problem?"

And they'll say it looks like I must have knocked the fence down. And that's why they got in there. They never should have been in there in the first place. They weren't trespassers.

How does an advocate, an activist, do anything about that? What am I supposed to do? Call my congressman? Well, that's not gonna work. My legislator? That's not gonna work.

So who do I go to? Who do I talk to?

The only solution I've come up with is just to keep hammering on the BLM, the superintendent, and maybe some of their staff members, and keep bothering them.

CHAPTER 12

Using FOIA to obtain information about livestock management

I have been to the Grand Staircase-Escalante National Monument, Escalante Office, which controls the grazing in this million acres.[22] Let's say I've been there twenty times. Nineteen times there's never been anybody in the office who knows anything about grazing. The grazing staff is always out working.

I have tried to make appointments with them when I can come see records—when I can come talk to them about situations. And I get nowhere. They're too busy to talk to me. I'm just a "public."

The only way I get any information whatsoever out of the Grand Staircase-Escalante is to basically FOIA them—Freedom of Information Act. But then you better make sure you cross all the t's and dot the i's. You better ask exactly what documents you're looking for. And if you don't know how they're organizing their files, then you're not gonna get much. You're gonna get a lot of copies of letters where they talk about "Rancher Bob needs to pay for his allotment. He's gonna be allotted eighty-eight cow head and three hundred AUMs. And we need that money next week." That kind of stuff. It's information that's not worth the paper it's printed on.

As far as trend analysis, or utilization studies, or how many cows they actually put on, you're not gonna get much of anything because they don't have it.

And they don't want you to know that they don't have it. It's not like there's a tally sheet that says, "For this allotment, we have seventeen pieces of paper. Six pieces of paper are trend analysis. Six are something else." It's not broken down like that. You get a pile of paper. Most of it's garbage. And most of the stuff is old anyway.

The Forest Service here has gotten really bad over the last five years. Their policy now for the Dixie National Forest is that on Freedom of Information requests, you cannot personally come to an office and view documents. They have a policy "not to view."

Their adopted policy is to put you through a couple of months worth of rigmarole. You have to be a bona fide environmental organization. You have to then answer a bunch of questions as to why you want these documents. Otherwise you pay. And you pay hundreds of dollars to get the documents. And if you don't know exactly what you're looking for, and you're just trying to do research, it could cost you tons of money. I've been billed thousands of dollars by these people. I've never paid them. I just turn around and find somebody else to break it down into smaller requests. And then we'd send different requests in, and help each other that way. But eventually, after several months, you can get a pile of documents.

What's in that pile? See if you can find something. Occasionally I find stuff. And when I do, I come to them and I say, "Here's the six pieces of paper out of the four hundred pages that I got from you. I want to know answers about why this is like this and that."

And they say, "We don't know."

"That guy moved on. He's got a new job."

"We're not sure what happened on that."

"That piece of paper really doesn't mean what it says. It means something different."

CHAPTER 13

Dishonesty of the Forest Service

Out of one side of the agency's mouth, they're talking about the problems. But they're blaming it not on the cows. In most cases, it sounds crazy, but they'll blame it on the environmentalists—"These aspen trees need to be timbered and cut down. We need to burn the forest. It's the environmentalists who won't let us do these programs."

But that's not the truth. The truth is it's cattle grazing. It's sheep grazing. That's where the problem is. They haven't got the courage to stand up and do the right thing. And even when they do their own science right here in our own forest—a draft analysis of the vegetation—they said it's cattle grazing.[23] It was very clear.

When they do an environmental assessment for some project that usually includes timbering and other actions—burning, underburning, that kind of stuff—we bring up their draft document. And we say, "Look, this is what your own staff decided—that it's cattle grazing that's causing this problem."

And they say, "Yeah, we did, but so what? We're not going by that."

What are you gonna do? Sue them?

We could file a lawsuit about it. And there are certain groups that do file lawsuits here in the state of Utah. But they end up losing in local court. Then they have to appeal to Denver, Colorado—the Tenth Circuit—where they sometimes win, and it comes back. Then the agency will adjust their EA so they still do essentially what they were gonna do in the first place. It's like little technical things, when it's the overall project, and grazing, that's the problem.

It's very sad. And lawsuit's take years. You're looking at five years. Meanwhile they've implemented the program and already done it. It's hard to get a stay, and stop them from going ahead, even when you're on appeal to the Tenth Circuit.

It's very frustrating and it costs a lot of money, but it's a tool that people have to use.

And this is the Tenth Circuit. This is not the Ninth Circuit. This is Intermountain Region. These judges are highly political, and probably we're not going to win most things. Then if you set a bad precedent in the Tenth Circuit, it can affect the rest of the country. So it's a very dangerous situation to be filing these lawsuits.

The best solution is to solve some of these problems at the very lowest level, and to implement some things. Like when they come in and they clearcut these forty-acre pieces of aspen, and we get them to put fences around it, and leave the fences up for five years. If that regenerates the forest in that area, good. That's what we all want. That's what everybody wants, except for the cattlemen. They want to get inside that enclosure, and they want to eat. But that's why they only do forty acres. They could be doing two thousand acres at a time. They could be doing these big pastures where they already have the fences. And they could just stop grazing in that whole pasture area, and not rotate through there for a year or two. It's that simple. But that would cause too much of a political outcry from the graziers—these handful of people who have their power.

CHAPTER 14

Why livestock should be removed from public lands

The best policy is to remove all livestock from all public lands. They're not suited for the public lands. And they don't belong there.

Why should one cowboy, or a couple of families, have a right to put their animals out on the public land, and I can't? If you're going to do that, let's put up a bid system. Let's see who will pay the most money to put the animals out there. They don't want to do that because they know that environmentalists like myself would bid on those allotments too.

Well, that brings up the problem of government. The government believes that under the Multiple Use Act,[24] they have to graze it—"It's good for the land to graze these animals." That's where the problem lies, because I don't think it's suitable for grazing. Under the Multiple Use Act there are lots of different things you can use an area for.

One of the things that they probably should use all of our lands for in southern Utah is recreation. Recreation should be the highest and best good. And cows do not work with recreation. Everyone that I see out hiking on the public land does not like the cows to be there. They ask, "Why in the Grand Staircase National Monument is there cattle grazing? We thought this was a monument." So people are not happy about it. Cows defecate in all the springs. And they destroy the streams from the top of the mountain all the way down through the monument. So there's real problems with having the cows here.

And what's the upside? How much money is somebody making on this? What good is it doing?

In the Grand Staircase environmental impact statement that's being produced, we'll be looking at exactly what reasons they say that they need to have cattle grazing on these lands. How much social/economic good are they doing?

I think if you total up the amount of money that these cowmen are making by having these cows out there, it's minuscule. We should just pay them every year for the next fifty years to not graze cows. The taxpayers would save money. And the public would be much happier about their recreation on these lands. And we wouldn't need to have any kind of monitoring program at all. We could get rid of the fences. We wouldn't have to fix the fences or put new fences in. It's a very simple solution. It makes total sense.

Why do a handful of people have all this power over everyone else? These are federal lands. They belong to millions and millions of people. They don't belong to just a handful of cowboys who have had control over them in the past. Things are different now. It's time for them to get a real job.

So why do they continue to do it? Because the political power in Washington, DC, and in every one of these western state capitals, is in the hands of a handful of people who are grazing advocates. They want to have the grazing.

CHAPTER 15

Persecution of environmental advocates in the rural West

I am the only person that I know that's a pro-environmental advocate who's lasted the last twenty years here.

There are a couple other people who have been here that long, but they've become pro-grazing. They've got their own horses, and they believe in the grazing principle. They've become outfitters to try to make their living.

There have been other environmentalists—the Southern Utah Wilderness Alliance began in this town. There's three founders; two of them are gone. They didn't last very long.

You don't last very long when you get knives stuck in your face. And you get your car's windows smashed in. And you get threats. And if you have loved ones, or family, they won't want to live here. You can't send your kids to school here. They will be destroyed.

There are a lot of pro-environmental people who come to the area. They visit. They make maps. They do studies. And then they leave. A lot of people are working on this area. It's not like they don't care about it. But very few people dare live here. And I don't blame them.

I hate to be very negative about it, but it's the truth. And people need to understand. If you want to fight the cattle grazing, you need to think twice. Maybe you better just go for a hike and look at the cow shit and walk past it. That's if you want to take care of your personal life. If you've got family, you better do that. Maybe just don't go hiking out there at all. Maybe just stay home, and see pictures on the computer of the Grand Staircase. The younger generation—that seems to make them happy.

But a person like myself, I'm stuck here. I made my life here, and I'm gonna be here probably until I die. I will continue to complain, and I will continue trying to make changes, but I have no illusions about this. The government's out to get you. If you make complaints, you're making their life miserable. They just want to go to work, collect their check, and go home. You're making trouble when you bring up problems about grazing, or monitoring, or anything else out there.

So that's the reality. But like I say, some of us are committed to do it. We'll keep trying to do what we can.

Steven G. Herman

Steve Herman, raised first in Pennsylvania and then in California, traces his passion for experiencing and protecting the natural environment to his childhood hunting and fishing adventures with his father. Later, as a college student, Herman's views on environmental conservation were strongly influenced by professors Richard Mewaldt of San José State University and Aldo Starker Leopold of the University of California at Berkeley. Herman received the PhD in zoology from the University of California at Davis in 1973 and has taught courses in ecology, natural history, and animal behavior at the Evergreen State College since 1971. Over the years, he has closely followed the management of livestock grazing at Malheur National Wildlife Refuge, Hart Mountain National Antelope Refuge (both in Oregon), and Sheldon National Wildlife Refuge in Nevada. He has also fought to protect the habitat of pygmy rabbits in Washington state.

Steve Herman made his remarks on the 18th of August 2004 in Olympia, Washington.

Chapters
1. Herman's early life, education, and career
2. Cattle grazing at Malheur NWR
3. Cattle grazing at Hart Mountain NAR
4. The last years of cattle grazing at Hart Mountain NAR
5. Management of Hart Mountain NAR after removal of cattle
6. Politics behind efforts to kill coyotes at Hart Mountain NAR
7. Cattle versus pygmy rabbits in Washington state
8. Federal agencies ignore public comments about land management
9. Public indifference to the decline of rangeland health
10. Tactics to reduce livestock grazing on public lands

CHAPTER 1

Herman's early life, education, and career

I was born in Elyria, Ohio, 1936. Spent the early years of my life mostly in northwestern Pennsylvania in Meadville and Conneaut Lake. I was the first of three sons that my parents had.

My father was a hunter and a fisherman. From a very early age I remember him carrying me out squirrel hunting and just planting me on a stump, while he called squirrels and then shot them. He had been a forestry major at the University of Pennsylvania, but was unable to complete his college education after his mother died unexpectedly.

When the Second World War started, he went to Buffalo, New York, I believe, where he signed up for the Navy, then came back having not told my mother that. It wasn't too long before he was sent out to Oakland, California, where he was stationed during the war directing materials for transport to the Pacific Theater.

I don't remember exactly how long, but it was more than a year before my youngest brother and I, and my mother set out to join my dad on the West Coast.

We drove across country, more or less in the winter, under circumstances of great rationing of everything from butter to tires. Got as far as Elko, Nevada, where my mother decided she wasn't going to drive another inch. She called my father and said, "You better come get us."

He arrived on the train and drove us the rest of the way out to the Bay Area. We lived for a short time in Oakland before moving to San Francisco, where we lived for several years—a time when there was still some wildness around the edges.

I remember finding a raccoon in a jaw trap not far from Lake Merced.[1] I was so outraged I called the SPCA.[2] When they arrived I took them to the willow thicket where the trap was. All that was left in the trap was the raccoon's right front leg. He'd chewed it off.

That might have had an influence on me. But I think the greatest influence was my father taking me hunting and fishing, and imparting to me what he knew about natural history, which was considerable. My grandfather—his father—was a big hunter, and that's where the tradition came to him.

When we left San Francisco I was about twelve, and we went to Redwood City. It was shortly after that move that I became interested in what was known as bird watching. It's now known as birding. (All nouns will eventually become verbs.) I joined the Boy Scouts, and the only merit badge I ever got was for bird watching, which was a tough one. I think you had to identify something like a hundred species of birds.

I went my first two years of high school in Redwood City. Then we moved to Hayward across the bay, where I graduated from Hayward Union High School. By that time I had a strong interest in falconry—training wild raptorial birds to hunt. My interest in herpetology, the study of reptiles and amphibians, was also very high—I was active in something called the "Northern California Herpetological Society."

After I graduated from high school, which I think was 1953, I went off to college at San José State, where I was enrolled as a biology major. I began studying wildlife management and wildlife science under a man named William Graf. And I met another professor, a big bird bander, Richard Mewaldt. He went on to help found Point Reyes Bird Observatory in Marin County, California. I came upon him many years later, in the 1970s, on Hart Mountain National Antelope Refuge, where he had a banding operation. But that gets a little ahead of my story.

After a year at San José, I flunked out because I spent all my time in the library reading bird journals. I was then not terribly popular with my parents.

So with an $18.75 war bond left over from the Second World War, I bought a bus ticket to Denver, Colorado. Didn't know anybody there except for a high school buddy, who had come by a different route at a different time. In Denver I took a young prairie falcon out of its nest and spent the next several months flying that falcon at things like pheasants and magpies—getting out in the country during the day and working at a bakery called the Curtsy Cake Company at night.

The following February, I decided to come home. On the way the car's engine blew up. So I did not return triumphant.

I then went to a couple of community colleges in the mid-1950s, after which I enrolled in the University of California at Berkeley, where I majored in what was then called "wildlife conservation." My major advisor was Aldo Starker Leopold,[3] Aldo Leopold's[4] senior son. That was quite a turning point for me. I took a lot of natural history courses. One of them, the Natural History of the Vertebrates, is where I learned a system for keeping a field journal in natural history. I also took conservation courses from Leopold and went in the field with him.

Then I worked as a technician with the University of California, Department of Biological Control, where I learned a lot about insects. I also learned a lot about ecology, about ecological relationships, predator-prey relationships, parasite-host relationships, and so on. I ended up working with forest insects—particularly the natural enemies of forest insects.

Meanwhile, I was going to school, but I didn't like chemistry very much. The first time I took organic chemistry, I spent the day of the first final at a golden eagle's nest. And I've never regretted that. I eventually got through organic chemistry. Later on, when I was involved in the fight to ban DDT, that chemistry came in quite handy.

I studied mountain chickadees, mostly in the Inyo National Forest south of Mono Lake in California (east of Yosemite). There was an infestation of a native insect in Yosemite, in Tuolumne Meadows, and Yosemite as well. My boss and I, and my technician colleague got kicked out of there, because we were not very secretive about our opposition to using pesticides to control this native insect, the lodgepole pine needle miner,[5] that had a clear and positive role in the ecological processes of the lodgepole pine forest.

Also, it was while working in the Inyo that I first became acquainted with the grazing issue. One day I came across a crew of people putting up a barbed-wire fence.[6] Their pickup truck had the insignia of the US Forest Service on it, which I thought was a little strange. So I went up and asked them about that.

They said, sure they were the Forest Service.

I said, "Aren't these cows privately owned?"

"Oh yeah," he said.

I said, "Well, what are you doing putting up fences for these cows?"

I imagine they told me. I don't remember what the answer was. But I got very upset about this. And I ended up going into Lee Vining, which is right at the base of Tioga Pass at Mono Lake, and looking up the district ranger, whose name was Jack Reveal.

I marched into his office, and I said to him, "I'm outraged by what's going on out in the forest, and I am interested in the extent to which a public servant will go to serve private interests."

I called my boss the next day, and he said, "Well, you go in and apologize to Mr. Reveal or you're out of a job."

I suppose I did apologize to Mr. Reveal, but I don't remember the details of my apology. I'm sure it couldn't be characterized as effusive.

I already knew about the damage these cows did in those mountain meadows. This was at eighty-five hundred feet in what was a fairly arid part of the forest with broad meadows, which were trashed in those days.

The Forest Service had wire cones, about three feet in diameter at the base, that were used to measure the extent of the grazing. If you were driving through the forest and you saw one, you could see very easily how much damage the grazing had caused, or how much grass had been taken. It wasn't too many years ago that some Forest Service statistician decided that those cones were not statistically significant, and all of them, or virtually all of them, were removed from our national forests.

So that was my introduction to livestock on public lands.

I finished my time at Berkeley in the mid-1960s, and moved then to the University of California at Davis to take up a PhD program. While I was finishing my degree in zoology, I worked on a contract with the US Fish and Wildlife program on western grebes[7] at Clear Lake, a hundred miles north of San Francisco, where these birds had been involved in a big die-off a few years earlier. I studied their reproduction and their relation to a material called DDD. By that means, and also because I was interested in peregrine falcons, I got involved in the DDT controversy, which was significant.

I was at Berkeley when Rachel Carson's *Silent Spring* was first run serially in the *New Yorker* magazine.[8] Because I was working in the department that was favorable to naturally occurring predacious and parasitic insects, I was pretty close to the people we call nozzle heads—the spray boys. Consequently, I had an unusually close perspective on the effort to beat back Rachel Carson's absolutely splendid book. A book which, by the way, holds up very well today, forty years after the original publication.

That was an environmental-engagement battle, if you wish, that was a learning experience for me. I found out what the "other side" could do—what dishonesty could get them.

Then in 1970 I did the first survey of nesting peregrine falcons in North America after that species had become virtually extinct over much of the continent. I looked at a situation in California, where there had been a hundred nests before 1945. I looked at sixty-eight of those nest sites, and found only two that had breeding birds present. That was another part of that whole battle: demonstrating conclusively that DDT was the prime cause of the eggshell thinning that led to the serious decline.

That's all positive history now, a virtual miracle that once DDT was banned, these birds came back. Of course, that's a small part of the story, but the whole battle impressed me with what I saw then as the value of science in environmental battles. I thought that science was going to win a lot of environmental battles. And it is science on which I've tried to rely in the grazing situation. But that has had mixed results. I'm a little cynical these days about the value of science in fighting environmental battles, especially under the current political regime. But I believe strongly in truth, and science usually qualifies in that arena.

After I finished the research and course work for my PhD at the University of California at Davis, I went to the University of California at Santa Barbara, where I taught for a brief time. Then I was invited to interview for a teaching/research position at Evergreen State College in Olympia, Washington. I was accepted there and have been doing that since 1971—teaching natural history, ecology, animal behavior, and related courses and programs at this small, state-supported school.

CHAPTER 2

Cattle grazing at Malheur NWR

My second real encounter with the grazing problem was when I started going to Malheur National Wildlife Refuge.[9] This was after I had met Denzel Ferguson,[10] another figure, a much more prominent one, in the anti-DDT fight. He and I ended up testifying on that issue in Washington, DC, together in December of 1972. I had already visited the Malheur Field Station with students, where he was then the director.[11] There was a big issue, as there is today, with grazing at Malheur. And we began fighting it.

In those days it was still useful to write letters. And Denzel wrote some beauties. On several occasions, we were able to confront the US Fish and Wildlife Service people, who were responsible for allowing this abusive grazing on that refuge. We were also able to involve the regional office in Portland. There was some promise at that time that some progress would be made. And some progress was made.[12]

Impacts of cattle grazing on birds

The livestock grazing at Malheur occurs in winter, but there were still cows out on Harney Lake in the spring.[13] I had experiences as late as 1981 with those. The first snowy plover nests I found on that lakeshore contained eggs that had been crushed by a cow. When I protested this, the refuge manager said, "Oh, don't worry about that Steve, they're just trespass cows," as if somehow, if the cows were trespassing, they had some sort of lease that would absolve them from any responsibility. Not so, I thought.

The refuge in many areas at the end of the winter and early spring is just a wallow. One sees the effects of cattle trampling—the compression of soil, the destruction of vegetation. In some cases, even nesting birds are impacted.

There's a significant population of greater sandhill cranes on the refuge, and their eggs are eaten by predators of various kinds. The problem for the birds, at least in part, is that they are forced to nest in areas that have been grazed off during the winter. Consequently, if both adults leave the nest for a short time to feed, the eggs are made very vulnerable by being in a place where there is little cover.[14] Of course, the solution of the manager in this case is to control the predators, not the cows.[15]

Cattle grazing as a management tool
John Scharff[16] was the refuge manager at Malheur in 1966, and he was still there in the early 1970s. He was a great believer in the management value of livestock grazing. But on the other hand, he was also a good naturalist. So in some respects, I would forgive him his view, because it was a very common one at that time.

But his successors have no justification to believe in the management value of grazing. The current refuge manager apparently is going to install some public display that will describe the values of grazing to wildlife. As nearly as I can tell, there are none in that situation. Yet they'll go on claiming it.

There are myths galore. One of them that the very distinguished scientist, Joy Belsky,[17] refuted expertly was the idea that grazing or browsing a plant stimulates it to grow.[18] That's one of the favorites, especially for the "deer people"—the belief that cows browse bitterbrush and stimulate leader growth, and then that those leaders, the year's growth, feed the deer in the winter. Certainly, plants respond, but they respond defensively to that kind of attack. And the value of it is, at best, questionable.

There are several major myths that George Wuerthner has outlined that we see in all sorts of publications.[19] Most of them are more or less easily refuted, but very popular. They just cycle and recycle as time goes by. The idea that grazing is a management tool has begun to fade, especially with younger agency personnel. I think it's just become an embarrassment to claim this.

CHAPTER 3

Cattle grazing at Hart Mountain NAR
My main focus, personally, over the decades was on trying to get privately owned cattle removed from Hart Mountain National Antelope Refuge in

south-central Oregon near the California and Nevada borders. I began teaching natural history there, usually bird courses, in 1976. The grazing there was so outrageous, so blatant, so destructive, that I began to be quite vocal to the authorities—the Fish and Wildlife Service people—about the necessity to get the cows off.

Of course, I was told the usual lies: this is a management tool; this reduces fuel for fires—the whole litany of stuff.

The refuge manager at that time was not too far from retirement, and was a rather engaging fellow who would talk to my students[20] and sometimes convince them, in the face of overwhelming visual and even olfactory evidence, that these cows were all right. He was the one who told me that he was managing for horned larks. Marv Kaschke was his name.

Eventually, the US Fish and Wildlife Service was sued to remove the cows.[21] That's what got them off. And since that time, there's been an enormous amount of recovery, not only in the vegetation, but also in the wildlife. There are more pronghorn antelope there now than have ever been counted before. That may well be true of sage-grouse as well. In any case, sage-grouse have come back from being a very seldom-encountered bird to a species that, on the refuge, is encountered with great frequency. Numbers of male sage-grouse on leks have increased dramatically since the cows were removed.

I teach a summer bird course there, and on July 28th of this year my dog put up fifty-six sage-grouse on a little walk just outside camp.

CHAPTER 4

The last years of cattle grazing at Hart Mountain NAR

From the earliest days, even those refuge managers who supported grazing also maintained a fence around a sizable piece of landscape called Buck Pasture. It was sort of a showplace. There's a creek that runs through it—Willow Creek. On the upstream side, when the cows were there, the water would be flowing over bare rock until it went under the fence. Beyond the fence another twenty feet, you couldn't see the stream, because it was a nicely covered slot, as a natural stream would be.

At the downstream end, where Willow Creek came out, there was a fifteen-foot hunk of erosion and lots of evidence of grazing.

I was there with an Oregon State University range ecologist probably about 1991, when the effort to get the cows off was heating up. We were standing at this site, and I said to him, "Look at that headcut. And look at the pasture above that. What do you think caused that headcut."

He looked at me and his voice went down a little ways. And he said, "Steve, I don't know. I wasn't here."

The other thing that happened there was at my camp in 1985. The refuge let a prescribed burn get out of control, and the resulting fire burned 13,500 acres. It came rushing through my little aspen grove. But just before the fire got there, my friend on the refuge staff had a water drop right where we had our kitchen. So that area was saved.

Because of my constant "bleating," it was arranged by the Fish and Wildlife Service that that area would be free of cattle from that time on, beginning in 1986. They put fences on the paralleling ridge tops. Previous to that, when I started working there in 1981, cows were everywhere. We would get up in the morning and my kids and my dogs and my wife and my students would chase the cows out, so we could put up our nets to catch and band birds. Cowpies were everywhere. Willows, aspens browsed down.

When cows were excluded in the spring of 1986, there were old-growth aspen present, but virtually no young ones.

I said something to my friend in the Fish and Wildlife Service about this—"Cows eat all the suckers that come up."

"Oh, no, Steve, that's not a cow problem. That's a human problem."

I said, "Really? Tell me about that."

He said, "In the fall when hunters come in here, they use the downed aspens for firewood. And those downed aspen logs would protect those aspen suckers from browsing by the cows if they didn't do that."

Here was a man who could tie his shoes with great skill. I'm sure he had other motor skills as well. He was a smart guy. And even he believed that crap.

When the time came to get serious about taking cows off, which was about 1992, these aspens were significant thickness and height. And so they provided a preview of what the entire refuge would look like in those situations if the cows were removed.

The refuge manager, Barry Reiswig, kicked my students and I out for a period of time so they could have workshops. They brought the cowboys in. They brought the managers in. They brought the environmentalists in. All they had to do was go like this—and point. Then take them to the next draw that was still grazed—and point. They didn't have to say a word. It was blatantly clear.

And so that was one of the precursors to the removal of cows from that 269,000-acre refuge.

CHAPTER 5

Management of Hart Mountain NAR after removal of cattle

There's a huge controversy over the role of cattle grazing in the decline or sustenance of sage-grouse. There are a couple of papers that demonstrate that one of the main determinants of good reproduction in sage-grouse is the presence of what's called "residual cover"—a nice little euphemistic term that just means "grass."[22]

When the sage-grouse hens build a nest on the ground under sagebrush—if they have a good batch of cover around that nest, predators are not nearly as effective in preying upon the eggs. Good cover is a big determining factor. There's another study with fake nests, which I don't believe in very much, where they just dumped out a bunch of eggs and checked if something found it. It's like dumping a newborn baby on a mall parking lot, and wondering when someone will run over it.

There was also an effort to control coyotes for pronghorn protection a few years ago, perhaps five years ago. It went through the process. And at the last minute, one of the animal rights groups—Predator Defense Institute—intervened and got it stopped.[23]

The next spring there was a higher survival of more fawns than ever before. Close to that, anyway. If they had gone ahead with that control program, they would have claimed that it was responsible for that reproductive success. Now there are more pronghorn antelope on that refuge than had ever been counted before.[24] And they're everywhere. Absolutely gorgeous, every single individual.

CHAPTER 6

Politics behind efforts to kill coyotes at Hart Mountain NAR

Support for killing coyotes comes mostly from the local community of Lakeview, and people who are convinced that in predator control lies the salvation of the world. They're very vocal. They believe in it. They like to kill things, whether they be coyotes with rifles, or ravens with poisoned eggs.

Or golden eagles with whatever. The amount of passion that's expended in hating problem animals or predators is enormous. And you see it manifest all over the place.

The people of Lakeview—what our current president[25] calls local control—the people who represent a very small percentage of the American population, which is what really owns those refuges, have an enormous amount of control. And they have traditionally dictated policy.

What needs to be understood about all of these resource managers is that while they may have gone to any number of very excellent schools, they're embedded with the local community. They are obligated to get along with the local people. If they have children, this is especially a barrier to their objectivity. Because if their kids are going to school with the cowboy's kids or the "timber beast" kids, those kids are vulnerable. And they're often singled out and attacked physically if their daddy, now, I suppose, their mommy too, is advocating a policy that the local populace disagrees with. Very powerful, very powerful factor.

CHAPTER 7

Cattle versus pygmy rabbits in Washington state
I've also fought grazing on public lands in central Washington, mainly in a place called "Sagebrush Flat" in Douglas County, where cows were grazed with the last Washington population of the pygmy rabbit.[26] This year, for the first time, pygmy rabbits have not been encountered in the wild in the state of Washington.

That was a major battle, which I engaged almost alone. And I had to endure the usual absurdities. The most paramount of which was a Washington Department of Fish and Wildlife man who leaned forward in his chair and said, "Steve, we're afraid if we lose the cows, we'll lose the rabbits." The rationale being that the cows, by eating the grass, increased the density of sagebrush, which was what the pygmy rabbits ate.

Stupid. Stupid, in part, because that's not all the pygmy rabbits ate. They required perennial bunchgrasses and forbs during the spring—during the reproductive season. And furthermore, sagebrush, in an environment like that, could never be limiting.[27] So by increasing the density of it, you weren't doing any favors for the rabbits.

The government went the length of putting together an elaborate research plan to study whether the cows and the rabbits were compatible. They also claimed that they were compatible.

Well, I pointed out in one document that people often have cancer for long periods of time. And using similar reasoning, you could say that the cancer and the body that the cancer was in were compatible. But notably missing from this research plan, which had to be approved by the University of Washington statisticians, was mention or recognition of the fact that here was the only fossorial (the only burrowing) rabbit in North America. Well, there's one more, the volcano rabbit in Mexico. But this is the only burrowing rabbit that weighs less than a pound, essentially pitted against sixteen-, eighteen-hundred-pound animals.

When they did the experiment, as usual they had an area that would be ungrazed, an area that would be grazed to a moderate extent, and an area that would be heavily grazed. But there were a few parts of this—I forget how big, but a couple square miles—where they took places that were far enough from water that they hadn't been completely devastated by cattle. And they put those into the areas that were to be grazed most heavily. They brought water in for the cows. And by this means they were able to essentially wipe out the last of the real shrubsteppe—"real shrubsteppe" being communities that are characterized by one or more layers of perennial grasses above which there rises a conspicuous, but discontinuous layer of shrubs. The cows, of course, ate the grass. So this is sort of like a clearcut in reverse. Instead of removing the overstory of trees and leaving the understory, the cows get the understory of grass and leave the overstory of shrubs.

At the end, I got a desperate call from a Fish and Wildlife Department employee, who had just been on the property and found three rabbit burrows that had been caved in by the hooves of cows. And he said, "Dr. Herman, Dr. Herman what are we going to do? They claim this has no effect, but they're not looking at this, and they're not looking at other things."

Eventually, when the rabbits were down to virtually nothing, there was an emergency listing and the cows came off in 2001, I believe. Now the place is without cows, but certainly not without the damage that they did.

I want to be clear that I'm not claiming that the cattle grazing alone resulted in the extirpation of this limited population of pygmy rabbits. But the question is the extent to which it contributed.

Other factors may have been involved. They really weren't studying the matter. They just waited until they got down to under twenty rabbits and then decided to put them all in captivity.

That captive breeding effort, of course, was very popular. The first time a little bunny was born to one of these captive rabbits at the headquarters of the Washington Department of Fish and Wildlife here in Olympia, they sent out for carrot cake and they had a little celebration.

But it turns out that those animals have not bred successfully. They're getting some captive breeding, but only by bringing in animals from Idaho and elsewhere. So the Columbia Basin pygmy rabbit is essentially extinct. And it was a unique form, a unique subspecies.[28]

The Douglas County site was the last big hunk of deep-soiled big sagebrush in the state of Washington. When the settlers came in a hundred and fifty years ago, they made choices as to where they were going to homestead based on the height of the sagebrush. The taller the sagebrush, the deeper and better the soil. So those deep-soiled sites that had loose, friable soil that could be utilized by a little animal that has to live in a burrow were gone early on. This relict landscape survived only because of some family feud.

When I got hold of the problem in 1993, the area was owned and managed (so-called) by the Washington Department of Natural Resources. They had a program that would allow places like this to become natural area preserves. There was a board of very distinguished men and women who made these decisions. And when Sagebrush Flat was brought up to them as a possible natural area preserve, they voted unanimously to make it one. Then they went beyond that and voted unanimously to have the cows taken off immediately.

Their recommendation would normally be followed by the commissioner of the Department of Natural Resources, who was widely alleged to be an environmentalist. And she vetoed it. The site was subsequently sold by the Washington Department of Natural Resources to the Washington Department of Fish and Wildlife, which immediately increased the grazing. The whole thing was just insane. It remains insane.

The plan now is to breed the rabbits in captivity and release them in good habitat. Recently I heard about a site that they have purchased.[29] The fellow in charge of the pygmy rabbit recovery said, "We're getting a fence up around that site."

I said, "Why would you want a fence around it?"

"Oh," he said, "to keep the cows out."

I said, "Gosh, why would you want to keep the cows out? You told me for a decade that the cows and the rabbits were compatible."

And he laughed, because I caught him.

Then he said, "Well, let's just say that the present owner doesn't want cows there, Steve."

These people have no guile; they just have careers.

CHAPTER 8

Federal agencies ignore public comments about land management

My theory is that the agencies learned from Nixon's[30] downfall how to behave. It's called stonewalling. You just tough it out.

There was a time into the early 1970s when it was meaningful to write a letter to the government. You wrote a letter, and you received a response. If enough people wrote letters, there was often some sort of action taken.

When NEPA, the National Environmental Policy Act, first came out, and there were environmental impact statements, those were, for a while, taken seriously. One could spend a few hours with an environmental impact statement and reply to the agency, and feel that he or she was being treated with some fairness. I think that all broke down about the mid-'70s.

In 1980 I was driving over the Ruby Mountains with a friend studying snowy plovers in western Nevada. We came to a place near the summit, where I had gone in 1957 to fish, and I now got out of the car for a better look. The area was devastated by cows. There were aspens down, and the dead aspens had been pounded into powder. There were virtually no willows along the stream. The stream was muddy. It was full of emergent vegetation that clearly was the result of nitrification—the application of cow poop.

I drove down to the next ranger station, and very politely I struck up a conversation with one of the employees. I said, "You know, I was on the allotment at the top of the hill. And there were an awful lot of cows there. I wonder if you think that might be a little overstocked and overgrazed."

The Forest Service employee looked at me and said, "That may look overgrazed, but it's just beat down."

A little later he said, "Why don't you write me a letter? I don't have anything to do all winter, and I love to write letters to professors."

CHAPTER 9

Public indifference to the decline of rangeland health
One of the things that really troubles me is that the American public doesn't seem to care much about what we call rangelands—a horrible term. But it's impossible, at least so far, to come up with an alternative one. Somebody's called them sagelands, and that's pretty good for part of it.

But people don't care about these areas. They find them mundane, boring, tedious, dry, and so forth. And instead, they tend to favor, and adore, and defend things like old-growth timber, which we used to call virgin timber. I am certainly a great defender of virgin forests, and that sort of thing. But it is a little bit galling that they don't catch on to the extraordinary biodiversity, and beauty of these landscapes that have grass and shrubs on them, whether those shrubs be sagebrush or greasewood or bitterbrush or whatever.

Teaching as I have for many years at a college in western Washington, where we went through the 1980s and '90s defending old-growth forests, I've seen a lot of activism among students, and among plain citizens who are motivated to do all sorts of things to defend these woods. It's very difficult to find a constituency that will defend shrubsteppe.

So I have to say that American conservation will have matured when hippies are chaining themselves to old-growth sagebrush. I'd like to see that day, but I'm not holding my breath.

I think that the effort to educate people about the beauty of that landscape is really very, very important. I've got a series of photographs of myself at the place where I teach students to band birds on Hart Mountain. One place, taken in the early '80s, has me sitting at a card table by a willow tree. All of me is visible. The next photo is after the aspen started coming up and the willow tree, which is unique and recognizable, began to leaf out more. I have a third photo where you kind of have to look to see me.

Then I have another series taken this year, just less than two weeks ago, where you'll have to look around to see me, because I'm virtually surrounded by thick aspen trees. That's all cows. That's the only difference—the absence of cows plus time.

I don't know how optimistic I am in the long-run. I think that the permit buyout program is a great hope.[31] And I certainly support it now, although I was very reluctant at the outset. I'm just hoping that I will live to see cows removed from a lot of public lands acreage. Maybe I will. Maybe you will.

CHAPTER 10

Tactics to reduce livestock grazing on public lands

I think that the tactic of conservation organizations bidding on grazing leases is a really good one. That happens a little bit here in the state of Washington. But that's often difficult, because the scales are weighted in favor of the cowboys.

I don't have very much faith in education alone, although I think it's a swell idea. As I get older, I'm much more concerned with beauty than I am with data. And I feel that we should elevate our argument to include a consideration of the aesthetics of landscapes. It's my understanding that there's something in the National Environmental Policy Act that would be receptive to that approach.[32]

When I was at a workshop on Hart Mountain in the run-up to the removal of the cows, when all stakeholders were present, a guy named Doc Hatfield,[33] a rather loquacious Oregon rancher, kind of took over the opposition. He got up and said to all of us, "Tell me, what do you want here? What is it you want?"

And I said, "A cow-free landscape."

"Oh, no, no, no, no, Steve. What is it you *really* want?"

I said, "A cow-free landscape."

And he couldn't deal with that.

So I think that it would be useful to experiment with looking at that level of argument. That's much of what we do with our slideshows anyway. Certainly bare ground—that's deleterious to some wildlife. What we're showing, for the most part, is damage to landscapes. Visual damage. Aesthetic damage. It's just a waste of all that beauty to deface it with cows or sheep.

Steve Johnson

Steve Johnson arrived in the American West from Virginia as a youth in the mid-1940s. He earned bachelor's and master's degrees from the University of Arizona, then went on to teach junior high, and high school biology until 1980. An interest in photography and a love of nature led to Johnson's awareness of the pervasive environmental impacts of livestock grazing on public lands. From 1972 until 1989 Johnson served as the southwestern field representative for Defenders of Wildlife. In recent years, he has consulted for several environmental organizations about the impacts of livestock production on public lands and endangered species.

Steve Johnson made his remarks on the 15th of September 2003 in Tucson, Arizona.

Chapters
1. Johnson's initial experiences with public lands ranching (1960s)
2. Johnson's activities as a school teacher (1968–72)
3. Johnson's activities at Defenders of Wildlife (1972–89)
4. Johnson learns of the banking connection to federal grazing permits (1978)
5. Livestock grazing degrades the southwestern landscape
6. Stripped of vegetation by cattle, streambanks quickly erode
7. BLM's failure to protect the desert tortoise
8. Sources of ranchers' political influence
9. Subsidies for public lands ranchers
10. Economic insignificance of cattle ranching on western public lands
11. Lawsuits improve cattle management on public lands

CHAPTER 1

Johnson's initial experiences with public lands ranching (1960s)

I'm a former biology teacher. And have lived here in the West since the third grade. From Virginia before that. I've always been an outdoors kind of person. Did a lot of hiking in connection with the teaching of biology. Took my kids on overnight hikes during the year. Had ecology clubs. All the things that in the mid-to-late '60s were just starting out. The Earth Day had not yet happened,[1] but I was already doing activities which later would be appropriate for it.

So being outside. Watching. Seeing the evidence of cattle. It was just kind of a background. The cattle shit was everywhere. I just assumed, never thinking about it, that we must produce a lot of cattle on western public lands, because I saw so many signs of the cattle, if not the cattle themselves. It took me a long time to realize that really very little was produced in terms of beef by the cattle that I saw. And yet I was increasingly aware of damage done by the cattle to everything that, to me, was most important. That is, the natural settings, the plants, the native wildlife. It took a long time, though, because I was never conscious of it as an issue until I began to really look at it.

My early visits to the Bureau of Land Management offices, in particular, were eye opening. To begin with, it was my impression in the mid-to-late 1960s that they'd never seen anybody in their office that wasn't a rancher. They were puzzled as to why anyone would care. Equally puzzled as to really what their job seemed to be. They didn't know how to respond to questions about the public lands themselves. It was my very strong impression that they rarely thought of them as public lands, because they were so into being the mouthpieces for the ranchers. They were there to serve the ranchers, and they did not know how to deal with someone who was not a rancher. They particularly did not know how to deal with someone who was asking questions that they obviously had never been asked, and for which they had no answers.

And so that was puzzling to me, and it was the beginning of what became a years-long process of gradual awareness and research. From then it grew until my activism really speeded up a great deal around 1972 to '73.

CHAPTER 2

Johnson's activities as a school teacher (1968–72)

Back in 1968 or so, I was teaching 8th grade—I preferred the junior-high age to high school, which I've also taught. More enthusiasm. More problems. But ultimately, I thought, more rewards too, in terms of being a teacher.

At the beginning of each year, I had to restock my room. This time I had to buy some rats to breed, so that I could keep my snakes going. I had a lot of snakes. And so I was visiting a friend of mine at a local pet store here in Tucson. I noticed that he had in a cage, near the back of the establishment, an alligator—about a three-foot-long alligator. I went to look at it to determine if it was, in fact, an alligator, which at that time was federally protected.[2]

It was an alligator. And the cage was filled with feathers and various detritus left over from the chickens my friend was feeding it.

I told him, which he already knew, that this was an animal he wasn't even supposed to have in his possession.

"Yeah," he said. "A couple of kids banged on my door and had it in a gunny sack. They asked if I'd take it."

Now, he didn't know what to do with it.

I contacted the Fish and Wildlife Service and told them the situation. I then decided to try raising money to send it back to protected habitat in Florida. That became my school project for the year.

To raise money, we collected aluminum cans. And the kids really got into it. The whole school smelled like a brewery from all the empty beer cans in the halls. The first prize was to be a ten-speed bicycle for the person who brought in the most cans. Second prize was a backpack, I believe.

It was lots of fun. We had alligator-naming parties. All the things that you do to get kids interested and keep them going. And the kids were bringing me things to feed the gator—beef steaks, roasts from the home freezer. Whatever they had. And the gator did very well.

We not only raised enough money to fly the alligator to Everglades National Park, but we had so much money left over that the ecology club, on its own, chose four different environmental organizations to give the excess money to. Since I was then chair of the Arizona Chapter of the Sierra Club, I disqualified them as a recipient. So the students chose four others. One of them was Defenders of Wildlife.

There happened to be a board member of the organization who lived here in Tucson, Ted Steele, who has since passed away. Quite a man. He showed up at the awards assembly as a representative of Defenders of Wildlife to receive the donation, which was four or five hundred dollars. He was really overcome by the whole thing. And he offered me a job to work for Defenders of Wildlife, which meant I'd have to take a leave of absence from teaching for a year. This was in 1972.

So that's how it started. I finally resigned from teaching for good in 1980. And I worked for Defenders on and off until 1989, when they closed the Southwest Office.

After that, I began consulting for a large number of national groups: Humane Society of the United States, Southern Utah Wilderness Alliance, National Audubon Society, the Desert Tortoise Council. But now my only client is the Humane Society of the United States. I represent them on public lands and endangered species issues.

CHAPTER 3

Johnson's activities at Defenders of Wildlife (1972–89)
I did a lot of writing for their magazine and a lot of field investigations. At that time, Compound 1080 was a big issue. It was the time, the early '70s, when there was a rancher in Wyoming, forget his name now,[3] who poisoned large numbers of eagles. Due to the fuss brought up by the killing of the eagles, this man was mainly responsible for Nixon's executive order restricting the use of poisons to kill predators.[4]

In doing some work on that, I found myself getting more and more into, not the grazing issue per se, but the animal damage control that was perceived as necessary to keep sheep and cattle safe on the public lands. And since sheep and cattle grazing was in itself a colorless, nearly invisible issue at that time, and the killing of mountain lions and grizzly bears and eagles and coyotes anything but invisible, it was a good way to bring attention to grazing.

"The winning of the West"—that hackneyed phrase really was the losing of many of the things that make the West most interesting to me: the grizzly, the wolf, the jaguar, the black-footed ferret, any number of creatures that were wiped out early on. And we had no voice in that. As I grew to realize that, I became less dispassionate about the issue, and more and more angry. I also had a lot of foreboding about how difficult it would be to fight.

Grazing was here in the West before the Forest Service; before the BLM. Those organizations came into existence, making room for what was already present. Whenever that happens, you're going to have difficulty getting any regulations that really address the issue. Mostly we got regulations that go toward establishing and keeping the grazing going. That's basically been true until about the last ten years. I would say that we've now begun to have so many people questioning this widespread use and abuse of the lands that change is coming. In order to see accomplishments you have to take a long-term view, but I can clearly state that things are moving in the right direction, in my opinion.

CHAPTER 4

Johnson learns of the banking connection to federal grazing permits (1978)
Reducing cattle numbers on public lands was a primary goal for many of us in the environmental community in the 1970s. But it was almost impossible to do. And I realized why when I discovered "permit value."

It was in the late '70s—'78, I believe. A banker was at a symposium I was attending in Texas. I was there to give presentations about grazing numbers and perhaps Animal Damage Control. But he was talking to me privately about how many of the ranchers were financed by his bank, which would actually loan money to a rancher based on the numbers of cattle he had under federal permit—not on the value of the private land on which the ranch house stood, but on the numbers of cattle that could be grazed on federal lands.

Well, that was a whole new thing to me. I suddenly realized that if you had a four-hundred-cow ranch, and you had a loan based on four hundred cows, then any attempt to reduce the numbers of cattle that you had, would not only affect the rancher, and his ability to pay it back, but the bank would be affected too, because its loan would be at risk. And this was another powerful element in the political system that would fight any reductions of cattle for the most fundamental of reasons. It also helped to explain why US senators often intervened in BLM and Forest Service grazing decisions and why so many agency employees were transferred.

I knew that ranches were financed, but I had always thought that the main value of a ranch was in the private land—the base property. Well, I've discovered since then that base property can be nothing more than water. If you have a well that produces water, that's base property to the BLM, and sometimes to the Forest Service. You don't even have to own any place to put a house on. So learning about permit value explained a lot of things I didn't understand before about the difficulty of the issue. And it gave us some new ways to work on it.

CHAPTER 5

Livestock grazing degrades the southwestern landscape

It's now thought that the Southwest was sort of balanced on a knife edge, environmentally speaking, at the time when cattle were first brought here.[5] There was a relict grassland possibly left over from the Pleistocene[6] when more rains fell than now. And the grass was doing okay. At least it looked liked it was. No one really knew what grass needed then to survive. But it had no real ungulate grazing pressure. The native wildlife that was here at that time in Arizona and most of New Mexico did not include the bison. The grazers were the pronghorns and deer, which are mostly browsers. Elk, which graze much like cattle, do not occur in the desertlands. So the grasses were not very heavily utilized at that time.

Anyway, the cattle came in, ate the grass that was barely hanging on, but doing okay, and then the grass just didn't survive. For the most part, our native grasses are gone in Arizona. They've been replaced by exotic species that came in through the droppings of cattle.[7] Or, in many cases, were introduced on purpose by the Department of Transportation to create roadside berms for erosion control.[8]

As the Pleistocene passed, and aridity really began to take hold on the West, the riparian areas—the few places that had running water—assumed a huge role, because they were the only places that water could be found almost any time of the year. And they'd never been more than perhaps 3 percent of the landscape. But more likely 1 to 2 percent.

As the Pleistocene receded and dryness took over, wildlife species increasingly moved closer to these green ribbons of life. Perhaps 75 percent of all vertebrate species were dependent on these narrow ribbons of green.

And, of course, cattle live in them. Unlike bison and other wildlife, cattle don't like to wander very far from water. About two miles is the maximum. And in the hot summers a mile is about the maximum that cattle like to get away from permanent water.

So with the introduction of cattle, the stage was set, even then, not only for the destruction of the dry-land grasses, but also for the drying up of most of our streams due to the cows' persistent eating of the plants that depend on water—cottonwoods, and so forth.[9]

CHAPTER 6

Stripped of vegetation by cattle, streambanks quickly erode

The destruction of the uplands, the grasses, and other plants that formed the watershed, of course, set into motion a greater runoff when the rains came.

There was a well-documented case in 1892 here in Arizona. There'd been a prolonged drought. I don't like the word "drought." To me, we have occasional mitigations of moisture. Drought is what is here all the time.

But in 1892, it was drier than usual. Governor Hunt of Arizona[10] recalled that one could throw a rock from one dead cow to another over much of southern Arizona. The die-off was huge. And these were not soft white-faced Hereford cattle. These were tough longhorns. Cattle that could survive nearly anything. And *they* died.

Then the rains finally came. And the Santa Cruz River, which then still flowed, which then still had beaver and some fish in it, in about two to three weeks incised its banks twenty to thirty feet down from the heavy rains, and from the grasses that were no longer there to hold the waters back.

Then the booms and busts that had been a normal part of the desert here increased in severity. The booms were greater because of greater runoff. The busts were more severe because the water table had dropped. It dropped so much that even the deep-rooted mesquite trees along the banks of the Santa Cruz—away from where the cottonwood trees were—began to die. That meant the water table had really dropped, because they have a really deep taproot.

Anyway, that's what happened, not just here, but wherever annual rainfall is less than ten inches. This just happens to be one of the best documented cases of the death of a river. The cottonwoods collapsed into the deep channel. Water tables dropped.

Yuma used to be flooded almost every year by the Gila River. It ran all across the desert down to Yuma, where it joined the Colorado. Hard to believe today. Gila Bend used to be a place where the Gila bent as it flowed by. But no more. It's just a dry place there with nothing to show that it ever was alive.

People today, who want to know about the West that they lost, have to learn to miss what they never knew. And you do that only by reading the accounts of people that were here at the time. That's what all of us have to do. If we ever want to get back any vestige of what has been lost, we have to live with an awareness of what was, not just with what we see now.

CHAPTER 7

BLM's failure to protect the desert tortoise

The desert tortoise is the one species of wildlife that most opened my eyes to the failure of the Bureau of Land Management to do any management for the benefit of anything but ranchers.[11] Consider the BLM's ephemeral-grazing policy. Ephemeral plants are flowers, and all the things, that make the desert beautiful in those rare years when there is really a lot of rain that occurs at the right times. And at these times, BLM permits ranchers to apply for extra grazing. So the ranchers are able to flood the desert with more cattle when the desert has that unusual, and very fleeting, abundance of plants.

Domestic sheep can create even more of an environmental problem, since water can be hauled to them, keeping them out longer, and spread over larger areas.

Ephemeral grazing, therefore, ignores the entire survival mechanisms of a whole range of desert creatures. The desert tortoise is probably the best known. It's been the most studied. It's an animal that lives perhaps a hundred years. Certainly, it has a breeding period of at least sixty years. And it can have more than one clutch of eggs during a year if conditions are good. But for most years, conditions aren't good. They may have a clutch, but they may have no survivors.

A desert tortoise, in order for a stable population to be maintained in sixty years of breeding effort, has to produce only one young that survives to successfully reproduce. A pair has to produce two tortoises. Thinking about that for just a minute would tell anybody with half a brain that life is tough out there—that those times of rare abundance in the desert are needed by the chuckwalla, and all the other creatures, because they have so many summers when they don't have anything, particularly when the cattle and sheep are here, as they have been for more than a century.

So you have an animal like the tortoise, who is underground for about 95 percent of its life. It's the closest to a "pet rock" that you can think of. It needs almost nothing. It eats less in a year—about seventeen to twenty-three pounds of plants—than a cow eats in a single day. And if you have an animal like this dying of starvation, do we need anything else to tell us what we've done to the capacity of the Mojave and Sonoran deserts to support life?

CHAPTER 8

Sources of ranchers' political influence
Ranchers have been here "forever," and many of their families are in state legislatures. They're in the Congress. They're senators. They're US representatives. They are on the Supreme Court, in the case of Sandra Day O'Connor,[12] who is the daughter of the family that has the biggest BLM ranch holdings in Arizona. Ranchers are in positions of power at almost every level. That comes from being here a long time. And that's one of their greatest sources of power.

Then there's the permit value that I referred to earlier. The fact that they are financed, based on the numbers of cattle that they can grow on land that belongs to you and me, is another source.

But inertia is the main reason. They have power based on the inertia of their vanished past, when it was generally, not just assumed, but *known*, that they gave us a lot of food. Today we know they don't. But like an arrow shot from a bow, the energy of the past still infuses their current situation. And very few people have caught up to what the reality of it is.

In the winter of 1988, I presented a paper at a conference of the Society for Range Management in Las Vegas. And "range management" is probably the ultimate oxymoron. But I told them that growing beef in the arid West was like growing tomatoes on the moon. You could do it if you were prepared to expend a lot of effort and work really hard. You *can* grow tomatoes on the moon; you *can* grow beef in Arizona. But they're gonna be very expensive tomatoes, and very expensive beef.

Missouri produces 8 percent of the beef in the US. All the western federal public lands combined produce 2, perhaps 3 percent. But more likely 2 percent of the nation's beef. Florida, eastern Texas—anywhere it rains—produces a lot more cattle with far less ecological cost.

So it's kind of like debating the number of angels that can dance on the head of a pin. We're having this discussion about grazing as if grazing mattered to the economic scheme of things on the public lands. It doesn't matter. It only matters to about twenty-six thousand people who are lucky enough to have permits on our federal land, and who are subsidized by all of us to do what doesn't make any sense to begin with.

CHAPTER 9

Subsidies for public lands ranchers

If the animal unit month cost is now $1.35,[13] which I believe it is, that's the lowest it can go, because of Reagan's executive order.[14] If he hadn't had that in, who knows how low it would be. But they pay $1.35 per cow, and usually a calf, per month. And that works out to $16.20 a year.

A cow eats roughly twelve thousand pounds of plants during that year. How long could you feed a horse or a cow if you bought all the hay yourself for $16.20? The price of a bale of hay is about $8.00 now. So that would be about two bales of hay. It would be a pretty hungry horse by the time those two bales were gone, and there was no more money.

It's a joke. It's a bad joke, but it *is* a joke. And that's just the most obvious subsidy.

The real subsidy is that which is paid by all of us who live here and depend on water—who depend on an aquifer that is healthy, depend on floods not coming more frequently than they need to—because we've destroyed all the watersheds throughout the West with cattle.

We build dams to impound water so we can live here. And yet those dams are silted up much more rapidly because of the overgrazing and the sheet erosion that happens after every rain because the grasses are gone.

We subsidize at such levels that it's almost inconceivable. So the grazing fee—that's just money. But the real subsidy is what it costs all of us who want to see something alive on the land besides a cow. That's what it really costs us. And that's the subsidy that makes me furious.

CHAPTER 10

Economic insignificance of cattle ranching on western public lands

The environmental impact statements (EISs) on grazing, which were done as a result of Johanna Wald's NRDC lawsuit forcing the BLM to prepare these impact statements, were a great boon to people like myself, who like to mine these statistics that the BLM is so reluctant to release,[15] or that they don't even know. But as a result of the lawsuit, they had to make an effort to obtain the information.

Almost all of those EISs have a socio-economic section in them. And in no place that I'm aware of in the EISs that I have read, would ranching, if it disappeared, be missed more than a little bit. It would not be missed nearly as much as other parts of the economy.[16]

Many ranchers are part-timers. They work somewhere else to support their "cattle habit."[17] They have jobs as teachers or other things. They really don't have enough to keep them busy if they're just grazing cattle.

You can drive through any small town in New Mexico or Arizona almost any time of the day, and you'll find a sea of hats and boots in the cafés. Ranchers are sitting there visiting, because really, in my opinion, it's almost an unrelieved coffee break to raise cattle in the West. The cattle have got salt. They've got water. The ranchers know that the cattle will come to the salt and water eventually. And then they can round 'em up. But that's it. They don't even need very many people for that.

And so the economic influence of these folks is just nil. We could get along without them very well. Any economic analysis that's honest will show that very clearly.

CHAPTER 11

Lawsuits improve cattle management on public lands

It's been really exciting for me to see the Center for Biological Diversity, based here in Tucson, and their willingness to sue and sue and sue. I really support that completely. I think it's too bad that you have to resort to litigation. But when you have a system that is so wrong, for so long too, something has to be done with that, even though it irritates a lot of people. There's just nothing else to do.

I've lobbied. I've gone to DC a lot. Worked very hard on these kinds of issues. We've tried to do something about grazing fees. All manner of things have been tried over the last twenty, thirty years. Progress has been very slow. So slow, in fact, that we can't really afford it.

Ralph Maughan

Although raised in the midst of Utah's livestock-degraded landscapes, Ralph Maughan did not take a great interest in environmental conservation until he returned to the West after completing a PhD in political science at the University of Wisconsin–Madison. Since that time he has been involved with several environmental organizations. Maughan was a co-founder of the Greater Yellowstone Coalition, and he has held leadership positions in the Sierra Club, including that of chair of the Northern Rockies Chapter. Currently, he is a professor of political science at Idaho State University, and is the president of the Wolf Recovery Foundation, an organization dedicated to the recovery of wolves in the northern Rockies.

Ralph Maughan made his remarks in Pocatello, Idaho, on the 25th of August 2003.

Chapters
1. Maughan develops an interest in environmental conservation
2. The Wolf Recovery Foundation
3. The politics of wolf reintroduction in the West
4. Turning over wolf management to states hostile to wolves
5. Difficulties of wolf recovery in the American Southwest
6. The politics of grizzly bear recovery
7. Brucellosis as political smokescreen in managing Yellowstone bison
8. Expanding the range for Yellowstone bison
9. The bison management plan (2000)
10. Montana initiates bison hunts
11. Many conservation organizations ignore public lands ranching

CHAPTER 1

Maughan develops an interest in environmental conservation

I grew up in Idaho and Utah. I've lived here all of my life except while going to graduate school, which was a real revelation to me because I got out of the West for the first time. And I came to understand that the West was a place which, while it was beautiful, had been degraded.

And so when I came back from graduate school, over thirty years ago, I immediately became an environmentalist. I began to join conservation organizations. The Idaho Environmental Council was the first. And later it was the Sierra Club. There were not a lot of Sierra Club activists at first, so I moved up quite rapidly in the club organization. Before long I was chapter chair for the Northern Rockies.[1] I did that for about three years and was always active in the Sierra Club. Held several offices for over twenty years, including a stint as regional vice president for the Pacific Northwest.

In the meantime, I was one of the founders of the Greater Yellowstone Coalition. I was primarily active in wilderness issues at that time.

My interest in livestock grazing came a little bit later, when I realized that in many cases we were protecting a landscape that, while it might not have roads, was severely degraded because grazing is allowed in wilderness areas. And so for about the last five years, I've been a lot more interested in grazing issues.

And I've also been very interested in the recovery of the large carnivores in the northern Rockies, particularly wolves, but also grizzly bears.

CHAPTER 2

The Wolf Recovery Foundation

The Wolf Recovery Foundation was established in the mid-1980s to facilitate the recovery of wolves in the northern Rockies. Wolves began to re-inhabit northwest Montana on their own, coming from Alberta and British Columbia in the early 1980s.

Over the years, many of the foundation's members got involved in other organizations. Then about 1999 the foundation was reinvigorated, and before long I was president. In some ways the centerpiece of the organization has been my webpage, which features reports on the wolves in the northern Rockies—detailed reports on how they're doing—plus a lot of other things, which I see related.[2] Things like the problems with bison not being able to legally leave Yellowstone Park in Montana. Problems with the recovery of grizzly bears and lynx. And problems we have with grazing.

And so the centerpiece of this rather small organization has been my webpage for about the last two years. Although, we also do other things, such as give grants to individuals for wolf education. And we've even given a grant to the Park Service to study the interaction of grizzly bears, wolves, and bison in Yellowstone Park in the early spring.

CHAPTER 3

The politics of wolf reintroduction in the West

I'm a political scientist by training. And I know that there are lots of different types of political conflicts. There are conflicts, for example, over economic issues. And there are conflicts over values issues. Conflicts over values issues are very difficult to resolve, because it's hard to split the difference when you're standing up for some kind of a principle.

After the first few years of studying the wolf reintroduction, I became convinced that this is almost entirely a values, a cultural issue. It's not an economic issue. Those issues are raised, but they're not the real issues.

The real issues are about lifestyle. They're issues about what the West is all about. They're issues about the changing West and the demise of a lot of the old extractive occupations.

And so the wolf issue is really part of a larger issue. It's not an issue apart. It's related to issues like timber cutting, grazing, the use of off-road vehicles, the management of our national forests and our national parks. And for that reason it's very difficult to work with people who are opposed to you, because there isn't any real economic incentive you can give them—that's not what they're really complaining about. They're complaining about the fact that wolves have been restored. And that's an assault on their values. So it's extremely divisive. And I think it will remain that way for a long time.

The opponents of wolf restoration have a rather simple set of arguments that they use. And they differ a little bit between the two groups of anti-wolf people. There's the farmer-ranchers, and then there's the hunters.

The farmer-ranchers will use the argument that wolves are going to kill large numbers of livestock. They raise an economic issue, but it's not really significant, because their losses are largely compensated by Defenders of Wildlife or by other funds in some other states. The actual amount of predation on livestock by wolves has been very low. Something like six hundred sheep since 1987, and about four hundred cows. And most of the sheep are lambs and ewes, and most of the cows are actually calves. Not very many full-grown cows have been attacked.

You can tell this is a cultural issue by the way it's treated by the news media. Wolf attacks on cattle, especially the first ones, but even today, oftentimes get more coverage in local news media than a homicide does. This shows you that this is not just about economics. It's something much bigger.

Another example is that very much larger losses of sheep and cattle from other causes don't get nearly as much media coverage. For example, recently three hundred sheep were poisoned near some of the phosphate strip mines near Soda Springs, Idaho, by selenium uptake of curlycup gumweed.[3] It hadn't been known that this plant would concentrate selenium. And that was a one-day story. Yet the loss of three lambs to a wolf might be a several-day story. So I think that shows that wolf reintroduction is a cultural matter.

Then there are the hunters. Their argument is that wolf populations will grow rapidly and will continue to grow rapidly, basically forever, until all of the big game is wiped out. And when that happens, "The wolves," as one of the leading spokesmen said, "will kill all the carnivores, and then they'll cannibalize themselves." He calls them land piranhas.

So there's the belief that wolves don't obey the same natural forces as other animals, whose populations reach a peak and then fluctuate. Wolf populations, it's erroneously believed, grow essentially to infinity and then crash.

Then there's the argument that the wolves that were reintroduced were not native to Idaho or Wyoming, but were Canadian gray wolves, which are a different subspecies.

There's absolutely no biological evidence for this claim. You won't find any biologist that will argue that. And yet that argument is used again and again.

An additional argument is used that the wolves were dumped on the people of Idaho and Wyoming. In reality, there were hundreds of public meetings. The public education, the public input went on for years and years. Most people thought the reintroduction would never happen. And yet within a year after reintroduction began, this idea that the wolves simply had been dumped on the people became current.

Outfitters like to argue that they're losing clients. But I don't see any evidence of that in general. Wolves are a convenient excuse, though, for individual outfitters that lose clients.

My late father-in-law was an outfitter, and he knew that one of the most important things about an outfitter was his personality. And the truth is, some outfitters have a bad personality, and they lose their clients.

It is also true that wolves change the distribution of elk and other big game. So elk that were on the meadow for the last ten years are still around, but they're somewhere else. Consequently, hunting tactics have to change and people are slow to do that. Again, the wolf is basically a convenient scapegoat.

But there's almost no evidence for an economic impact from wolves. Almost all the evidence is that the impact, though modest, has been positive and largely confined to the area around the northern range of Yellowstone Park, where you're almost guaranteed to see a wolf. Communities like Gardner and Cooke City, Montana, now have a lot more visitors on the shoulder seasons and in the wintertime, because that's one of the best times to observe wolves.

CHAPTER 4

Turning over wolf management to states hostile to wolves
Until recently the wolf was classified as an endangered species in most of the lower forty-eight states. The wolves that were reintroduced in Idaho and in the Greater Yellowstone Area had a special classification from the Endangered Species Act called "experimental, nonessential." The wolves in Minnesota, which have spread to Michigan and Wisconsin, were classified as an endangered species, and they have now been downlisted to a threatened species.

The wolves in the experimental population area, that is, Idaho, Wyoming, and Montana, are slated for state management, because it's believed that the population has biologically recovered—the recovery has been successful. And so when each state adopts a conservation plan for the wolf and the federal government decides that the state will protect the wolf—stop it from going back on the endangered and threatened species list—then the state will gain management authority.

Now, all three states have to produce management plans that are deemed acceptable by the US Fish and Wildlife Service before they can go into effect, because the wolf populations are tied together. It's really one meta-population. And so the only thing right now which is holding up the delisting of the wolf is the adoption of these plans. Two states have adopted plans. And Montana just issued its final plan. After review by the Fish and Wildlife Service as to their adequacy, management authority can then be given over to the relevant state agency, which is usually the fish and game agency.

Wolf supporters are concerned about turning wolf management over to the states, because in all three—Montana, Idaho, and Wyoming—the politicians, with very few exceptions, have expressed extreme hostility to the wolves. And so people wonder how a state, where almost all of the "political elites" (as political scientists have called them) are against wolves, is going to adequately manage them.

Most hostile have been the officials in the state of Wyoming, where they have classified the wolf as a "trophy game species" in a small portion of the state, and a "predator" in the rest of the state, where they could be shot for any reason.

If the US Fish and Wildlife Service does not grant the authority to the states, it will probably be because of the defects in the Wyoming wolf plan.

The Idaho plan and the Montana plan look much better on paper. Wolves would be allowed to wander where they want. And they would be judged on their behavior, not by their location. They'd be treated, at least in principle, much like black bear or cougar or other animals of that class.

But the management is what really counts, not what's on paper. And many of us suspect that, in fact, the wolf populations will decline in these states. In fact, will decline rather rapidly. There won't be enforcement of the laws. And we'll be petitioning to put the wolves back on the threatened or endangered species list pretty fast.

Things look a lot better for the states in the upper Midwest (Minnesota, Wisconsin, and Michigan), which have more enlightened fish and game commissions and where there is more experience with wolves, because they never totally disappeared, for example, in Minnesota.

CHAPTER 5

Difficulties of wolf recovery in the American Southwest

The restoration of wolves in the Southwest has been very difficult for two reasons. One reason is that there were no wild wolves, unlike in the northern Rockies, where there were wild wolves north in Canada that you could bring south. These wolves knew what elk were, and they knew what deer were. And they knew how to chase them down.

These Mexican wolves were essentially—. I won't call them tame wolves, because officials tried to keep them away from people—but they did not have hunting skills. So that was the first difficulty.

The second difficulty is that they were limited to small areas where they could roam. And if they roamed outside of these areas, they were captured and put back. So the southwestern Mexican wolves have to obey artificial boundaries, which don't make any sense certainly to the wolves. And the boundaries really don't make any sense even to biologists.

The wolves in the northern Rockies were allowed to go wherever they wanted. Although they were reintroduced in central Idaho and in Yellowstone Park, they've since spread out, and they're scattered throughout three states. There have been wolves that have even gone to Washington, Oregon, and Utah. And they've been protected there as well. So that recovery has been more successful than the one in the Southwest, because wolves are judged by their behavior, not by their location under current federal management.

CHAPTER 6

The politics of grizzly bear recovery

I've been interested in grizzly bears since I took my first hike in grizzly bear country over thirty years ago. And that's longer than I've been interested in wolves.

There was a plan to reintroduce bears to central Idaho, which was shot down by Bush and Norton.[4] But other than that, there haven't been any efforts to reintroduce grizzlies where they've become extinct.

So most of the grizzly bear recovery is really the recovery of existing populations in Yellowstone Park, the surrounding area, and in what's called the "Northern Continental Divide Ecosystem," which is Glacier National Park, the Bob Marshall Wilderness, and other wilderness and nearby lands.

That has been proceeding quite well in terms of the population being restored and a lot of the biggest insults being eliminated, such as large grazing allotments, especially of sheep. Although a few of those still remain.

There have been two recovery plans now, which have been in force. And many, many things have been done. In areas, which are inhabited by grizzly bears, all of the trash cans, including in communities like West Yellowstone, have been replaced with bear-proof garbage cans.

Roadside feeding of bears in Yellowstone Park has been absolutely prohibited. And people are subject to heavy fines if they do that.

All of the dumps have been closed in the concentrated grizzly bear territory.

There are a lot of threats to the future population of grizzly bear,[5] but in the meantime, the grizzly bear population has expanded.

CHAPTER 7

Brucellosis as political smokescreen in managing Yellowstone bison
While bison and wolves might seem to be quite different, the politics of the issues are very much the same. Management of bison is primarily, in my view, a cultural issue. And so the arguments that are being stated are not the real arguments that drive the policy.

Bison, which are a success story of recovery of an endangered species from ninety years ago, are by Montana state law confined to the boundaries of Yellowstone Park. It's been Montana policy that if bison leave the park boundaries, which are generally not set on any topographic feature—they're simply straight lines in the forest—then they'll be shot or hazed back into the park or various other things. And this controversy's been going on for about twenty years.

The ostensible reason why this is done is that a certain percentage of the bison—and there's controversy over what percentage—are infected with brucellosis. Brucellosis is not a native disease. It's a disease which is very costly to the cattle industry, because it causes pregnant cows to abort their first calf. That's the ostensible reason why they won't let bison leave the park—they're afraid they'll give the cows brucellosis.

There's been an effort of about forty years and over a billion dollars spent trying to eliminate brucellosis from cattle herds in the United States. And except for one or two herds, which still have brucellosis, it's been completely eliminated. So the argument is that the bison are a threat to the brucellosis-free status of almost every state.

Now, the brucellosis-free status is an actual designation given to a state. If a state doesn't have that, its cattle have to be quarantined before they can be transported to another state, which can be very costly. And so the Animal Plant Health Inspection Service, that is, APHIS, of the federal government every once in a while has said, "Well, if we don't do something about the brucellosis in the bison in Yellowstone Park, Montana could lose its brucellosis-free status."

But it's an idle threat, because people always focus on bison only in the state of Montana. There are also bison in Wyoming immediately south of Yellowstone Park. And almost 100 percent of them are infected with brucellosis, and they go wherever they want. They mingle with cattle. And APHIS takes a completely different approach there. And Wyoming does too. Wyoming does not shoot the bison. They let them wander basically where they want. Instead, they vaccinate their cattle against brucellosis.

Another reason why bison are not a threat to cattle in Montana is that essentially all of the cattle immediately adjacent to Yellowstone Park in Montana have now been eliminated. So there are no cattle for the bison to give brucellosis to.

And furthermore, it's very difficult to transmit brucellosis to cattle and vice versa. (Bison in Yellowstone got brucellosis when they were trying to recover the herd back around 1917, when they brought dairy cows into the park. They fed supplementary cow's milk to the buffalo calves to make them grow faster and survive. Unfortunately, the cattle herd they brought in was infected by brucellosis. So the bison actually got it from infected cattle. There's a real irony to that.)

And finally, brucellosis can be transmitted to species other than cattle and bison. It can be transmitted to elk. Many of the elk in Yellowstone Park are infected by brucellosis. And in Wyoming, in the Jackson Hole area, an even larger percentage of the elk are infected. But, unlike the bison, the elk are allowed to wander wherever they want. They're allowed to wander outside of the park. In fact, hunters expect them to wander outside of the park. They count on the elk hunt every winter outside of Yellowstone Park on its northern range.[6]

And elk are equally efficient at transmitting brucellosis to cattle as bison are. In fact, the only instance of transmission of brucellosis from wild animals to cattle in the northern Rockies took place, I believe, two winters ago in Idaho where infected Wyoming elk were being fed by people who were also feeding cattle. What probably happened was that an aborted elk fetus was dropped onto the feed line and was then licked by some cows. And then, that cattle herd had to be eliminated.

Yet all the prejudice is against the bison. That's why I say it's an issue like the wolf. The elk is highly valued in the western states by most people, including a lot of them who hate the wolf. The bison is not valued for various reasons. For one, it hasn't been a big game animal. There seems to be a prejudice against it, particularly in Montana. Oddly enough, there doesn't seem to be a prejudice in Wyoming. So it's probably a prejudice of a few important decision makers in certain agencies in Montana.

What appears to be an economic issue, namely the grave danger that brucellosis will infect the cattle of Montana, I think, on close examination turns out to be purely a cultural issue. It's the Montana Department of Livestock, and certain politicians, pushing us around and showing us their power by killing the bison that leave Yellowstone. And so it's a clash of cultural values. They kill Yellowstone bison to show who's really in charge in the area.

CHAPTER 8

Expanding the range for Yellowstone bison

Historically, bison numbered millions, and they roamed about the Great Plains and into the Rocky Mountains. They didn't inhabit every part of the Rocky Mountains. For example, there never were bison in Nevada, and there were very few bison in southern Idaho. But there certainly were a lot of bison in Montana and in Wyoming.

After the great slaughter[7] the only surviving bison were in a few private herds and in Yellowstone National Park.

I think most conservationists would like to see bison be able to roam outside of Yellowstone Park like every other animal—grizzly bears, elk, wolves. And, in fact, there is room for them to roam outside of Yellowstone Park.

There's not a lot of room. Bison are big animals, and they walk through fences. They're potentially dangerous. And so they couldn't go everywhere. They couldn't become as numerous or be in as many places as elk or deer. But there is a lot of country to the immediate west of Yellowstone Park that is almost completely cattle free now—the Madison Valley. And it has natural boundaries where there could be a choke point that would prohibit bison from going farther.

And if some livestock grazing allotments were shut down, bison could be allowed to expand farther along the Idaho/Montana border. So I think there are great opportunities for the expansion of free-roaming bison to the west of Yellowstone Park.

To the south of the park, we already have free-roaming bison. They roam all over Grand Teton National Park. And they are also found in the adjacent national forest.

And that was done without a plan. It was an escaped herd. They were not Yellowstone bison. And, in fact, the bison population there needs to be controlled, because they're not subject to limitations of winter range, like the Yellowstone bison are, which keeps their numbers down. They go out on the National Elk Refuge and eat alfalfa pellets all winter long. So their only constraint is the amount of summer range. Consequently, I'm strongly in favor of a bison hunt in the Jackson Hole area, because the herd does not obey natural constraints there.

There is a little bit of opportunity for bison expansion to the north of Yellowstone Park. Most of that region is extremely rugged and is not bison habitat. But along the Yellowstone River up to Yankee Jim Canyon, about twenty miles north of the park where there is another natural choke point, bison could roam freely.

CHAPTER 9

The bison management plan (2000)

The bison management plan requires the National Park Service to work with the Montana Department of Livestock to capture bison near the boundaries of Yellowstone Park and to test them for brucellosis. Bison with brucellosis will be slaughtered. Those which test negative will have paint slapped on their side, and they will be allowed to roam outside of the park, at least during the wintertime.

And so park rangers are actually involved now in what amounts to a bison-slaughter campaign. It's not being done willingly, but it's something that was coerced upon them by a bison management plan, which was adopted late during the Clinton administration.

But now we have a situation where the Park Service is actively participating in the capture and the slaughter of Yellowstone bison inside—not far inside—but inside of Yellowstone National Park. And there's a potential for great controversy if there is a severe winter and a large bison die-off, which, of course, could happen at any time.

The current policy regarding bison which leave the park in Montana is a little bit complicated. The way it's supposed to work is that those bison which test free of brucellosis will be released. And those which are positive will be taken to slaughter. However, Montana Department of Livestock has been shifting its rationale now, which indicates that animosity to bison is the real goal, not brucellosis protection. The new management plan, which I referred to, says that the population of bison in Yellowstone Park shall not exceed three thousand. Well, there are currently about four thousand bison in Yellowstone. The view of most conservationists is that the bison population should be able to fluctuate according to natural conditions. If five thousand bison are in Yellowstone, that's fine. If the population drops to eight hundred bison, that's fine too. But anyway, the new management plan says three thousand is the maximum. So late last winter, Montana Department of Livestock changed their rationale and said, "Well, the population is over the limit, so we're just going to kill all of them we can capture."

And so two hundred some odd bison were killed last winter,[8] which was the most in quite a while. However, it didn't make any real dent in the population. It was more of just a public insult.

There was an amendment offered to the appropriations bill recently in Congress to withdraw all money for National Park Service cooperation in bison slaughtering. It lost by about twenty votes,[9] but that's been the first time there's been a vote on it. And that's not a big loss.

So a lot of people are optimistic that with more publicity, and sadly enough, if quite a few bison are slaughtered, that this amendment will eventually pass. And the Park Service, at least, will be allowed to withdraw from this bison-slaughter campaign, which is now meaningless for Montana. A reason why it's meaningless from the standpoint they argue is that there are no longer any cattle adjacent to Yellowstone Park in Montana. There used to be one large grazing allotment to the west in Montana. That was recently purchased, and the cattle were moved to a grazing allotment well in eastern Idaho.[10]

And so the only cattle remaining are a few that are in fenced pastures. There are also a few cattle in the northern end which are in fenced pastures. But the cattle, which the Montana Department of Livestock has been saying they're trying to protect, are all gone now due to the closure of the Horse Butte Allotment. Meanwhile, down in Wyoming there are cattle running every day with bison at their side.

CHAPTER 10

Montana initiates bison hunts

Montana's dealt with bison leaving Yellowstone Park in various ways over the last twenty years. For a while they had a hunt, or a so-called hunt.[11] And that was very controversial, because it wasn't really a hunt. The Department of Livestock would find a bison outside the park. And then they'd get a hunter, and they'd say, "Shoot that bison." And so it wasn't a hunt at all.

There were protests against that. Extreme protests.

So instead, a new plan was created, which eliminated the hunt, and whereby the bison would be kept inside of the park. And if they left the park they'd be shot, or they'd be herded into corrals and taken away to slaughter. That was extremely controversial too.

In the severe winter of 1996–1997, when a lot of the park grass froze over— it was basically a January thaw followed by a freeze—about a thousand bison were killed and another thousand bison starved inside of the park. And the bison population was greatly reduced.

The Buffalo Field Campaign was organized about that time. And they've been extremely active every winter monitoring the Montana Department of Livestock and harassing them in some sorts of ways. The result of it's been that the bison herd has increased to its former size and is nearing record size. And so public outrage has helped the bison population recover.

Many politicians in Montana realize that Montana has really gotten a black eye because of all the publicity. And some people have also realized that the reason that the brucellosis-infected elk are treated gently, but the brucellosis-infected bison are not, is that elk are economically valuable. The elk hunt is a big thing.

And so the last session of the Montana Legislature passed a bill restoring the bison hunt outside of the park. However, it won't be like the old bison hunt, at least it may not be like the old bison hunt. The Montana Department of Livestock won't say, "Here's a bison. We found it for you. Shoot it."

The hunter will find the bison on his or her own and shoot it that way. The details are still quite vague. We don't know what's going to happen. We don't know if the Department of Livestock will still be slaughtering them and be hazing them back into the park, while other people are shooting them. But it is potentially something that could make bison seen as something valuable, rather than just something inside of the park which is a potential problem.[12]

CHAPTER 11

Many conservation organizations ignore public lands ranching
There are a number of western conservation organizations that are reluctant to address the grazing issue. And there is a reason for that reluctance. It's because ranching is a cultural issue. To a considerable degree, ranchers

are cultural icons in the West. And for that reason, groups that are working on "wilderness" or "fisheries" or maybe something else are reluctant to go up against the icon of the rancher. And so I've seen a lot of groups not do anything about grazing.

I also know about growing up in the West. When I grew up in Utah, I didn't understand the degradation that was going on. I thought it was just natural that streams out in the deserts ran down straight washes. That was just something about the geology of the area. I didn't realize that was caused by overgrazing.

And so it's been a slow process for conservation organizations to take up the cause of restoring our lands from abusive grazing and eliminating grazing where that's the best. And that's in an awful lot of places. For example, in places where precipitation is less than twelve inches a year.

So it's a difficult thing politically for many conservation groups to take up. And, in as much as a lot of people don't see the damage that grazing does in many cases, they've also overlooked it.

Some of the groups are starting to come around, however. There have been considerable efforts recently to buy out grazing allotments. Money from the Bonneville Power Administration was used by some of the Indian tribes in Idaho to close grazing in the Bear Valley Creek area of central Idaho. It's been closed for two years now. And the restoration of the salmon spawning streams and the increase in the number of elk is simply astonishing. The area is just utterly beautiful.

Recently, the National Wildlife Federation and other groups provided monies to buy up the Horse Butte Allotment[13] west of Yellowstone Park. And even if bison aren't allowed to roam there, still the area's going to improve, and there's going to be elk and other wildlife there.

And similar groups also recently bought out the Walton allotment, which is near Grand Teton National Park.[14] This allotment had been a real sore point with grizzly bears for many, many years.

And so these are rather cautious groups, and they are starting to move into dealing with the livestock issue in an indirect sort of way.

But, on the other hand, there are groups like the Western Watersheds Project, which are really "out there"—active. And they've probably done more to raise the issue, and probably done more actual change than these other groups.

Even though there aren't very many traditional ranchers left, ranchers are held in high esteem by politicians, because ranching is a cultural icon in the West. And so ranchers really have kind of a political hold on the land.[15]

One of my colleagues, who's a Republican, really got it though. I was saying, "Public lands ranching is like socialism."

"No," he said, "it's like feudalism—it's a pre-capitalist sort of thing, where certain people have ties to certain politicians, and they mutually support one another."

Even though the number of ranchers is growing smaller and smaller, politicians like to have that cachet of the ranch. What do they do when they gain power? What did George Bush[16] do when he gained power? What did Ronald Reagan[17] do when he gained power? They bought what they called "ranches." That just shows the hold that the ranching mystique has on the United States even though there are very few traditional ranchers left.

Basically, cows get to trample our public land stream banks and defecate in the streams because they are politically protected animals, and because their owners are politically protected.

Bobbi Royle

As civilization has encroached on the western landscape, water sources and migratory ranges of wild horses have became blocked. And horses, once viewed as living symbols of the Old West, have become increasingly regarded as pests by people whose economic interests compete with them for land and water.

Bobbi Royle began rescuing persecuted wild horses in 1989 and four years later co-founded, with Betty Kelly, Wild Horse Spirit, Ltd., a nonprofit wild horse rescue, advocacy organization, and sanctuary.

Bobbi Royle made her remarks on the 11th of August 2004 in Carson City, Nevada.

Chapters
1. Royle's early experiences with wild horses
2. Wild horses in North America
3. Wild horses slaughtered for profit
4. Ranchers' attitudes toward wild horses
5. Conflicting attitudes about wild horses
6. Number of wild horses not known
7. Wild horse management under the G. W. Bush administration favors ranchers
8. Adopted wild horses sent to slaughter
9. Deficiencies of the government's wild horse adoption program
10. Horse slaughter in the US
11. Violence committed against wild horses (December 1998)
12. Efforts to make the wild horse Nevada's second state animal (Feb.–June 2001)
13. Improving the conditions for wild horses on public lands

CHAPTER 1

Royle's early experiences with wild horses

I'm a native New Yorker. When my folks came out West, I came with them as a child and just fell in love with it here. But after moving around a great many times, we finally went back East. I went to art school and got a BA. Worked as an artist most of my life.

When I came back West again, I moved into a neighborhood where horses came down right to our backyards. And not only that, but they were breaking everybody's sprinklers. It was like a banquet for them, all the green grass.

Of course, everybody got upset because the horses were trampling their rosebushes, and so on. But the reason they were trampling was because the people were chasing 'em. If they had just left them alone and moved them off very slowly, the horses probably wouldn't have done any damage. But people, as people, need to be in control.

The county was threatening to round up these horses. And when a county or any government wants to do something, it's usually bad for whatever is the brunt of it. So we and another group got together, and we rounded up the horses and started adopting them out.

We kept gathering horses and finally we decided to form our own group.[1] Then we started rescuing injured horses or ones that were so abused that they never would be trusting of humans. We decided to keep those that weren't really fit to be adopted out.

Nobody was really in control, so we were able to pretty much rescue and do as we wanted. But the battle began when the state and the county wanted to get in on it. Then things got much worse. Finally the Ag Department went ahead and took control of all the horses that weren't covered by the federal government—by the BLM.

So now it's a split. The Virginia Range, which is in back of Reno, Nevada, is controlled by the state. Whereas the BLM is more in the outlying districts all around the state. That's basically what we deal with.

CHAPTER 2

Wild horses in North America

The wild horse is an indigenous species to North America. They were here before they came over with the Spaniards who brought horses.[2]

Ranchers talk about the wildlife that was here. Well, the bighorn sheep is not indigenous to North America.[3] But the horses are. And that's the big hew and cry of the ranchers who say, "Well, why should we have them because they're nonindigenous to the state?" But they are and they belong here. And it's only in the last forty years or so that development has moved in. People encroached upon their habitat.

So BLM, I think, started with about 302 herd management areas. Which meant that the horse supposedly was safe in those areas. And those horses that were found in those areas were to remain there.

But then people started putting up fences. That stopped the migratory pattern of the horses, and they couldn't get the water that they needed. And they didn't have enough food.

If you fence off a horse and he has to walk twenty miles around to get to water, then the horse is eating all that twenty miles around. So the ranchers are saying the horse is eating all the food. No, they're not.

If you put cows down somewhere and you put water out, the cows will only stay within a mile of that water. Now, suppose you move the water half a mile up. The cows will all go up half a mile. But the horses graze high in the summer; low in the winter. And they eat a lot of forage that the cows don't. Also, they don't ruin the water holes. They don't trash 'em and wallow in it like a cow does.

CHAPTER 3

Wild horses slaughtered for profit

A perfect example is up in Storey County, Nevada. Almost all of Storey County is private land except maybe a very minimal portion of checkerboard which is BLM. Just very small. And what happened is BLM had horses up there and was protecting 'em. I think it was in the early '80s, if I'm not mistaken, that they finally said, "Well, we're gonna zero out the horses here. We have no more interest in the horses. There aren't any horses up here."

Well, that was a blatant lie, because there were a lot of horses up there. And there were a lot of mustangers[4] up there. And those mustangers would put a stallion out with the horses, or let the horses breed. And then they would come and say, "Those horses over there are the progeny of our horses, and we're entitled to 'em." And that could be two hundred, three hundred horses.

In late December 1990, there was a mustanger up there name of Nick Mansfield,[5] who gathered up 391 horses and put 'em into horse trailers and trucked 'em off to slaughter. No ifs, ands, or buts. Now, the Ag Department is responsible for giving brand certificates to anybody who rounds up estray horses[6]—anybody except BLM. Those horses went to slaughter with three pages of brand certificates!

A brand certificate is issued so that you should be able to find your horse if it's lost. The brand certificate tells the age, the sex, and the color of the horse. Now, when you have a page that says "thirty-four mares," you can't possibly find your horse in that thirty-four mare statement.

So a whole bunch of us got onto the Ag Department. And they went, "Oops, we made a boo boo here."

So then they passed a law that said they need to post a notice in the paper when they're rounding up horses and what horses they're going to round up.

So it kinda shut down the mustangers. And there had been a tremendous amount of 'em because it was just a feast for them.

Then the Ag Department took over. And as with any agency, it wasn't much better than the mustangers except that we decided that the nonprofit groups would adopt out the horses that were rounded up. So that kept them from going directly to slaughter—directly to killer buyers.

But there again, they were rounding 'em up so quickly that we couldn't keep up with it. They got all the money. And we did all the paperwork. Reported back to them. And it wasn't a very wonderful relationship for any of us. It's still going on because we're still cognizant of the fact that if we don't continue, the horses will go directly from the gather site to the auction site. And, of course, the killer buyers are waiting there like that fellow that gathered the 391 horses and then made about $60,000. So it's very lucrative.

CHAPTER 4

Ranchers' attitudes toward wild horses

I think that a rancher who says, "This horse is competing with my cattle for forage" is also really mad because they lost their cash crop. They can't legally round up these horses and take 'em anymore, because they don't belong to them. The horses belong to the American people. The BLM has the responsibility for managing them.

And that's what one of the big fights is about. Ranchers always say, "Oh, they're telling me how to run my cattle." Or "They're telling me to take my cattle off." BLM is making them pull their cattle because of drought, so they're turning the water off and the horse has no water anymore. The horses are dying of thirst. That happened a couple months ago when about seventeen horses died. They put 'em on the wrong side of the fence. Gone. And if we trespass and go out there and cut fences, of course, we're going to go to jail.

There are good ranchers. There are a lot of small ranchers who let the wild horses roam on their land. They rotate their stock. Some are good stewards of the land. And they've been there a long time. But there's a lot of very large ranchers, which are bankers and corporations of a lot of different make-ups. And it's a tax write-off for them.

But the West really wasn't meant for cattle grazing. We're a high desert here. We don't have oodles of grass. We're not in California. We're not in the Midwest where there's a grass belt.

Yet the ranchers get paid for the cattle that wander onto highways and are killed by motorists. They get paid for putting in wells. They get paid for growing alfalfa. And so it has become a way of life—a welfare life.

It's the "cowboy mystique" that's running it. And it's the lifestyle. "My grandfather and my father had this ranch. And I want my kid to have it," the ranchers say.

Sure they like to live this life. Who wouldn't? It's great. You're being subsidized. You get to live out on the land. You don't have to go and be a plumber and crawl under houses and things like that. You get to do nice things. But it's coming to the point where it's not feasible anymore. There's cattle coming in from other countries. There's beef being grown back East. And feedlots and so forth. And the West supplies very, very few cattle to the beef-buying public.

CHAPTER 5

Conflicting attitudes about wild horses

The advocates regard wild horses as wildlife. As free. As belonging to you, me, and everybody else who pays taxes. They were there. They need to belong there.

As far as the ranchers are concerned, they're livestock.

I have something written by a guy named Abbey.[7] And if you'd like, I'll quote it to you.[8] I think it's very interesting about the rancher. And this is exactly what the rancher does. He thinks that everything belongs to him. That he can control everything. That he can kill everything he wants to. The coyote—poison it, shoot it. You know the old story.

And so he regards the horse the same way. "Well, if it's in my water hole, I'll shoot it."

That's the way he regards the horse. Very few ranchers say, "Well, I take a few horses a year. I cull 'em out and make good ranch horses out of 'em."

That's the way it used to be, too. And it may still be in certain places, but not very frequently. Instead he can go out and buy horses. Trade horses. Do whatever.

But the bottom line is that the wild horse is genetically the most wonderful horse in the world. It has thicker feet. A better immune system. Stronger bones. It has stamina. The stallion who selects the mares and creates the band is much smarter than any rancher who says, "We'll cull and keep all the paints," which are probably the weakest. Or "Gee, they're pretty. We'll keep those instead of the brown mare."

When man comes and starts to manage things, they start to destroy it. And this is what's happening to wild horses.

CHAPTER 6

Number of wild horses not known

The BLM is saying that there's something like 21,000 wild horses in Nevada. I'll bet you dollars to donuts there aren't any more than 10,000 horses, if that.

BLM puts out gather plans. And they say we're gonna gather six hundred horses. And we're gonna put two hundred back or one hundred back or sixty back.

They go out there, and even some of the BLM employees will tell you—they go out there for six hundred horses; there aren't three hundred horses out there to gather.

So according to my arithmetic, if you say you're going out for six hundred horses and you gather three hundred, there are no horses left. It's a big zero.

Now BLM can say, "Well, there's 21,000 horses out there. And we gathered 9,000 this year and we gathered 10,000 last year." And so forth and so on. But they'll never tell you that they did a count and that this is what the numbers are.

They are in the numbers game, but they're not giving us the right numbers. They're juggling numbers around. So the public will never know when the last wild horse is out there.

CHAPTER 7

Wild horse management under the G. W. Bush administration favors ranchers

I think that the BLM, the Bush administration, Interior Department, the whole regime in power now is giving the rancher carte blanc for pretty much anything he wants to do. And I think we're going toward "sale authorization" for the wild horse.[9] Up 'til now it's been pretty much that all of us have demanded that they be adopted out and saved. Now, I think what they want is sale authority to speed up the process of getting rid of these horses. And they are appointing a federal advisory board that is leaning toward this. And most of 'em are ranchers. They're suppose to be made from the public in general. But what they want is anti-horse people. And then they can manipulate these people into saying, "Oh, this is what we need to do."

And I also think they're putting together edicts and laws. Once something is done, it becomes kind of a thing that goes on. This has happened a lot in the West. "Oh, let's try this," they'll say. And then, "Well, that's the way it is."

So I think it's a purposeful annihilation of the wild horse. And it's going very, very quickly. We're on a downhill slide. And if we don't get a new administration, I think the wild horse is gonna be in a lot of trouble.

Responses to government actions against wild horses

You could write and protest to IBLA—the Interior Board of Land Appeals. You can do that, but it's not gonna do you any good anymore.

So we don't really have any recourse.

What happens, media wise? Let me say a little bit about this, which is very powerful. A lot of people don't think so, but I do, because it just sweetens everything.

Media wise, the BLM has a choke hold on most of the newspapers and magazines. The minute we come out and say something, there's a big spread in the newspapers with BLM saying, "Oh no. *This* is what we're gonna do, and *this* is how we feel about it."

Another thing that BLM is doing is that they want to disband their sanctuaries and let the ranchers take care of these horses.

Already there's a rancher in Fallon who gets two-something a day, and he has maybe fifteen hundred horses on there. Well, the guy's making maybe a million bucks a year.

And then the BLM keeps screaming, "We need more money."

Well, sure if you're givin' all your ranchers a million dollars a year. That's a big drain.

And then BLM's screaming, "We need more money for gathers. All these horses are costing us a lot of money because we're keeping 'em in Palomino Valley. We're feeding them. We're inoculating them."

It's this snivel and cry and hew—we need more money all the time. But if they didn't gather 'em up, they wouldn't need this money. And that's what the advocates are saying, "Leave 'em on the range and stop listening to all these whining ranchers that say, 'We need. We want. We must have.'"

CHAPTER 8

Adopted wild horses sent to slaughter
During the 1980s they had a holding facility out in Lovelock. And it wasn't working. I don't know if there was disease out there or too many horses or what. And what was happening is they were adopting out from ten to a hundred horses to a rancher, say. And he was immediately taking them to slaughter. There was not even a year waiting period.

I think if the American public knew what was going on, they would cry out and say, "Let's save these horses." There's a lot of apathy simply because they don't understand what's going on. They think that the American cowboy is great. And the horses are gettin' real good homes. And everything's goin' real fine here. It's like anything else if you don't know the true facts.

CHAPTER 9

Deficiencies of the government's wild horse adoption program
The BLM is adopting horses out to people who don't have a clue, number one. They don't know how to feed a horse; how to train a horse. And most of 'em get it because it's a cheap horse. It's only a hundred and a quarter. And therefore, they're not gonna hire a trainer at $450 a month to train the horse.

So three months down the line, when they go out in the pen and the horse is still draggin' a rope, they go, "Here horsee, would you like a carrot?"

And the horse says, "Take a hike."

It takes a lot of finesse and a lot of knowledge and a lot of work to train a horse—domestic or wild. And especially wild, because you have to make friends with a wild horse before you can train it. Most people need help. Very few people can just innately take a horse and do this. And if you give 'em two or three horses, it's just impossible.

So the horse ends up at the neighbors. The horse ends up goin' to auction. The horse ends up in an accident—killing it or maiming it. Or it starving to death. Because they go, "Well, geez, I spent a hundred and a quarter, and now I got to spend a hundred a month?" So it's not a real feasible thing.

CHAPTER 10

Horse slaughter in the US
There are only, I think, three slaughter plants in the United States right now. Two in Texas and one in Illinois. There's Beltex and another one in Texas.[10] And the one in Illinois that was burned down. They're rebuilding it larger. Its owned by a company from the EU.[11]

They're just horrible, horrible places.

That's why we have this sanctuary. A lot of these horses have had two or three homes. If we hadn't taken 'em, they'd end up at slaughter.

We got one horse a few hours from goin' down to the auction house. A woman had sent some money to one of the killer buyers and said, "Please hold it." Then she asked us, "Would you go down on Sunday morning and pick it up." And so we went down there and picked up this beautiful papered mare and brought her here. She was so thin it took us three or four months to build her up, so that she could go on a transport down to Ocala, Florida, to a sanctuary for domestic horses.

So we not only take care of the wild ones. The idea is to help the horse. That's the bottom line here.

CHAPTER 11

Violence committed against wild horses (December 1998)
Well, for the past probably twenty years, horses have been shot. Since it's way out in the boonies, nobody can get a handle on it.

There was a horse massacre around Christmas of '98, slopping over into New Year of '99, where thirty-four horses were shot to death up here in Devils Canyon, which is a little northeast of Reno. We were called on the scene one Sunday afternoon by Reno Animal Control and the sheriff's department.

A mare was shot—a baby. We were called up there because a hiker had found this horse who was still alive. And they thought perhaps we could help. So we went up there. But she'd been shot in the spine, and she had to be euthanized because she couldn't have made it.

Then we started to look around, and we saw all these other horses that were dead. Betty[12] started to video it. If she hadn't, this may have not come out, because we were approaching Storey County, which is notorious for living back in the 1800s.

And so Betty started videoing this, and then we went down and the police got in on it again. And then Storey County got in on it. Washoe County got in on it. And more and more horses were found.

Betty's video went on CBS News in New York City that evening. And we were in communication with people from Hawaii and all over. People were just outraged at what had happened.

It came out that a guy was bragging at a party that he shoots wild horses. He shot wild horses as a kid. He used to live in my neighborhood.

And the two buddies of his were Marines. They were eventually let go by the Marines, 'cause the Marines didn't want anything to do with this. This was a hot potato.

They shot one up there. Then they spray painted 'em. And there was a picture in the evidence file of this guy with his foot on a horse like a trophy. So you know it was a real kick for 'em.

And so the trial began. And the evidence was lost. And the scene wasn't protected. And it went on and on and on.

The trial was just a sham. There were three really high-powered lawyers against Storey County DA and assistant DA. And they weren't prepared for it. And like I said, the evidence was boogered up.

So the ultimate punishment after about three years was each one of these guys was fined, and two of them spent a month in jail.[13]

So there again, the people weren't adequately punished. It was a bad scene and people were mad about it. But what are you gonna do?

CHAPTER 12

Efforts to make the wild horse Nevada's second state animal (Feb.–June 2001)
There were a group of students in a 4th-grade class from Henderson, Nevada—down by Vegas. And they proposed that the Mustang be Nevada's state animal.[14] Well, Dean Rhodes, who is one of the worst horse haters and a state senator from Nevada, was up in arms because that would give 'em a special thing—you can't shoot 'em. And as some of the ranchers call 'em, they're

"shitters" and they're "jugheads" and a lot of derogatory terms. And so he didn't want this animal elevated to the state animal, because then they couldn't be maligned.

The kids all testified for this. They were excellent. Wonderful. And they were so disappointed that this didn't pass. And it didn't pass because of one vote. And that was a swing vote from Storey County: Mark Amodei. Other senators said, "Oh, well, I was absent that day."

It's the same old shuffle. And that's why nothing gets done in this state.

CHAPTER 13

Improving the conditions for wild horses on public lands

Well, there's always talk of privatization—of the people taking over the management of the wild horses. I don't know if that would be any better. Immediately it would go to the people who knew horses, of course. And those people would be the ranchers.

Somebody asked me one time, "If you had fifty million dollars what would you do?"

And I said I'd buy the BLM and run it. I was kidding. But maybe I wasn't kidding. Maybe we should have horse advocates running something for the horses.

But then again, as in anything else—groups, clubs—everybody's up there bucking for chief.

I think that there can be no freedom for the wild horse, because we've not allocated any room for it. The only way you could get room is to buy it. Buy up a state. Buy up a corridor. Buy something. But the minute you put something in a limited space, you have to take care of it. And this "taking care of it" brings out a lot of different ideas.

Mike Sauber

Before moving to New Mexico in 1978, Mike Sauber had given little thought to livestock grazing. His interest markedly increased, though, when he found that New Mexico's open-range law left his market garden vulnerable to invasion by hungry bovines. Years later, when the Forest Service planed to construct earthen water tanks in nearby wilderness areas to benefit a cattle rancher, Sauber and his partner, Susan Schock, formed Gila Watch to oppose the project. Sauber gives a personal account of the ensuing struggle between environmental advocates and a broad assortment of pro-ranching entities that included bankers, news media, law enforcement, and the Forest Service.

Mike Sauber made his remarks on the 30th and 31st of July 2003 in Silver City, New Mexico.

Chapters
1. Sauber's early experiences in New Mexico
2. Susan Schock and Mike Sauber form Gila Watch (1992)
3. Banking industry influence perverts management of public lands ranching
4. Proposals to reintroduce Gila trout on the Diamond Bar Allotment
5. Susan Schock is harassed by a sheriff
6. Permittee Kit Laney applies political pressure to the Forest Service (1995)
7. Kit Laney sues the Forest Service (1996)
8. Sauber at the Quivira Coalition meeting in Silver City (8–9 June 2001)

CHAPTER 1

Sauber's early experiences in New Mexico

Out of high school I joined the Air Force, where I was an electronic repair technician on secure communication gear. After that, I got a degree in automotive technology at a junior college in northern California, then worked for a few years at an automatic transmission shop rebuilding transmissions. I then returned to school on the GI Bill at Cal Poly Technical University on the central coast of California. I took classes in mechanical engineering and in mechanized agriculture.

I left before getting a degree, and moved to the Mimbres Valley of New Mexico in the spring of 1978, where I worked at the local Ford garage for two years, then opened my own automotive repair garage. When we got here, I really wanted to be a part of the community, so I got to know people and joined the rural volunteer fire department. That was kind of important initially. It's a small valley, and as long as I didn't talk politics, I got along very well with everybody.

I also spent a lot of time outdoors. We had bottomland on which we grew some crops—mainly sweet corn—that we sold at a farmer's market in Silver City. Most everyone on our side of the river was growing some crops on their irrigated bottomland.

There was one fence between all of us and one person to the north that raised cattle. That fence was continually down, and his cattle were continually getting into our crops. Much to our surprise, the law said it was up to us to repair the fence that held his cattle in his pasture. That was the beginning of the realization that laws relating to cattle were really outdated. The open-range law, in essence, states that a rancher has no responsibility to keep cattle on his own land.

We were constantly calling the brand inspector to try and get the cows out of our crops, but since it was our responsibility to fence the cows out, we didn't get very far. It was a very frustrating experience having to maintain this terribly old fence for one person who wanted to make profit from raising cattle.

I've always thought it should be a cost of doing business for him to do that—to maintain his fences. It was a very trying experience. In fact, the one neighbor who was closest to the rancher's fence, had the biggest crops and was the one that was hit the hardest, eventually moved due to the frustration of putting everything they had into a beautiful crop for the farmer's market and then having it eaten. You wake up one morning, and it's gone. That's pretty frustrating.

CHAPTER 2

Susan Schock and Mike Sauber form Gila Watch (1992)
In the early spring of 1992, we heard from Rex Johnson, a cartographer for the Forest Service, that the Forest Service was formulating a plan to use bulldozers and dynamite to create thirty-three new earthen stock tanks for cattle

to drink out of in the Gila and the Aldo Leopold wildernesses. Susan Schock, my partner at that time, realized that was not legal and would set a very bad precedent. She wanted to have something for her daughter to see that was still wilderness when she grew up. So Susan was really a lot of the impetus for starting what would be a very long battle with this public agency.

In March of 1992, we formed the group Gila Watch and told the Forest Service that their proposed actions were illegal.

They said, "No, you're interpreting the law incorrectly. We're doing it in order to protect the resource."

We asked, "Exactly what resource are you protecting by taking bulldozers and dynamite into the wilderness and creating a bunch more fetid mud pits for nonnative cattle to drink out of?"

And they said, "The riparian areas."

The riparian areas are in a degraded state from continued use by cattle. So the Forest Service was saying they could go into the wilderness with motorized intrusions and do these things if its purpose is to protect the resource. But they were proposing to protect the riparian areas by damaging the uplands—merely transferring the damage from one place to another.

Initially, the Forest Service was just going to do an environmental assessment for this project. And we said, "No, this is a very significant action on a wilderness resource. There's sufficient public controversy. You need to do an EIS."

We had no power at that time. We were just an unknown group without a lawyer. They would not have listened to us. But at the same time, the Wilderness Society was saying the same thing to them. And they were saying they'd sue them if they didn't do an EIS.

The Forest Service waited as long as they could. And then the day before the EA was going to be released, they said, "Okay, we'll do an EIS." But they were using fifteen- or seventeen-year-old range data in order to do this EIS. And we said, "No, that's meaningless. You need to do new transects and analysis of the range to have something to compare to what it was seventeen years ago. Is range condition improving? Is it declining? What's the actual condition of the range?"

It took them almost a year to complete the range and transect analysis. And then they were picking and choosing what information to use out of that data. So we did a Freedom of Information Act request to get all of their data. They then illegally claimed they could refuse our request since they weren't done yet.

Finally, the day came that the data was supposed to be in our hands, because we had turned in a FOIA request. And then they said, "Okay, we'll give you the data."

The National Wildlife Federation lawyer who was working with us said, "We don't have the information in our hands."

Gerry Engel, the district ranger overseeing the management of the Diamond Bar Allotment, promised to have it the following day.

And the lawyer said, "You've got a FOIA violation. You didn't produce it in time."

Engel really pleaded, but it was no good.

So with that FOIA violation and with lawyers backing us at that time, we started to get a little bit more information out of the Forest Service, but it was still a real struggle.

Once we had a university student go to the Forest Service to request information that Susan and I had tried to get. Right in front of him, someone from the Forest Service said, "Where's that other stuff that we took out when Michael and Susan were here?" They admitted to an illegal act!

After a while we had a lawyer donate his time to us. When Susan went up to Albuquerque to give a presentation about the endangered Gila trout to a trout group, someone there said, "We have a lawyer that's interested in helping out. And I'll get him in contact with you."[1]

And so that lawyer donated his time. He wasn't an environmental lawyer, but he knew the court system in Albuquerque and was a very good lawyer. I'm sure he sent chills up the spines of the federal lawyers that he was going up against.

Of course, before we could file a lawsuit, we needed to exhaust all administrative options, which we had not yet done. The Forest Service finished their final EIS, which said that they were going to create only fifteen stock tanks instead of thirty-three.[2] As if fifteen was more legal.

We appealed their final decision, and we were overruled in the Albuquerque Regional Office. That's when we filed an intent to sue.

That's when Jack Ward Thomas, who was in charge of the Forest Service,[3] did a discretionary review of the case. And he told the Gila Forest personnel to go back to the drawing boards—to start over.[4] They were instructed to put out a new EIS with no motorized intrusions, no development of the wilderness. Construction of stock tanks was prohibited even if they wanted to do it with primitive means. And finally, Thomas declared that grazing is not a historic use. He made it clear that this is not a balancing act. They're not balancing wilderness values against grazing privileges. If grazing can take place without affecting wilderness values, so be it. If it can't, then they need to modify the grazing or eliminate it in order to maintain wilderness values. So we set a precedent for wilderness management throughout the nation through Forest Service policy.

CHAPTER 3

Banking industry influence perverts management of public lands ranching
When a rancher wants to get a permit to graze federal public land—a privilege, not a right, as so many newspapers and media often call it—a rancher buys a base property. And for that base property he's paying an inflated amount. The difference between what the base property costs—the true value of the base property—and that inflated value is based on the number of cattle that he can run on his adjacent federal allotment, among other things. Basically, a bank loaning money for this has no security except for the ability of a rancher to run cattle on his public land allotment. If the Forest Service attempts to reduce numbers of cattle for resource protection, that's when the banks really freak out, because the permittee will no longer have the ability to make sufficient payments to the bank.

There is a memorandum of understanding (MOU) between the Forest Service and the banking industry[5] that allows the Forest Service to act as an escrow agent, holding the grazing permit for the bank and the permittee until the loan is paid in full. If the permittee defaults on the loan, the Forest Service is to give the bank another chance at reselling the base property with the permit attached to it, so the bank won't lose out on the deal.

In the case we were dealing with, the former permittee on the Diamond Bar had gone bankrupt.[6] The Federal Intermediate Credit Bank of Texas that held the mortgage on the base property with the permit attached to it, then sold it to the Laneys.[7] The Forest Service was in the process at that time of lowering the allowable number of grazing cattle on the allotment from 1,188 down to 833. There was a letter to the Forest Service from the bank that threatened to play hardball if the numbers were reduced.

The Forest Service is required to notify a bank of any reductions in cattle numbers. And through the MOU with the bank and the Forest Service, the Forest Service cannot lower numbers anything more than 20 percent per year after notification of reduction. If there's a really severe drought and the numbers need to be reduced drastically, according to this MOU, that can't happen.

We decided to investigate the bank that held the mortgage on the Laneys' grazing permit. With the money they raised, they donated to many congressmen and senators, amongst them a significant amount to Senator Domenici and even some to Democratic Senator Jeff Bingaman, also from New Mexico.

This is why we have such difficulty getting good honest management of our federal lands. Senator Domenici, who was the chair of the Budget Committee in the US Senate at that time, held a lot of sway over decisions on the Forest Service and utilized his power to the detriment of wilderness and wildlife values.

People typically assume that ranchers have a lot of power, but it's not really true. It's the banks holding the estimated $2 billion that's loaned out on grazing permits on western public lands that have power. Our public lands are being used as collateral for bank loans. Our wilderness areas, archaeological sites, watersheds, wildlife habitat is being mortgaged—used as collateral for bank loans of ranchers that are buying base properties with grazing permits attached to them.

CHAPTER 4

Proposals to reintroduce Gila trout on the Diamond Bar Allotment

Prior to the Laneys being issued the grazing permit for the Diamond Bar grazing allotment,[8] the US Fish and Wildlife Service said that Black Canyon was a prime re-introduction site for Gila trout. But as soon as there was a big bank mortgage that needed to be paid on this allotment, things changed. They needed to run as many cattle as possible in every possible location. Suddenly there were now stumbling blocks as to why the Gila trout should not be placed in Black Canyon. We have the same physical layout, but different politics. Suddenly it was too close to a road where people might put in nonnative trout. Suddenly there was no place to put in the fish barrier. And so throughout our battle, we were never going to see Gila trout reintroduced into Black Canyon.

It wasn't until after we won,[9] and Black Canyon was mostly closed to grazing, that it once again became a viable reintroduction site. In the upper reaches of Black Canyon there was a big forest fire. And after that there was a lot of rain. When the rain filled the creek, the ash from the fire came with it and killed all of the fish in the stream, which created on opportunity to restock the stream with Gila trout without the possibility of cross breeding with other species.

In the summer of 1998, the Fish and Wildlife Service, along with El Paso Sierra Club, Gila Watch, Trout Unlimited, and possibly a few other groups, created a rock dam barrier in Black Canyon. With the barrier in place and a clean stream to put them in, the US Fish and Wildlife Service finally reintroduced the Gila trout back into upper Black Canyon on the east side of Forest Road 150. Gila trout have also been reintroduced in the uppermost reaches of Main Diamond and upper South Diamond creeks, as well as on the Diamond Bar Allotment.

CHAPTER 5

Susan Schock is harassed by a sheriff

After we had made a name for ourselves in the region, there was an incident out in the Mimbres Valley where two women went hiking. They parked their little truck on the side of the road and took off on a hike, apparently through the private property of a local rancher.

The rancher saw their vehicle and, for whatever reason, thought that it was Susan Schock's. Assuming she was taking pictures of his land, the rancher called the sheriff, and together they went searching for the women.

When the women finally made it back to the truck, it turned out that one woman was a local that they knew, and the other lived right down the road, but they didn't know her. The deputy sheriff evidently hounded the woman, claiming he knew she was that Susan Schock woman by the license plate he ran on her. Finally, she produced identification that proved she wasn't Susan.

This woman later said, "I don't know who in the hell Susan Schock is, but she had really better be careful if she ever comes out here. These people wanted blood. They would not believe—they would not take no for an answer—that I was not Susan Schock."

It appears more than coincidental that a mere two weeks later when Susan showed up at the Gila Watch office in downtown Silver City, thirty miles away from that incident, that the same deputy sheriff was across the street apparently waiting for Susan to arrive in her vehicle. During recent construction, a handicapped-parking sign had been installed, but no markings on the pavement had been laid out yet. Non-handicapped people parked there all the time. When Susan parked where she normally did, the front of the vehicle was a few feet in front of the sign that said "handicapped zone."

The sheriff was at her door before she got out and was writing her a ticket.

She said, "Excuse me, what's going on?"

He said, "I'm writing you a ticket for parking in a handicapped zone."

And she said, "Oh, I'll back up." And she backed up a few feet.

He still said he was writing her up.

She said, "What for? I'm not in it."

But he insisted on writing her up.

So she said, "This is too weird. I'm gonna get a camera."

And he said, "No you're not. You're staying right here."

She said, "I am." And went up, got the camera, came back down and started taking pictures of the sheriff and of the vehicle.

She asked the sheriff where exactly the handicapped zone was, since there was no way to know.

He indicated that it was everything behind the sign.

Susan said to the sheriff, "So, all of these other vehicles are illegally parked too? And you're going to write all of these people tickets?"

And he said no.

At that time, a local policeman came up and said, "Oh, you don't have to worry. We don't ticket people here because it's not designated on the road yet."

When the time came, we went to court with all the documentation. The sheriff never showed up, and the case was dismissed. I tried to talk Susan into filing harassment and stalking charges against the sheriff, but she was too busy doing other things.

CHAPTER 6

Permittee Kit Laney applies political pressure to the Forest Service (1995)

Kit and Sherry Laney, the most recent permittees on the Diamond Bar grazing allotment, started their business with a few hundred head of cattle and slowly built up their herd year by year. As the number of cattle rose, the ecological health of the land started to deteriorate. In 1995, Laney had thirty head of cattle die in the Squaw Creek area, because his cattle had consumed all the vegetation, and the only things left were the toxic plants that abort fetuses or kill the cattle outright.

Just about that time, Sue Kozacek, who was the wilderness district ranger, had the annual permit renewal meeting with Laney to determine how many head of cattle he could have and in which pastures he would have them. At that meeting she said to Laney, "What am I supposed to do? Your cattle are staggering around, bony and starving. You just had thirty head die. The whole allotment is overutilized. You can keep a hundred head on the allotment, but you'll have to move the rest to the T-Bar Allotment."

According to Ranger Kozacek, Laney refused. He said, "If I move 'em to the T Bar, you'll never let 'em back on the Diamond Bar. And they're too weak to move. I'm gonna lose too many if I move 'em right now."

So Laney called Domenici.[10] He called Skeen.[11] And he called the governor's office. We ended up having a big three-day meeting out on the ground the week of the Oklahoma City bombing.[12] The neighboring permittees were there, as well as members of the Range Improvement Task Force from NMSU, officials from the governor's, congressman's and senators' offices, and Forest Service personnel from the Gila Forest, the regional office in Albuquerque, and staffers from Washington, DC, Office. Our experts were there as well: Dr. Bob Ohmart,[13] our riparian expert, and Bill Worf,[14] our uplands expert. I believe at least one lawyer who was helping us came out too, as well as Susan and I.

It was during this meeting, after we had heard that some deranged person or persons had bombed a federal building, that Laney made the threat: "If you come to move my cattle off, you're gonna meet a hundred people with guns. It's stupid, really stupid, but it's gonna happen."[15]

Those threats are the kind of things that make these agencies back off from what's necessary in doing their job and lowering cattle numbers the way they're supposed to.

After the three days out on the allotment, nobody was willing to say that the Diamond Bar Allotment looked good, or that there was any place to put more than a hundred head of cattle. But lo and behold, a few weeks later a new decision came out of the Forest Service. Rather than a hundred head being allowed on the Diamond Bar, there was a decision to allow four hundred head there. Laney had six hundred head on the ground at the time, and nobody ever talked about where those other two hundred would be put.

When our local *Silver City Daily Press* covered the story, they implied that the Forest Service was trying to put Laney out of business, when in reality they were bending over backwards offering him a brand new grassy allotment after this 145,000 acres of wilderness was so cow burnt that his cattle were dying. The paper failed to mention that this was all precipitated by thirty head of cattle dying, that the records showed that the cattle were bony and starving and staggering, and that Laney was offered a whole new allotment to put all of his cattle on, except for the hundred head that he could keep there. It didn't mention any of his threats. And it didn't mention that their new decision was an *increase* in number over the one hundred head.

That's when Susan contacted the *Daily Press* and said, "Excuse me, you don't have all the facts here." She then provided them with information so they could run another story and correct a lot of the really strong bias that the first article had.

They never did run a second article, and so Susan wrote out the chronological series of events that took place in a letter to the editor.

And they refused that.

Then we decided to run an ad. We gave the newspaper $200 to run an ad of the series of events to correct the misinformation that they were putting out. And they refused and returned our money and wouldn't do the ad.

Unfortunately, that's not the end of the story. The *Daily Press* then wrote the most vicious editorial I'd ever seen from the newspaper blasting Gila Watch as being a divisive group that was trying to destroy this rural way of life and the local economy.

So those were very difficult times for us. Susan specifically was really vilified throughout this whole process, and it's one of the reasons she doesn't live here anymore.

CHAPTER 7

Kit Laney sues the Forest Service (1996)

When we won on our points in the EIS for no development in the wilderness, we got a commensurate lowering of the number of cattle allowed on the allotment as well. Since the whole purpose of developing the wilderness with "range improvements" was to increase cattle numbers by disbursing the cattle, without those new stock tanks the cattle numbers came down to three hundred head instead of the proposed 1,188.[16] When the numbers came down to three hundred head, Kit Laney refused to sign his permit and filed suit against the Forest Service.[17]

Laney lost his case, and because of his threats about the Forest Service "meeting a hundred people with guns" if anyone came to remove his cattle, Judge Bratton stipulated that in order to appeal his decision, Laney would have to remove his own cattle from the Diamond Bar Allotment. By stipulating that, it avoided a potentially violent situation.

It took Laney a year and two months to remove his cattle. Why? Because his penalty for the continued trespass was a doubling of his grazing fees. Let's see, $1.35[18] times two is $2.70 per cow-calf unit per month. Fair market value for leasing similar private land for grazing is $10.00 per month. So he was being "penalized" by charging him one-fourth of fair market value. After Laney dragged his feet for months, and claimed he couldn't get the cattle out because the roads were snowed in, we went up in two-wheel drive and documented that the road condition was passable. Instead of costing Laney money, he was making money during his trespass.

Then Laney appealed the case to the Tenth Circuit Court in Denver and lost there as well.[19]

CHAPTER 8

Sauber at the Quivira Coalition meeting in Silver City (8–9 June 2001)
I'm co-owner of a bicycle and outdoor equipment store in Silver City, and I see a lot of people come through hiking the Continental Divide from Mexico to Canada. One day a guy came in the store who was hiking the Continental Divide Trail from south to north, and he described his efforts at trying to get badly needed drinking water in the Burro Mountains just south of town. He was shocked and disgusted that cattle had trampled the small streambed so heavily and turned the little seep of water into a cow-crap-and-mud slurry, which would have plugged his water filter immediately. It strained his ability to go on without that badly needed water.

I said, "You're describing the damaged riparian area really well. Where is that?" And so he showed me on a map.

I went out there, and I took a bunch of pictures,[20] which I enlarged to 8-and-a-half-by-11-inch color photos.

The Quivira Coalition held a meeting in Silver City during June of 2001. And on the evening of the first day, after their slide presentation was done and the lights were turned back on, I went over to Courtney White, one of the founders of the Quivira Coalition, and I showed him the pictures I had taken the year before.

After mulling over the photos of the trampled riparian areas, I believe he said something on the order of "This person really needs some management," or something to that effect. And then he asked where the pictures were taken.

I replied, "This is where you're taking us tomorrow, your role-model rancher for the region." And he just shut up. He didn't know what to say.

That's when Dan Daggett[21] came up and said, "What's wrong with these pictures? Did you go back there months later? What did it look like a few months later?" Which I understood him to mean that cows are good for the land; that it will be better a few months after being grazed—better than if the cows hadn't been there.

I said, "It doesn't matter what it looked like a few months later. Where's the habitat right now? Where are the species that normally would be living in this area? They're dead under the hooves of all these cattle."

This is public land that is supposed to be managed for watershed values, wildlife, recreation, hunting and fishing, as well as grazing. Multiple use does not mean grazing multiple cows. The federal multiple-use mandate means the public land in question could be used for many types of legitimate uses, and no single use should take away from any of the others.

The next day we went out to the allotment.[22] And I was being quiet because I wanted to see how far they would go without my input—without my criticism.

At one point, they did ask for our comments in an area that was supposedly ungrazed and in "bad shape." They did admit, at one point, that for a number of years the fence wasn't too great, and cattle were in there on a regular basis. So it wasn't really ungrazed for the long time period they had originally claimed. The point that they were trying to make, though, was that all the grass there was dead, and that dead grass is bad. And it means the place needs to be grazed to bring it back to health.

To me, looking at all the dead grass meant there was great habitat for ground-nesting birds, rodents, and insects. You don't need to be a rocket scientist to realize that even if the tops of the plants are dead (or even if the plant is indeed dead) that there's still value in it for preventing erosion and keeping the sun from drying out the soil.

They were looking at it purely as a forage-capability issue—what value is it as forage for livestock? They talk very little, if any, about habitat needs of wildlife species. It was really a bovine-centric presentation.

Later in the day, we went down to the Gila River area,[23] where the permittee didn't let the cattle graze very heavily. And the talk there was about why there are so many Southwest willow flycatchers in that area.[24]

When the day was almost done, and we were back at the cars getting ready to leave, I just quietly started passing out my large color photographs to this new group of people.

One of the people standing there said, "Wow, where were these pictures taken?"

And the permittee spoke up and said, "This is my place here."

And, I think, I corrected him, "His public land allotment. This is public land. And perhaps this is why the Southwest willow flycatcher is doing a little bit better here in the bottomland, because he is 'bombing' the upland areas with his cattle."

Nobody could look at these pictures without having a real visceral reaction. The pictures were so graphic you could smell the putrid mess, and the few willows remaining were broken and eaten down to twigs.

Anyway, the permittee said, "I invited everybody here in good faith, and I don't really appreciate Mike showing these pictures."

I responded that, first of all, this is public land, and I didn't expect we would be able to go out and see the whole allotment. So I simply brought some pictures so we would all have a good idea of what is going on up there where we didn't visit.

I then mentioned that when these pictures were taken, there were two Forest Service range personnel assigned to this region to oversee its management, and now there is only one. I asked, "Will there be twice as many places this bad in the years to come?"

I have to hand it to Steve Libby, the top range person for the Gila Forest,[25] who said it was good for me to speak up, and that it was good for the permittee to respond, because we both remained calm and we all know a little more now.

What I know is that the Quivira Coalition probably won't be going to areas where knowledgeable grazing critics are likely to attend anymore. It makes them look bad, and their role-model permittees don't like it either.

Todd Shuman

In the 1980s, beer maker conglomerate Anheuser-Busch purchased the Cabin Bar Ranch in California's Sierra Nevada Mountains to access underground water for use at its Van Nuys brewery in case of drought. Also associated with the ranch were permits to graze cattle on two allotments within the Golden Trout Wilderness—home to California's state fish, the golden trout.

The Golden Trout Wilderness had been grazed by cattle for several decades before the arrival of animals owned by Anheuser-Busch. Consequently, by the late 1980s, environmental studies began showing damage to the trout habitat from this long-term grazing.

While hiking through the Golden Trout Wilderness in 1995, Todd Shuman witnessed this environmental degradation. Over the next year, Shuman further investigated the environmental conditions of the region and then began organizing Sierra Club members to oppose the continued grazing of the Anheuser-Busch cattle. He also worked to build a coalition of groups that included Trout Unlimited and California Trout for that same purpose.

Shuman provides a personal perspective on the ensuing five-year campaign that concluded in February 2001 with the Forest Service's decision to remove the Anheuser-Busch cattle.

Todd Shuman made his remarks on the 9th of August 2003 in Tehachapi, California.

Chapters
1. Shuman's formal education and youthful activism
2. Shuman learns about public lands ranching (early 1990s)
3. Shuman learns about the Golden Trout Wilderness (early 1990s)
4. Collaboration among organizations (1996–2000)
5. Energizing the Sierra Club (1996–2000)
6. Drawing the news media's attention
7. Increasing the opposition to Anheuser-Busch
8. The Golden Trout Wilderness is protected (2 February 2001)
9. Shuman reflects on the success of the campaign

CHAPTER 1

Shuman's formal education and youthful activism

I was kind of a child of the '60s, so I was steeped in the values of that time. As I got older I was naturally drawn to activities that would implement those values.

In 1979 I became involved with the UC Nuclear Weapons Lab Conversion Project.[1] There was a lot of anti-nuclear activity at that time. So that's how I got my first involvement with things.

Before I went to college, I did some peace and justice work with a religious group called "Clergy and Laity Concerned." I actually worked there for my future mother-in-law. We worked on the MX Missile issue, Salvadoran intervention issues, veteran's justice issues. Stuff like that.

Later I went to college at UC Santa Cruz, where I got a degree in sociology. During this time I worked with a local group opposing US intervention in El Salvador, but I also continued working on high-technology issues.

I remained in Santa Cruz during the 1980s. During the latter years of that decade and into the '90s, I was petitioning on behalf of some progressive environmental initiatives being circulated in California. That was probably my first real experience with environmental issues other than those of nuclear weapons.

Around that time I met some people who were hunt saboteurs. They would go out and sabotage bighorn sheep hunts, bear hunts, mule deer hunts in California. Through my involvement with them and participating in hunt sabotage, I became involved with the EarthFirst! network's other environmental issues.

In the early '90s I worked in association with Tom Skeele of the Predator Project trying to gather information on the status of the wolverine in California.

Through that activism, I was exposed to some of the issues concerning public lands livestock grazing.

CHAPTER 2

Shuman learns about public lands ranching (early 1990s)

During the early 1990s, I read some articles by Lynn Jacobs concerning public lands livestock grazing that he was publishing in animal rights magazines.[2]

Plus a lot of the hunt saboteurs were vegetarians and vegans. And they were involved with some of these issues already.

One of my good friends, Lee Desseaux, was doing lots of banner hangings for the anti-grazing movement at that time, such as at the Denver Livestock Show.

So I gained some exposure about these issues through my involvement with hunt saboteurs and, increasingly, just through the EarthFirst! movement.

Also, I was involved with some litigation concerning logging of residual old-growth forest in the early '90s. We actually entered into settlements with a logging company to protect trees that would constitute habitat for the marbled murrelet,[3] and to limit cutting in ephemeral drainages that would promote erosion that could impact steelhead trout spawning waters farther downstream.

CHAPTER 3

Shuman learns about the Golden Trout Wilderness (early 1990s)

My first exposure to grazing issues in the Golden Trout Wilderness occurred in the early 1990s when I was traveling with my girlfriend to the Ancient Bristlecone Pine Forest in the Inyo National Forest. A beautiful area.

Up there they have a University of California Research Lab Station. And in their library, when I was thumbing through some publications, I came across one by Odion, Dudley, and someone else. They had done some research during the 1980s concerning vegetation and grazing impacts on the Golden Trout Wilderness on the Kern Plateau.[4]

John Muir[5] had been a shepherd up there. And his famous comment about the hoofed locusts and sheep grazing, I think, referred to his experiences on the Kern Plateau.[6]

So when I saw this publication, I became interested in pursuing the issues further.

In late 1994 I left Santa Cruz to work in Mammoth Lakes—in the ski resort. And at that time I began focusing some of my energies into activism on public lands livestock grazing. At that time a court decision concerning the Comb Wash Allotment[7] was issued that we thought would require the agencies to conduct NEPA reviews for all grazing allotment permit renewals or permit issuances.

Then in 1995 I became aware of the NEPA review for the Monache Allotment on the Kern Plateau. I knew that a review would also be occurring for the Mulkey Allotment. And although I wasn't sure when reviews would be completed for the other allotments up there, I became very interested in visiting that area.

That summer I phoned Brett Matzke, the public lands director of CalTrout.[8] And I told him that I was going up there with my brother who was a fly fisherman. I asked Brett if there were particular areas that we should look at. Brett suggested some, and we went to some others.

That first time I went with my brother, we hiked up the South Fork of the Kern River into the Templeton Allotment, where we did a lot of photo documentation. And from there we continued hiking northwest into the Whitney Allotment, where we did additional photo documentation.

Later Brett told me that the Whitney and Templeton Allotments were controlled by Anheuser-Busch,[9] which had purchased a ranch during the 1980s that had a number of grazing permits associated with it. And so they were actually the permittee for two of the four allotments in the area.

I later made a solo trip, where I did photo documentation in the Mulkey and the Whitney Allotments as well. The Mulkey Allotment was not grazed by Anheuser-Busch, but by the Hunter family. Roy Hunter controls the permit. And his son and his daughter actually manage the cattle on the allotment.

At that time I was already getting information from the Forest Service through Freedom of Information Act requests. I was receiving allotment management plans. I was receiving the permits. I was receiving analytic documents that had been produced in the past. I was receiving annual operating instructions. So I was getting a sense of how the allotments were being managed.

They were also sending me utilization review reports, and trampling and chiseling surveys. I got watershed inventory assessments a little bit later.

I was gathering a file on these allotments all during the summer of 1995—doing my own photo documentation; trying to document where there was apparent damage to soil and damage to vegetation, degradation of riparian area, lack of willow cover. Things like that.

That's how I got started in 1995. At that time I was working closely with Dano McGinn, who was the chair of the California Grazing Reform Alliance and the California Mule Deer Association. But I was also in contact with Brett Matzke of CalTrout. And I was increasingly making contacts with conservation chairs of fly fisher clubs throughout California.

I would focus a lot more on outreach and working with these other groups in 1996. But during 1995 I was focused more on gathering information and doing my own photo documentation. So that's where I began.

CHAPTER 4

Collaboration among organizations (1996–2000)

Over the years I would work with both California Trout and with Trout Unlimited. My brother and I would also give presentations to chapters and gatherings of Federation of Fly Fisher Clubs. We would get stuff published in some of the independent magazines that the fly fishing community would produce. So we had numerous dealings with these groups.

In 1996 I really started going on this. I had gathered a lot of information, and now I was writing up analyses of it and distributing it to journalists and to anyone who was interested.

There tended to be competition between the California Trout Unlimited Chapter and California Trout. California Trout, at one time, was the California Trout Unlimited Chapter. But at some point they had serious conflicts with the national organization, so they broke away and established their own organization. Then Trout Unlimited created another California Trout Unlimited Chapter and Council. So those two groups always have been competitive over issues.

California Trout, at that time, was interested in—how shall I say it? They wanted to try a consensus, collaborative approach with the Inyo National Forest and with the permittees and other interested parties to see if grazing reform could be implemented at a site-specific level.

I went to the first meeting where this process was initiated. But I didn't go to further meetings of this. It seemed to be exclusively CalTrout. They were the lead organization. The only other environmental group involved was Range Watch—Jane Baxter.[10] It also looked like the process was an attempt to muzzle people who wished to talk about the issues on the Kern Plateau with the news media.

At that time I knew some people in the local Sierra Club group—the Range of Light Group—that covered the jurisdictional area of the Owens Valley, where the Inyo National Forest was headquartered and where the Anheuser-Busch ranch was located. They were also kind of involved in this collaborative process.

The Forest Service was looking toward CalTrout and the ranchers to come up with solutions that the Forest Service could implement. The Forest Service really didn't want to be pressured into implementing its own laws. And we were concerned that this group would be developing policy and prescriptions in a semi-private setting—excluding the public. That, we thought, would be a violation of the Federal Advisory Committee Act (FACA), which regulates how nongovernmental collaborative efforts like this relate to the government. We thought there were violations of this act, because this was a private body that was starting to develop policy concerning public land. And we pressured CalTrout to withdraw from this. So the collaborative effort never really went far.

There were other problems. The ranch manager of Anheuser-Busch and their lawyer were not very sympathetic to the collaborative process, probably because we were increasingly focusing on Anheuser-Busch. We wanted to get reform and grazing reductions on those allotments more so than on the other ones in the Golden Trout Wilderness.

So there were some divisions. There was an alliance between myself and the California Grazing Reform Association, and increasingly with Trout Unlimited, to take a harder line with respect to grazing reform efforts. And California Trout was taking a softer line in terms of what kind of prescriptions they wanted.

California Trout did put some money into monitoring. In 1996 the Inyo National Forest was having difficulty securing funds for monitoring. They had done very serious monitoring in 1995, but then they ran out of money, or they didn't get money allocated through the regional office for doing monitoring in 1996. So the only monitoring that occurred was basically financed by California Trout.

And it was pretty high quality monitoring that occurred there. In 1995 and 1996 we had extensive documentation of resource damage that had occurred during those grazing seasons. So I was able to get that information, write it up, and distribute it. And try to educate people about what was happening up there.

I thought I had a good issue because that area is the last native waters for genetically undiluted golden trout. It's also important habitat for the Monache mule deer herd. There's a number of riparian-dependent birds—the yellow warbler, the willow flycatcher—that have historically used those areas as habitat. And there was extensive documentation of livestock-related damage.

And we had Anheuser-Busch. We had a permittee that we could influence through potential threats of boycott.

So there were a number of factors that made it productive to focus on these allotments. We felt that we could actually impact and influence and advance grazing reform nationally by focusing on these allotments just because of the nature of the permittee and the resources involved.

CHAPTER 5

Energizing the Sierra Club (1996–2000)

In 1995 and 1996 I operated, along with my brother, pretty much independently. We had named ourselves the "Golden Trout Wilderness Protection League" just to give a formal name to our activities. But we worked pretty much as independent activists.

One person we increasingly worked with was Sally Miller, who had been associated with the Mono Lake battle.[11] Her husband, Roland Knapp, had conducted a great deal of very important research on the impacts of cattle grazing modifications to golden trout habitat in that area.

Sally had a lot of activist experience and connections. And because of her husband's involvement, she had concern about what was happening in the area. She had also worked closely with Joe Fontaine, who was a past president of the Sierra Club.[12] Joe had been active in actually establishing the Golden Trout Wilderness back in the '70s. So Sally suggested that I should talk to him and work with him closely on these issues as a way of involving the Sierra Club.

I had worked with club members earlier. I had worked with the conservation chair of the Toiyabe Chapter's Range of Light Group—a wonderful woman named Betty Goodrich. And I'd worked with other people who are on the board of that local group—Wilma and Bryce Wheeler.

And so they were interested in the issues, and Betty had been involved. But Sally encouraged me to also establish communications with Joe.

Joe and I kind of formed an alliance,[13] and we were able to craft some resolutions. Our first effort was to craft a Sierra Club resolution that called for resting this area from livestock grazing for an indefinite period because of resource damage.

Then we just kept working the committees. We went to the Southern California Forest Committee. We drafted the resolution there. We got that committee to pass it. Then we went to the Eastern Sierra Club Activists Committee gathering where we actually refined the initiative.

Rose Strickland[14] had come down from Reno. Rose and Sally Miller and Joe and I crafted a resolution that essentially called for the Sierra Club to adopt the position that this whole wilderness area should not be grazed. I think we maybe even put in for a twenty-five year period of rest due to resource damage.

With that resolution passed, we then went to the Southern California Regional Conservation Committee meeting in Joshua Tree in March 1997. We were working with Tim Frank, who was chairing that committee and who at that time was also the chair of the overall California-Nevada Regional Conservation Committee. We got that resolution passed.

And then we went, maybe a month later, to the San Luis Obispo convention gathering of the California-Nevada Regional Conservation Committee. And we got that resolution passed.

So that basically established Sierra Club policy for the area, which was "kick the cows out," because the Regional Conservation Committee had the authority—had been delegated the authority by the Sierra Club—to establish policy on its behalf for areas within its jurisdiction.

And so we continually used these resolutions to work things up within the club.

After we had this resolution, Tim Frank and I wrote an article for *The Planet*[15] about grazing in this area, Anheuser-Busch's involvement, and the degradation of resources.

So we were getting the word around through the Sierra Club and we were getting policy change.

At this time I was starting to cultivate relationships with a couple of influential people within Trout Unlimited—Jamie Hunter, who edited their California newsletter and who was their California Council chair, and Don Duff, who was the Forest Service–Trout Unlimited partnership coordinator. They both had acute interest in this matter because the California golden trout is the state fish. And so they had great interest in assuring that its habitat was preserved and that any activities that were threatening it with extinction or damaging its biological viability were stopped.

I was using the Sierra Club's passage of these resolutions, and publication of these articles, to try making the fish organizations more aggressive, so that we could form an alliance to coordinate tactics for pressuring both Anheuser-Busch and the Forest Service.

By 1998 California Trout concluded that they had not been making progress with the Forest Service and the Anheuser-Busch permittee concerning grazing reform on these allotments. And at that point, CalTrout decided to take a more aggressive stance.

Now CalTrout was talking more explicitly about the need to rest these areas for a longer time. By then the groundwork had been laid for more harmonious relations between the Sierra Club, California Trout, and Trout Unlimited. So we were able to conceive of building an alliance that would actually pressure Anheuser-Busch and the Forest Service.

I had known from the start that any perceived divisions among the fisheries community would probably be discerned and noticed by Anheuser-Busch. And that would undermine our efforts to force the Forest Service to implement some serious reforms.

By 1998 all the environmental organizations were much more aggressive. And so we were able to make significant progress. That year the Forest Service did substantial monitoring in parts of the Anheuser-Busch allotments. And there were a number of areas that they documented as trending downward where there was serious damage.

I did further documentation with Jamie Hunter and Don Duff of Trout Unlimited by performing some aquatic surveys.

And California Department of Fish and Game was becoming more adamant about ongoing resource damage.

By late 1998 and into 1999, we finally had some organizations that were willing to consider calling for a boycott of Anheuser-Busch products. First of all, we were trying to get into a meeting with some top-level Anheuser-Busch officials to discuss this stuff. At first they said maybe we'd have a meeting, but then they didn't want to meet with us after all. At that point, Brett Matzke of CalTrout asked me, "Can you get some Sierra Club support for a boycott or at least to consider it?"

Joe Fontaine and I then went to the Sierra Club chapters that have jurisdictional relations to these allotments. And there's a number of them. The actual area that's grazed is in the Kern-Kaweah Chapter. The ranch is located in the Range of Light Group of the Toiyabe Chapter. The reason the ranch exists—the reason Anheuser-Busch controls the ranch is to secure water rights so that they could supply their brewery in Los Angeles if there was a drought. Well, everything related to Los Angeles comes under the Angeles Chapter, which is the largest chapter in the Sierra Club.

And we also had some connections with people in the San Diego Chapter, since Sea World, which is owned by Anheuser-Busch, is a big facility down there.

So we were able to get some resolutions requesting the Sierra Club Board of Directors to consider adopting a stance of boycott and promote a boycott of Anheuser-Busch, because of the damage done to the golden trout and the resources in the Golden Trout Wilderness. And we were able to get the Range of Light Group to endorse the resolution. The Kern-Kaweah Chapter, the Angeles Chapter, and the San Diego Chapter also endorsed it.

With those resolutions in hand, Joe Fontaine and I went again to the California-Nevada Regional Conservation Committee to ask them to endorse the resolution, which they unanimously did.

About that same time, the Inyo National Forest was initiating a NEPA review for these two Anheuser-Busch allotments. They conducted scoping during the summer of 1999. By the summer of 2000, they had released an environmental assessment—a big thick environmental assessment. It looked more like an environmental impact statement.

We wanted to insure that Anheuser-Busch would not appeal any kind of positive Inyo National Forest decision to rest the area or cut back stocking rates. We wanted to let Anheuser-Busch know that if they did challenge this stuff and they tried to maintain the status quo up there that we were seriously considering a boycott.

It was then, during the summer of 2000, that we asked the Sierra Club's executive director, Carl Pope, to send a letter to August Busch,[16] basically discouraging Anheuser-Busch from aggressively supporting continued grazing up there.

So at this point it was CalTrout and the Sierra Club that were starting to threaten Anheuser-Busch with a boycott. And Anheuser-Busch did not like it. We received a letter back from their vice president for environmental affairs—a real nasty letter saying, we don't appreciate this being threatened with a boycott stuff.

And that was exactly the response we wanted from them. We wanted to piss them off. We wanted them to be on the defensive. We wanted them to know that there was some serious organizational muscle that could be mobilized against them. And that it could hurt them.

I was very happy about the way things were going.

Trout Unlimited did not want to participate in the boycott issue. They have some corporate connections. But they were willing to take the lead in developing an Endangered Species Act listing petition for the golden trout, which Roland Knapp and others of us felt was needed at that time.

So we now had California Trout and the Sierra Club and possibly even Natural Resources Defense Council acting in alliance on this threat of boycott. We were putting pressure on Anheuser-Busch directly that way. And at the same time, we were also putting pressure on Trout Unlimited.

I was working behind the scenes with Trout Unlimited to help write the listing petition. And through that petition, they were putting pressure on the government by potentially constraining whatever the government might want to do with regard to future management in this area.

So things were moving among the different organizations kind of in parallel at just about the right time.

CHAPTER 6

Drawing the news media's attention

At different times we'd been able to get journalistic interest in this area. Frank Clifford of the *Los Angeles Times* had been real interested in the issues in early 1996 and wrote the first major piece on the conflict between cattle grazing, and wildlife and fish resources in the area. I had sent him an essay that I had written based on all the recent Inyo National Forest information that I had. And he ran with it. And that piqued the interest of other California journalists in this matter.

Later on Paul Rogers picked it up. And he published in November of 1999, a set of stories in the *San Jose Mercury News* called "Cash Cows."[17] One of the side stories was about the grazing allotments that Anheuser-Busch grazed in that area.

So we had been able to get a number of stories that were favorable to us.

CHAPTER 7

Increasing the opposition to Anheuser-Busch

Anheuser-Busch was starting to get some bad publicity. And now the major environmental organizations that have resources were intimating threats of boycott against them. Plus, a listing petition was published, issued, and filed even before the comment deadline for the environmental assessment was up.[18]

So we were moving pretty well at that time. I never felt we would get a progressive decision from the Inyo National Forest or get Anheuser-Busch

332 / Nongovernmental Conservationists

to back off on this, unless all the major fish organizations and environmental organizations were in agreement, and were aggressively calling for major reductions in grazing or outright rest of the area. By fall of 2000 everyone was pushing in that direction.

CHAPTER 8

The Golden Trout Wilderness is protected (2 February 2001)

The environmental assessment was issued in August of 2000. And the comment period on the assessment ended October 31st. Then the Inyo National Forest made the decision to eliminate livestock grazing on the Whitney and Templeton Allotments (the allotments grazed by Anheuser-Busch) for approximately ten years. The decision was made by Luci McKee, who had recently become district ranger.[19]

I had gone on the Whitney Allotment Proper Functioning Condition Assessment with Luci back in 1998, and I had a good impression of her. For her to be a ranger, she had to be politically minded. But she was a hydrologist. She saw the damage up there. And she made a decision to protect the resource in the long run. Jeff Bailey, who was the supervisor for the Inyo National Forest, deferred to his ranger on that. She made the decision.

And we were surprised.

The decision was appealed by Janice Allen, who is the daughter of Roy Hunter who controls the Mulkey Allotment, which is adjacent to the Anheuser-Busch allotments on the Golden Trout Wilderness. Janice Allen had some Forest Service experience. She had been a vegetation analyst earlier in her life. She had been hired by Anheuser-Busch as a consultant. And she had done some of their monitoring in earlier years.

She filed an appeal on behalf of herself, not on behalf of Anheuser-Busch. And Anheuser-Busch did not appeal the decision.

The regional office denied the appeal.[20] They felt that the documentation of resource damage was sufficient to justify the Inyo National Forest decision.

They did say that the Inyo National Forest would be required to establish a monitoring plan that would compare rates of recovery on the ungrazed allotments with what was happening on the two remaining Golden Trout Wilderness allotments that are still being grazed.

The Inyo National Forest is still developing that monitoring plan. They came up with a proposal in 2002 that was roundly criticized by all the groups that have been following this issue. This year they developed a revision in the monitoring plan that is much better. And we're trying to improve it.

We also want to insure that the Inyo National Forest does not prematurely continue grazing on these allotments. We want to make sure that the Inyo does not even consider initiating a new NEPA review for the Whitney and Templeton Allotments until 2009 or 2010—until the area has actually had about ten years of rest from livestock grazing.

CHAPTER 9

Shuman reflects on the success of the campaign

What's most instructive and most important about what we did in this case is that we formed an alliance of prominent fish groups and prominent national environmental groups, such as Sierra Club and Natural Resources Defense Council. It was essential to construct this alliance in order to have an impact upon both the Forest Service and Anheuser-Busch.

For me to make progress within the Sierra Club—in getting the Sierra Club to look at this issue and actually stake out some positions on this—it was important that I formed an alliance with Joe Fontaine, who has long-time respect within the Sierra Club.

People who have been very active in the Sierra Club for a long time listen to Joe Fontaine. Whereas they might have challenged me and been reluctant to support anything that I had encouraged them to adopt, they were much more willing to defer to Joe. And that enabled the Sierra Club to establish a very strong position on what they wanted to happen in this area.

Once we had that support, we were then able to form a strong alliance with Trout Unlimited. They are sometimes a cautious organization, but because the Sierra Club had staked out a strong position and had backed it up using the Forest Service's own documentation in many cases, Trout Unlimited was willing, in an alliance with the Sierra Club, to pressure the permittee and the Forest Service.

Once that alliance had been established, California Trout was more willing to take a strong position, because it saw a national environmental organization and a sometimes competitor fish organization pressing hard on this issue. That enabled us to form an alliance in which all the fish organizations were agreed that strong actions were needed. That whole alliance then enabled us to bring other environmental organizations into this effort, such as Natural Resources Defense Council.

This alliance, in which groups sometimes worked together and sometimes worked in parallel, enabled us to achieve victory. We needed all that kind of pressure at all these different levels to make things happen. And if we'd had division, we would not have presented the face of a serious threat to the permittee and to the Forest Service.

So I would say for the future for any activist, if you're working within the Sierra Club, try to find some progressive-minded respected elders who you can work with. And do your best to convince them that there's resource damage taking place and that strong actions are required—that cattle need to be reduced or removed from an area. If you can do that, you can then establish strong alliances with fish organizations. And when you have an environmental–recreational fisher alliance, that's something that corporations and the agencies pay attention to. Then you can make progress in changing the status quo.

Charmaine White Face

Teacher, writer, and activist Charmaine White Face (Zumila Wobaga) is Oglala Tetuwan Oceti Sakowin, a member of the Oglala Sioux Tribe of the Great Sioux Nation. A political columnist since the early 1990s, she is the author of **Testimony for the Innocent** *(Brunswick, ME: Audenreed Press, 1998), a book about financial corruption in tribal government. White Face is the founder and coordinator of Defenders of the Black Hills, a volunteer group working to ensure that the United States government upholds the Fort Laramie Treaties of 1851 and 1868.*

Charmaine White Face made her remarks in Rapid City, South Dakota, on the 6th of September 2004.

Chapters
1. Relationship between the buffalo and the Oceti Sakowin
2. Ranching destroys grasslands and the Oceti Sakowin
3. Prairie dogs made scapegoats for the livestock industry
4. Livestock grazing desecrates the Black Hills
5. Legislation that would protect the national grasslands
6. Leasing of reservation lands
7. A call for studying the biological diversity of the Black Hills and the grasslands

CHAPTER 1

Relationship between the buffalo and the Oceti Sakowin

Buffalo are our big brothers. They take care of us. They provide everything for us. And so when they wanted to do away with us, they did away with the buffalo first.[1] The last twenty-three were found in Yellowstone. Now there is a herd in the Black Hills in Custer State Park. And the tribes now are building up their herds.

We're supposed to live with them. If we don't, then we're not who we're supposed to be. That's a concept that a lot of people can't understand.

It's like on the Northwest Coast—a lot of the tribes there say, "We're salmon people. We have to live with the salmon."

There's a tribe on the California coast that has to live with the shells—different kinds of sea shells.

And we're buffalo people. We have to live with buffalo. There's reasons. There's a lot of reasons. It's hard to explain.

CHAPTER 2

Ranching destroys grasslands and the Oceti Sakowin

It's not just about what happened in the past, but it's right now. My people are becoming extinct, because we're not allowed to live the way we're supposed to be. And cattle grazing—cattle ranching—in this area is part of what is causing the extinction of my people. We will be brown Americans running around with American values speaking the English language, but that is not what we are meant to be.

The other part of my stance on this is that the only sustainable economy in this grasslands area is the way that my people lived—with the buffalo. With nature.

In the 1868 Treaty of Fort Laramie there was a specific land area where we could live. And another specific land area where we could hunt—where the buffalo and all the wildlife (what is called wildlife) were supposed to be allowed to remain so we could remain—so we could become viable—living the way we were living for thousands of years as a nation.

And then gold was discovered in the Black Hills,[2] our sacred place—a very sacred place. The whole hills, not one place, but the whole mountain range is sacred for many reasons. When gold was discovered, then the US deliberately broke their treaty and allowed people to come in—to trespass. And that's what we still say, "They're trespassing."

They forced us into "prisoner of war camps." But they call them "reservations" because that's what was able to "sell" with the liberals on the East Coast in the mid-1800s.

People don't know that those designations are still there—that we still live on POW camps, but they're called reservations. And that our way of being has been destroyed for the purposes of cattle ranching, for the purposes of mining, oil development, and, of course, the gold mining in the Black Hills, tourism development. All of these things.

There were wolves in this area—where we live in western South Dakota. There were bears. There were a lot more bobcats. There were a lot more predators than there are now.

Where are the wolves? They're gone.

Where are the bears? They're not here anymore either.

Where is the buffalo?

Where are all the different kind of grasses?

Where's the tallgrass prairie?

Even at the time of my great-grandfather there was still tallgrass prairie—grasses that were taller than his horse. Grasses, of course, that would be taller than me if I stood up. Where are they?

There's only one place left where there is still tallgrass prairie and that's a little tiny national grassland in southeastern North Dakota.[3] And here I live in western South Dakota, and there's no tallgrass prairie out here. Yet this place used to be covered. This place was covered with buffalo. There was a way to live in this area.

The Oceti Sakowin—the Great Sioux Nation—we need the grasslands to continue as Oceti Sakowin. Without the grasslands, we just become brown Americans. We disappear as a people. Just like—. Where are the wolves? Where are the bears? Where's the buffalo? We're disappearing just like them.

Just because people think that we have acclimated or we have assimilated to the American culture—maybe we have assimilated to the American culture, but that means then that we're disappearing.

That's what the grasslands mean to me.

CHAPTER 3

Prairie dogs made scapegoats for the livestock industry

Ranchers are going out of business now because of the drought. And they want to kill off the prairie dogs, because they say that they eat the grass that their cattle eat. And that's malarkey.

They've already started poisoning the prairie dogs on the Pine Ridge Reservation,[4] just south of the Buffalo Gap National Grassland. The black-footed ferrets will become extinct without the prairie dogs. It's just appalling to me that they would poison them when they know that they are needed for the continuation of other species.[5]

When my son went out hunting here, he found a dead prairie dog. And not far away he found a dead eagle. So it's not just the black-footed ferrets that will be hurt if they poison these prairie dogs. It's gonna be a whole lot of others too.

If the ranchers would quit their overgrazing, then the prairie dogs wouldn't be so numerous. But they can't see that connection. And so to them it's just money. It's greed.

CHAPTER 4

Livestock grazing desecrates the Black Hills

If you go up into the Black Hills, in an area where the trees are allowed to grow naturally, you will find sacred sites where the trees do not grow inside of that circle. And cattle walk all over these sacred places. They walk all over our burial sites. They poop and pee and eat the grass. It's not their fault, but it is the fault of the human beings who put them there. And the human beings also then create fences and roads in these same places.

Yeah, the buffalo were there. Buffalo, antelope, deer, elk, bear. There was a lot of wildlife that lived in the Black Hills. But they didn't overgraze it. They didn't ruin it.

We have an awareness, a connection, a relationship with these other species. We were not allowed to hunt in the Black Hills. Anything that was in the Black Hills is like in a church. You don't kill something in a church. You don't kill them in the Black Hills. And so the people would wait until the migration outward in the spring happened.

Yet there are human beings who don't understand this, who have not been made aware of this. But then there are some who don't want to hear it either, this story that we say—this is what the Black Hills is. It's a sacred place. The Forest Service knows this. We told them over and over. They won't listen. They won't listen about the Black Hills. And they won't listen about the national grasslands.

I can't say all of them. It's the higher ups. I know scientists that work in the Forest Service. There's archaeologists also. We talk about these things. And they don't like what they're ordered to do. But yet, if they want to keep their job, then they do what they're ordered to do.

Some of them have enough integrity, enough trust in their own self, that they will say, "No, I can't do this anymore." And they'll resign or they'll leave some way or other. I know, I've had friends that are in the Forest Service.

I feel sorry for those that choose to do what they're told to do whenever it goes against their own learning, their own understanding, their own integrity. I feel sorry for them, because there are consequences. There will be consequences for the Black Hills and the grasslands. There will be consequences for the whole world, because this is a sacred place, a major sacred place.

CHAPTER 5

Legislation that would protect the national grasslands
The legislation that I wrote was specifically for the national grasslands.[6] There's the Thunder Basin Grassland in Wyoming. Buffalo Gap, Fort Pierre, and Grand River National Grasslands in South Dakota. And then there's the Sheyenne Grassland, the Little Missouri, and the Cedar River in North Dakota.

All these national grasslands, first of all, are within our treaty territory—the 1868 Fort Laramie Treaty territory. And I wrote the bill because there are rivers that are being polluted—strongly polluted. There are species within these rivers that are going to become extinct from pollution.

These rivers right now are being polluted by cattle. One of the things my legislation says is *no grazing*. There's grazing on the national grasslands. And it's this grazing that is really hurting them.

So my bill asks for no grazing by livestock, but for the restoration of buffalo. Buffalo won't overgraze if they're managed properly, which means the other predators must be allowed to come back in also. You can't have grasslands without allowing the other species. They go together. It's like Chokecherries. Chokecherries have to go through the bodies of birds in order to germinate. If you kill off all the birds, will there be any more chokecherries?

In the grasslands there are other things too that need all the species. And with only cattle allowed on grasslands, you're killing the grasslands. That's what's happening. So my bill asks for no grazing. No mining. No ORVs.

How can you have a wilderness and yet allow vehicles to go in there? You can't.

How can you have wilderness and allow fences and cows to be in there? You can't.

There's also mining—coal surface strip mining in Wyoming on the national grasslands. In North Dakota national grasslands there's oil wells. And, I imagine, eventually there will also be coal strip mining. I know there's uranium up there. So I'm just waiting for that to open up too.

They're also beginning to be polluted by coalbed methane wells. These are going to hurt not just the wildlife but also cattle that drink from that water.

These grasslands are all near to, or even border on, reservations. One of them, the Grand River, is actually on Standing Rock Reservation. Standing Rock Sioux Tribe has asked permission to co-manage this national grassland, which is within their borders. They've never been allowed to do that.

On these grasslands there are burial sites. There are massacre sites. There are sacred sites. And there are also petroglyphs and petroforms—they are sacred forms for us. They have what would be called "religious connotations." We don't have religion in our language, it's more a spiritual connection with these certain areas. And all of these grasslands have these, but most of them are not known because of the exploitation that would take place. The few sacred sites that are known, like in the Black Hills, are exploited for different things.

And so a lot of these sacred places all over—burial sites, massacre sites—we would rather keep as private as we can to protect them from exploitation. There are also medicines that grow there that we need. So I don't know why the Forest Service or the US government will not allow tribes to co-manage these areas.

The bill asks for studies of the water. All the riparian systems on all these national grasslands need to be studied, because these cattle are the ones that can kill off a little tiny stream that feeds into a larger one.

Wild and scenic rivers[7]—we need to create more of those. How many do we have left? All the rivers running through these national grasslands are polluted. At one time you could drink from them. Now you can't. They're polluted from a lot of things, not just cattle.

We never did a real active push on the bill. We just didn't have time. There's so many things going on out here. If human beings could realize that it's proactive for them also, then, I think, we would get more of a push for it to be passed through Congress. But people don't understand the whole picture. They're very narrow, especially when they want to use something just for what they call natural resources. Then they only see "natural resources equal money." They don't see the bigger, broader picture. And they have to, otherwise everything will become extinct. That's common sense. It doesn't take a genius to see that.

One percent of the grasslands in South Dakota, that's all I'm asking for. Thunder Basin Grassland is only 1 percent of the total acreage of Wyoming. The three grasslands in South Dakota are 1.78 percent of the total acreage in South Dakota. And the largest one is the one in North Dakota and that's only like 2.5 percent of the total acreage of North Dakota. So it's not very much.

And if this doesn't happen, then the prairie ecosystem is going to totally disappear. Right now, the only pristine prairie ecosystem that's left is a little teensy tiny part called "Red Shirt" within the Buffalo Gap National Grassland.

At one time my people covered fourteen states and three Canadian provinces. We were pushed back, pushed back, pushed back until the last treaty asked for half a state and pieces around it. And they couldn't even do that. They pushed us onto these little reservations—POW camps is what they really were.

And now I'm saying we've got to do these grasslands or it's all gonna be gone. That's all I'm asking. It's not much.

CHAPTER 6

Leasing of reservations lands

On Pine Ridge Reservation leasing is handled either by the Bureau of Indian Affairs, or privately by someone who has taken their land out of trust status. Trust status means that when these lands were first allocated to people on the reservations, that the US has a trusteeship. And the Cobell Case[8] is the most recent example of the mismanagement of that trust status that the US had.

In the Cobell Case, a woman was trying to find out why the lease money that one of her family members was supposed to receive wasn't coming in. And she found out that the US government had mismanaged billions of dollars in trust money for leases for native people all over the US.

The Cobell Case is still going. She won the case. And the difficulty that they're having is that the US government—the Department of the Interior and the Bureau of Indian Affairs—did not keep good records. And so people were not being paid for the leases on their lands.

That's only part of it. That's just the money part that the US government mismanaged. It wasn't just grazing lands. It was oil leases, mining leases, logging—a lot of different kinds of leasing.

As for grazing leases, the process is, if a person's land is still within trust status, then they sign a paper that says the Department of Interior, through the Bureau of Indian Affairs, leases their land and is supposed to watch over it. They have range managers that are supposed to insure that it's not overgrazed.

Now, I have two kinds of land. I have one that's in trust status, where the BIA actually does this. And I have another piece of land that I've taken out of trust. And so that one can be privately leased. If you live on a reservation and you privately lease your land, then it's your responsibility to watch the grazing. And you also have responsibility to keep up the fences.

If you have land that's in trust status, then BIA will lease it out for you. And they lease it out according to the amount per unit head that your land is capable of grazing. In a good year, if you had ten acres, maybe you could have two or three cows on it. In a bad year, when there's a drought like we've been having, maybe one cow.

They have range managers that go out and watch over these areas. And like everything else in government, there are some that are good and some that are bad.

It happened that my land was being overgrazed by an off-reservation leaser. The range manager that was to oversee my land was a good guy. But he didn't know that his boss, who was corrupt, had connections with the leaser. And so, when the range manager went out to talk to the rancher that was leasing my land, this rancher literally ran over him and tried to kill him, because he didn't want him to tell him, "No, you have to take some of your cattle off this land."

This is a proven fact. They went to court and everything. The manager was injured very badly.

The rancher left him out there for dead. But he didn't know that his partner was sitting in a pickup a mile or so away.

These are the kind of things that happen today on reservations. This is the kind of corruption that goes on.

So the overgrazing on reservation lands happens. At Pine Ridge Reservation the people have been asking for a land audit for a long time. They want to know what's happening with all the land on the reservation. Who is watching the grazing? Who actually owns the land? A lot of people are going into the land office and finding out that their land was sold. And they never did that. So who's selling it? Who's getting the money?

There's a whole lot of corruption going on with land on the reservations. And it's not just at Pine Ridge. It's on other reservations too.

CHAPTER 7

A call for studying the biological diversity of the Black Hills and the grasslands

There is a UN agency concerned with biological diversity. They need to come out here and conduct a study on the biological diversity of the grasslands and, of course, of the Black Hills. The Black Hills is a unique species area.

We need this study because the US has a vested interest in the environment only for the dollar. They don't really look at the value of biological diversity. And because of their policies and practices in the Great Plains area, they are destroying all of it. Why? Because they don't understand the prairie ecosystem, nor the unique speciation of the Black Hills.

And because we don't have the human population out here, thank God, that there is on the East Coast or the West Coast, we also don't have the political clout to say that something is wrong here.

So as an indigenous person from this area, I have asked the UN to please send out people to do studies on the uniqueness of the biological diversity of the Great Plains. And to also look at cattle grazing, and oil and strip coal mining, and housing development, and tourism in the Black Hills. And excessive logging. It would look at all that. I don't know if it will happen, but it needs to.

There need to be studies on what has disappeared since Lewis and Clark came here.[9] Go back and read those Lewis and Clark diaries. Look at the diversity of animals, plants, insects, birds that were here when they came up the Missouri River. And look at it now, what's left—a minuscule amount compared to what they originally saw.

I see it from a different perspective, because it wasn't just the plants and animals that disappeared. It was my people too. My ancestors.

So if anything is to be saved, we've got to get away from just looking at the dollar. You can't eat the dollar when there's no food, and can't breathe it when there's no air. Can't drink it when there's no water.

Notes

Preface
1. Ferguson and Ferguson (1983).
2. Jacobs (1991).
3. Donahue (1999).
4. Wuerthner and Matteson (2002).
5. Steiner (1980) is a compilation of transcripts from interviews conducted with ranchers during the 1970s.
6. Studs Terkel is the author of several books of oral history, including *Working* (New York: Pantheon Books, 1974), *American Dreams* (New York: Pantheon Books, 1974), and *Hope Dies Last* (New York: The New Press, 2003).

Introduction
1. Roosevelt (1888).
2. Savage (1979, 110).
3. Savage (1975, 6).
4. Jacobs (1991, 438).
5. Wuerthner (2002a, 27).
6. Ibid.
7. Government Accountability Office (2005, 15).
8. The number of "authorized to graze paid permits" issued in the seventeen western states by the US Forest Service is 6,608 (US Department of Agriculture, Forest Service 2006, 8, 81, 90). The number of grazing permits and leases on BLM lands in fifteen western states is 17,940 (US Department of the Interior 2005, Table 3-9c). The actual number of livestock operators may be less than the sum (24,548) of the permit/lease numbers reported, due to some operators holding permits with both agencies.
9. Government Accountability Office (1988b, 11).
10. US Environmental Protection Agency (1990, 5).
11. Government Accountability Office (1988a, 24).
12. US Department of the Interior, Bureau of Land Management (2005, Table 2-1).
13. Compiled from Wuerthner and Matteson (2002, 251–53).
14. Horning (1994, 1).
15. Government Accountability Office (2005, 7).
16. Power (2002, 264).
17. US Department of the Interior, Bureau of Land Management (1992, 2).
18. Donahue (2005, 749).
19. Sharman Apt Russell (1993, 12), in recognizing the deficiencies of the cowboy myth in contemporary society, called for "new images and new role models, ones that include heroines as well as heroes, urbanites as well as country folk, ecologists as well as individualists."

Douglas K. Barber
1. The Blue Range Primitive Area, established in 1933, encompasses 173,762 acres on the Apache-Sitgreaves NF in Arizona.

2. Barber means that when the Sandrock Allotment was closed in 1984 that most of the topsoil was gone and, over large areas, weathered rock was exposed.

3. The cattle on the Sandrock Allotment were rounded up at the time the allotment was closed in 1984. Doug Barber, e-mail message to author, 6 September 2005.

4. Barber (1998).

5. The Apache trout (*Oncorhynchus apache*), Arizona's state fish, was listed as endangered in 1967 under the Federal Endangered Species Preservation Act of 1966. In 1975 it was downlisted to threatened after re-evaluation of its status.

6. The West Fork Allotment is on the Alpine Ranger District, Apache-Sitgreaves NF, Arizona.

7. The development of the stream-fencing project took place during 1993–94, with construction of the fence in 1994. Doug Barber, statement to author, 10 June 2004. Barber (1998) states that twenty miles of fencing were constructed at a cost of $100,000.

8. Money from the rangeland improvement fund may be spent on building fences, but it may also be used for other improvements, such as water projects or reseeding.

9. See Note 4.

10. Aldo Leopold (1887–1948), forester, scientist, teacher, and philosopher was influential in the development of modern environmental ethics and was an early voice for wilderness preservation. He is best known for his book *A Sand County Almanac*.

11. In 1910, Aldo Leopold was forest assistant on the Apache NF, Arizona.

12. As of 1 March 2007 the fee to graze a cow and a calf per month on the national forests in the sixteen western states was $1.35. Private land grazing fees are typically six to nine times this amount.

13. "The grazing permits and leases the 10 federal agencies manage generated a total of about $21 million from fees charged in fiscal year 2004—or less than one-sixth of the expenditures to manage grazing." (Government Accountability Office 2005, 6)

14. In reality, the ranch owner holding a federal grazing permit is selling only the base property to a new owner. But added into this selling price is a hidden additional cost based on the forage production capability of the ranch's permitted federal grazing allotment(s).

15. For a discussion about the monetary relationship between ranch value and the forage-production capability of federal grazing allotments, see Stern (1998).

Clait E. Braun

1. One epithet applied to the Montana Department of Game and Fish is "Department of Game, Fish, and Garbage Cans." Clait Braun, e-mail message to author, 13 January 2006.

2. Blue grouse (*Dendragapus obscurus*).

3. White-tailed ptarmigan (*Lagopus leucurus*).

4. A severe winter would provide the opportunity to discover the Gunnison sage-grouse population's critical winter range. Clait Braun, e-mail message to author, 13 January 2006.

5. Young (1994).

6. Braun (1995).

7. Braun means that environmental advocates would petition the US Fish & Wildlife Service to list this newly identified species of sage-grouse as a threatened or endangered species.

8. Sisk-a-dee (2002).

9. The Uranium Mill Tailings Remedial Action Project is located in the Gunnison Basin (CO).

10. Critical habitat: a term defined and used in the Endangered Species Act. It refers to a specific geographic area (or areas) that contains features essential for the conservation of a threatened or endangered species, and that may require special management and protection. Critical habitat may include an area that is not currently occupied by the species, but that will be needed for its recovery. For more information, see US Fish & Wildlife Service (2005b).

11. Braun refers to Young et al. (2000), which was awarded the Edwards Prize for best paper published in the *Wilson Bulletin* during 2000.

12. The petition to list the Gunnison sage-grouse as "endangered or threatened, emergency listing, and designation of critical habitat" was filed on 25 January 2000 by several organizations, including Sinapu, American Lands Alliance, and Biodiversity Legal Foundation. For details, see Kritz (2005).

13. By December 1999 the US Fish & Wildlife Service (USFWS) knew that a petition to list the Gunnison sage-grouse as a threatened or endangered species was about to be received. Clait Braun was told by a person in the USFWS that a memo had been signed in late December 1999 approving the draft candidate listing of the bird. One might view this listing as an attempt to defuse the petition before it was received, in effect, showing that the USFWS was working in the best interests of Gunnison sage-grouse. The candidate listing was signed by Ralph O. Morganweck, regional director of the USFWS, on 18 January 2000. Clait Braun, e-mail message to author, 27 January 2006.

14. On 11 April 2006 the US Fish & Wildlife Service announced its determination that listing of the Gunnison sage-grouse is not warranted. For details, see US Fish & Wildlife Service (2006).

15. These Gunnison sage-grouse populations were located within Archuleta, Conejos, Costilla, and LaPlata counties. Clait Braun, e-mail message to author, 13 January 2006.

16. Gunnison sage-grouse have evolved several survival mechanisms to drought: (1) not commonly renesting in drought years, (2) long survival of hens, (3) ability to move long distances, (4) ability to find and use small habitat patches, (5) ability to live solely on the leaves of sagebrush, and (6) ability to derive moisture from plants. Clait Braun, e-mail message to author, 13 January 2006.

17. Braun et al. (2004).

18. Dolores County is located in southwestern Colorado.

19. Clait Braun and Norwin Smith applied to CDOW on 4 June 1998 for a grant to purchase property owned by Charles Hughes. Notice was received on 3 February 1999 that $2.5 million in funding would be available to purchase the property. After subsequent negotiations, Hughes accepted the proposal on or about 8 June 1999 with all documents finalized on 15 October 1999. The property sold by Hughes comprises 1,120 acres, including the main lek site and nesting areas for Gunnison sage-grouse in the Miramonte Basin. In Braun's opinion the purchase of the Hughes property will protect the population of Gunnison sage-grouse in the basin for the next ten to twenty years. Clait Braun, e-mail message to author, 15 October 2006.

20. The lawsuit was filed on 29 September 2000 by Sinapu, Biodiversity Legal Foundation, American Lands Alliance, and Grasslands Advocate for American Lands for the purpose of forcing the US Fish & Wildlife Service to consider the groups' petition to list the Gunnison sage-grouse as endangered under the federal Endangered Species Act. On 11 April 2006 the US Fish & Wildlife Service announced its determination that listing is not warranted. For details, see US Fish & Wildlife Service (2006).

21. Anderson (1960).

22. Clait Braun believes that to significantly benefit sage-grouse on the Uncompahgre Plateau that large portions of the plateau must be subjected to controlled burns and further restriction of livestock grazing. Even then, he notes, it would take at least twenty to thirty years for the area to develop into useful sage-grouse habitat. Clait Braun, e-mail message to author, 13 January 2006.

23. In support of his assertion that high-intensity livestock grazing is detrimental to many avian species, Braun cites the following sources: Baker et al. (1990), Bareiss et al. (1986), Bock & Bock (1999), Brown (1982), and Wilkins & Swank (1992). Clait Braun, e-mail message to author, 29 January 2006.

24. Connelly et al. (2004).

25. On 22 December 2003, twenty-one conservation organizations submitted a petition to the US Fish & Wildlife Service to list the greater sage-grouse as threatened or endangered under the Endangered Species Act.

26. Braun (1998).

27. By combining the Connelly et al. (2004) estimate of 50,556 males on leks in 2003 with an estimator for males and leks not counted, along with the presumed number of hens, the total number of sage-grouse could be more than 200,000. Clait Braun, e-mail message to author, 13 January 2006.

28. Braun (1998).

29. Haplotype: a contraction of the phrase "haploid genotype." It is the genotypic constitution of an individual chromosome. Multiple haplotypes indicates there is variation within the population that may be helpful in allowing members of the population to withstand a catastrophic event.

30. Braun's reference to "paper birds" denotes an estimation of the bird population derived by over-extrapolation from counts of males on leks. Clait Braun, e-mail message to author, 13 January 2006.

31. Braun (2004a).

32. Braun (2004b).

33. On 7 January 2005 the US Fish and Wildlife Service determined that the greater sage-grouse did not warrant protection under the Endangered Species Act. For more information, see US Fish & Wildlife Service (2005a).

34. On 20 July 2006 a coalition of environmental organizations issued a press release announcing their desire to see the "adoption of science-based sage grouse conservation measures drafted by grouse expert Clait Braun to increase sage grouse populations and range, in order to avoid a new Endangered Species listing petition." For more information, see Salvo et al. (2006).

35. Connelly et al. (2004).

36. Naugle et al. (2004).

Leon Fager

1. The fisheries biologist is Jim Cooper.

2. The botanist is Renee Galeano-Popp, whose remarks appear on pages 50–65 of this volume.

3. Fager (1998).

4. Congressman Steve Pearce in 2002 was first elected to represent the Second Congressional District of New Mexico.

5. The League of Conservation Voters (LCV) gave Congressman Steve Pearce's votes on environmental legislation in 2003 a 5 percent pro-environmental rating. For the year 2004, Pearce's LCV score was 0 percent.

6. From 1989 until selling the business in 2003, Steve Pearce was owner-operator (with wife Cynthia) of Lea Fishing Tools, an oil services company that does down-hole repairs in oil wells.

7. The Mexican spotted owl *(Strix occidentalis lucida)* was federally listed as a threatened species within its entire range on 16 March 1993.

8. In the early 1990s, Forest Guardians and the Forest Conservation Council brought a lawsuit against the US Forest Service under provisions of the Endangered Species Act charging that timber-cutting on Southwest national forests was bringing the endangered Mexican spotted owl to near-extinction. GS (1996) writes: "On August 24, 1995, District Judge Carl Muecke ruled in the environmentalists' favor and ordered an end to 'timber harvesting' on 11 National Forests in New Mexico and Arizona—pending an EPA-mandated region-wide study on logging's effects on spotted owl habitat."

9. Fortune 500: an annual ranking compiled and published by *Fortune* magazine of the top five hundred US corporations as measured by gross revenue.

10. Leon Fager served as the sensitive, threatened, and endangered species program manager for all national forests in Arizona and New Mexico from 1994 until his retirement at the end of 1997.

11. Fager recalls having written this letter sometime in the mid-1990s.

12. The "old days" to which Fager refers are the 1950s and 1960s.

13. Fager (1998).

14. See note 10.

15. Fager recalls that the study was conducted circa 1992.

16. The $10-per-cow cost of grazing administration and the $1.35-per-cow income are monthly figures.

17. For a thorough examination of federal grazing fees, see Government Accountability Office (2005). One major finding of the report (p. 47) is that in fiscal year 2004 the BLM and Forest Service together spent approximately $132.5 million on direct, indirect, and range improvement activities for

grazing programs, while collecting only about $17.5 million in grazing fees. In other words, receipts represented only about 13.2 percent of expenditures.

18. For a perspective on the economic costs of degrading ecosystems, see Balmford et al. (2002).

19. The full title of the legislation is the Forest and Rangeland Renewable Resources Planning Act. Enacted in 1974, this was the first federal statute to require forest planning. The legislation was reorganized, expanded, and amended by the National Forest Management Act of 1976.

20. Fager refers to Douglas Barber, deputy forest supervisor of the Apache-Sitgreaves NF from 1 January 1989 until 12 July 1994. See Barber's remarks on pages 2–7 of this volume.

21. Fager was a staff wildlife biologist on the Black Hills NF in South Dakota from 1978 until 1982.

22. Creosote bush (*Larrea tridentata*).

23. Fager refers to his time at the Albuquerque (Region 3) Office of the Forest Service from 1992 until his retirement in December 1997.

24. See Note 21.

25. George Armstrong Custer (1839–76), American military commander.

26. Fager refers to the American Southwest, specifically Arizona and New Mexico.

27. For an overview of how long-term livestock grazing can adversely change the dynamics of forest ecology, see Belsky and Blumenthal (1997); Wuerthner (2006a).

28. The Healthy Forests Initiative, formally known as the Healthy Forests Restoration Act of 2003 (HFRA) was signed by President George W. Bush in December 2003. According to a government website, HFRA "contains a variety of provisions to speed up hazardous-fuel reduction and forest-restoration projects on specific types of Federal land that are at risk of wildland fire and/or of insect and disease epidemics." (http://www.healthyforests.gov/initiative/legislation.html (accessed 3 May 2007)) Environmental advocates found much in HFRA to criticize. As a *New York Times* editorial stated about the legislation: "Its mandate is so broad that it practically invites commercial logging on millions of acres in remote areas of the national forests, where fires pose little or no threat to people or property but where the trees are the biggest and the opportunities for profit are the largest." (*New York Times* 2003)

29. For details of how long-term livestock grazing promotes catastrophic forest fires, see Belsky and Blumenthal (1997).

30. For a comprehensive view of fire's role in the ecology of western forests, see Wuerthner (2006b).

31. For details of long-term livestock grazing's effects on riparian systems, see Belsky et al. (1999).

32. Fager refers to Nick McDonough, forest supervisor of the Apache-Sitgreaves NF from the late 1980s until April 1990.

33. John Bedell became forest supervisor of the Apache-Sitgreaves NF in April 1990. (Ruch 1999)

34. The deputy regional forester to whom Fager refers is Forrest Carpenter, who apparently drew the ire of environmentalists for a variety of reasons. Case in point, on 30 November 1993 a coalition of thirteen environmental groups urged Forest Service Chief Jack Ward Thomas and his boss, Assistant Agriculture Secretary James Lyons, to immediately dismiss Apache-Sitgreaves NF Regional Forester Larry Henson and Carpenter, his deputy. (*High Country News* 1993) The *High Country News* article reports that "the groups say Henson and Carpenter punished agency biologists whose findings threatened logging, failed to halt timber theft and tried to head off the listing of the Mexican spotted owl as a threatened species by leaking confidential information to the timber industry."

35. Jeffrey Ruch, executive director of Public Employees for Environmental Responsibility (PEER), in 1999 requested that Forest Service Chief Michael Dombeck remove John Bedell from his position as Apache-Sitgreaves forest supervisor, the position Bedell had held since April 1990. Among the reasons cited by Ruch (1999) was the conflict over elk management on the forest: "Not all of Mr. Bedell's provocations are so public. Many involve behind the scenes attempts to undermine the reasonable wildlife management efforts of the AGFD. Typical is the role Mr. Bedell has been playing with respect to Apache County, which includes the Apache National Forest. Apache County is continually seeking help from the governor and the Legislature to 'significantly control AGFD management activities' and reduce elk populations. They have sent repeated letters to the governor and selected legislators complaining about the Department's management actions."

36. In arguing for removal of Forest Supervisor John Bedell, Ruch (1999) cites "questionable dealings with members of the Arizona Game and Fish Commission, the governor-appointed policy-making board over AGFD. For example, it is widely believed that Mr. Bedell played a role in the politically motivated transfer of an AGFD regional supervisor. These interactions fostered conflict and contributed to increasing political pressures on AGFD from the Arizona Legislature."

37. Fager was a wildlife biologist on the Apache-Sitgreaves NF from 1976 to 1978.

38. Fager refers to a workshop offered by the Savory Center (currently named "Holistic Management International") about a grazing management system developed by Allan Savory. Over the years the grazing management systems promoted by Allan Savory have been successively named "Savory grazing method," "holistic resource management," and most recently, "holistic management."

39. In a multi-year, multi-treatment study, the case of peak plant standing crops with prairie dogs alone was 24 percent higher than with cattle only. The case of peak plant standing crops with cattle plus prairie dogs was 13 percent higher than with cattle only. (Uresk and Bjugstad 1983)

40. Fager's involvement with the Animal Control Committee occurred during his tenure as a wildlife biologist on the Apache-Sitgreaves NF, Arizona, from 1976 to 1978.

41. The Roadless Area Conservation Rule is an administrative rule that was issued by the US Forest Service in January 2001 to protect the last remaining wildlands in the national forest system. It places about one-third of the national forest system's total acreage off-limits to virtually all road building and logging.

42. Desert topminnow (*Poeciliopsis lucida*).

43. Theodore Roosevelt, president of the US from 1901 to 1909. During his administrations there were designated 150 national forests, 51 federal bird reservations, 5 national parks, 18 national monuments, 4 national game preserves, and 24 reclamation projects.

44. Ronald Reagan, president of the US from 1981 to 1989.

Renee Galeano-Popp

1. Henry David Thoreau (1817–62): American essayist, poet, and practical philosopher, best-known for his autobiographical story of life in the woods, *Walden*.

2. Rachel Carson (1907–64): American biologist and ecologist, best known for writing *Silent Spring* (1962) in which she revealed the dangers of widespread use of pesticides.

3. Lincoln NF, New Mexico.

4. Galeano-Popp was a range conservationist on the Apache-Sitgreaves NF from 1985 to 1987.

5. As forest supervisor of the Lewis and Clark NF (Montana), Gloria Flora in October 1997 refused to approve gas or oil leasing on a significant portion of the forest. In November 1999, Flora, as supervisor of the Humboldt-Toiyabe NF, Nevada, resigned in protest over the lack of support provided by the agency in response to threats made against Forest Service employees.

6. The legislation is formally known as the Forest and Rangeland Renewable Resources Planning Act of 1974.

7. Galeano-Popp refers to Reserve, New Mexico.

8. For examples of social and political pressures put on federal land agency personnel whose management proposals have been disliked by ranchers, see Luoma (1986, 101–3).

9. Private land grazing fees are typically six to nine times those on federal public lands.

10. Lincoln NF, New Mexico.

11. The gray fox *(Urocyon cinereoargenteus)* is native to all four deserts of the American Southwest.

12. Galeano-Popp refers to agents of Animal Damage Control.

13. Region 3 is the US Forest Service's designation for the American Southwest, comprising the states of Arizona and New Mexico.

14. The Sacramento Allotment contains approximately 111,000 acres of Forest Service lands on the Lincoln NF in New Mexico.

15. Not up for decision: another way of stating that the grazing permit is not up for renewal.

16. Sacramento prickly poppy (*Argemone pleiacantha* ssp. *Pinnatisecta)* was federally listed as an endangered species on 24 August 1989.

17. Sacramento Mountains thistle (*Cirsium vinaceum*) was federally listed as a threatened species in its entire range on 16 June 1987.

18. Galeano-Popp refers to Sec. 9(a)(1)(B) of the ESA, which prohibits the "take" of federally listed fish or wildlife without a permit. No such prohibition of "taking" exists for plants. "Take" is defined in the ESA as "harass, harm, pursue, hunt, shoot, wound, kill, trap, capture, or collect, or to attempt to engage in any such conduct." Protection of federally listed plants must rely upon prohibition of "jeopardy" provisions in ESA Sec. 7(a)(2) and Sec. 7(a)(4).

19. Galeano-Popp refers to the Paragon Foundation, based in Alamogordo, NM.

20. Pete Domenici, first elected as US senator from New Mexico in 1972.

21. See my interview with Leon Fager on pages 30–49 of this volume.

22. As a result of a lawsuit filed by the Southwest Center for Biological Diversity (now Center for Biological Diversity), Forest Guardians, Maricopa Audubon Society, and Carson Forest Watch, defendant US Fish & Wildlife Service on 6 June 1995 established critical habitat for the Mexican Spotted Owl. In June 1996, in partial response to the critical habitat designation, the US Forest Service amended its forest plans for eleven national forests in Arizona and New Mexico by adding significant new protection for old-growth forests, grasslands, streamside areas, Northern goshawks, and Mexican spotted owls. On 28 September 1996, Forest Guardians and the Southwest Center for Biological Diversity filed a lawsuit (Case number: 96-2258-PGR) in the US District Court (Phoenix, AZ) against the US Forest Service alleging that existing logging and grazing projects violated the June 1996 forest plan amendments.

23. The goshawk guidelines were adopted in 1996. For details, see US Department of Agriculture, Forest Service (1996).

24. Galeano-Popp's meeting with the forest supervisor took place on 10 April 1998. (Galeano-Popp 1998)

25. Galeano-Popp worked for the BLM on the Ely District in Nevada, spring to spring 2002–4.

26. Sneezeweed (*Helenium autumnale*) is an invasive perennial forb.

27. Galeano-Popp notes that livestock grazing has facilitated the invasion of vegetation that, although visually appealing to many people, is often noxious or toxic to native wildlife. Invasive annuals also typically fail to stabilize soil to the extent of the native perennials they replace.

Steve Gallizioli

1. On 23 October 1953 the Crook NF was abolished and its land transferred to the Coronado, the Apache, and the Tonto National Forests.

2. J. Howard Pyle, not Jim Smith, won the Arizona gubernatorial election of 1951.

3. Coyote getter: formally known as the M-44.

4. Widespread misuse of Compound 1080 led to its US ban in 1972. President Ronald Reagan reinstated its use in 1982 but only in sheep collars.

5. Predators accounted for 3.6 percent of the total deaths of cows and calves in 2000. (US Department of Agriculture, APHIS 2006, 9) In 2004, 224,200 sheep and lambs were killed by predators in the US. (US Department of Agriculture, National Agricultural Statistics Service 2005, 1) As of 1 January 2005 there were 6,135,000 head of sheep and lambs in the US. (US Department of Agriculture, National Agricultural Statistics Service 2007, 1). Combining the statistic for sheep and lambs killed with the figure for the total sheep and lamb population suggests that the current annual loss of sheep and lambs to predators is approximately 3.5 percent.

6. Gallizioli (1976).

7. Bob Jantzen subsequently resigned from Arizona Game and Fish to become the director of the US Fish & Wildlife Service during the presidential administration of Ronald Reagan.

8. Bruce Babbitt was governor of Arizona from 1978 to 1986.

9. Gallizioli (1979).

10. Allan Savory is the developer of livestock management methods that have been successively named the "Savory grazing method," "holistic resource management," and most recently, "holistic management." For critical assessments of Savory's methods, see Brown (1994), Jacobs (1991, 525–35), and Wuerthner (2002b).

11. Holistic resource management is currently known as "holistic management."
12. The Arizona Cattlemen's Association was established in 1985 as the management organization for the Arizona Cattle Growers' Association, Arizona Cattle Feeders' Association, and Arizona Beef Council.
13. A similar sentiment from Allan Savory is reported by David E. Brown, who asked Savory whether there was a farm in Africa where SGM [Savory grazing method] had brought about a demonstrable improvement that could be compared to neighboring systems. Brown quotes Savory as stating, "No, they have all collapsed through lack of planning and failure to follow the model." (Brown 1994, 27)
14. Gallizioli visited the summit of Dutchwoman Butte in September 1997.
15. For more information about environmental conditions on Dutchwoman Butte, see Ambos et al. (2000).
16. Croxen (1926).

David Gilman

1. Alexander and Gilman (1994).

Martha Hahn

1. Hahn served as BLM's Colorado associate state director from May 1992 until December 1994.
2. Hahn refers to the period between the mid-1970s and early 1980s.
3. Hahn refers to the Natural Resources Defense Council's lawsuit, *NRDC v. Morton*, 388 F. Supp. 829 (DDC 1974), *aff'd per curiam*, 527 F.2d 1386 (DC Cir. 1976), *cert. denied*, 427 US 913 (1976). NRDC and other plaintiffs challenged the failure of BLM to comply with the National Environmental Policy Act (NEPA) in administering grazing on the public lands under its jurisdiction. The District Court rejected the agency's claim that a programmatic environmental impact statement (EIS) would suffice, and it ordered the preparation of NEPA documents that provided site-specific information about the impacts of current grazing practices and alternatives. The court required the BLM to produce 144 EISs for all of its grazing districts because the evidence had shown that grazing had done, and was doing, widespread damage to western ecosystems, especially riparian areas.
4. William Jefferson Clinton became president of the US on 20 January 1993.
5. Bruce Babbitt, secretary of the interior during the years 1993–2001.
6. Jim Baca resigned as BLM director on 3 February 1994 after nine months in that office. For more details about the events leading up to Baca's ouster, see Davis (1994a).
7. Gale Norton, secretary of the interior from January 2001 until 16 March 2006.
8. The proposed changes in BLM grazing regulations to which Hahn refers were finally released on 12 July 2006.
9. For a more in-depth and more critical view of Babbitt's tenure as interior secretary, see St. Clair (2007).
10. Resource advisory boards: often called "resource advisory councils." For more information about their structure and function, see Hinchman (1994).
11. Hahn refers to the presidential election of November 2000.
12. For a 5-year perspective on implementation of Rangeland Reform, see Carlson & Wald (2001).
13. Idaho Watersheds Project in February 2001 was renamed Western Watersheds Project.
14. For a thorough presentation of the controversy over the Air Force's plan to put a bombing range in the Idaho desert, see Nokkentved (2001).
15. Cecil D. Andrus, governor of Idaho during the years 1971–77 and 1987–95.
16. Bert Brackett, the rancher "displaced" by the bombing range, was compensated by the Air Force approximately $1 million, the full value of his operation. (Elliott 1998)
17. Larry Craig, first elected to the US Senate from Idaho in 1990.
18. Patrick A. Shea, a prominent Utah lawyer, educator, and businessman, as well as adjunct professor of political science at Brigham Young University Law School, was confirmed as BLM director by the US Senate on 31 July 1997.
19. George Herbert Walker Bush, US president from 1989 to 1993.

20. For details of subsequent changes to BLM's grazing regulations, see US Department of the Interior, Bureau of Land Management (2006).
21. Dirk Kempthorne was elected governor of Idaho in 1998.
22. Hahn was reassigned as executive director of the National Park Service New York Harbor operations in January 2002.
23. Hahn submitted her resignation to the BLM on 8 March 2002. Martha Hahn, e-mail message to author, 29 September 2006.
24. Kathleen B. Clarke became the director of the BLM on 2 January 2002.
25. K. Lynn Bennett replaced Hahn as BLM's Idaho state director. Bennett's career with the BLM from 1972 to 1993 included assignments as associate state director in Nevada, Shoshone district manager in Idaho, and chief, Branch of Range Management in BLM's Washington, DC, Office.

David A. Koehler

1. Koehler lived in Skokie, Illinois, during the 1940s and 1950s.
2. Koehler received his MS degree from the University of New Mexico in 1974.
3. See Koehler's master's thesis, Koehler (1974), and his preliminary report, Koehler (n.d.), for details about the environmental impacts of feral burros at Bandelier National Monument. For the National Park Service's historical overview of burro impacts on soil and archaeological sites at Bandelier, in addition to information about removal of burros from the monument, see National Park Service (2001).
4. *La Alianza de las Razas*: Spanish for "The Alliance of the Races."
5. Ronald Reagan, president of the US from 1981 to 1989.
6. Bruce King served as governor of New Mexico during the years 1971–75, 1979–83, and 1991–95.
7. Koehler and Thomas (2000).
8. Hoary cress (*Cardaria draba*), noxious perennial native to Europe.
9. "The importance of sagebrush communities as wildlife habitat is illustrated by the fact that at least 87 wildlife species use them as habitat in Intermountain States; several species are obligately tied to sagebrush habitats including sage grouse (*Centrocerus* [sic] *urophasianus*) and pygmy rabbits (*Brachylagus idahoensis*) ... Sagebrush provides nesting, hiding, and thermal cover for various animal species." (McArthur 1994, 349)
10. Populations of greater sage-grouse in the late 1960s and early 1970s were approximately two to three times higher than in 2004. (Connelly et al. 2004, 6-71) Sage-grouse expert Clait E. Braun wrote of this report: "habitat fragmentation and degradation would appear to be understated and implied current population levels would appear to be inflated." (Braun 2004a, 7)
11. Robert F. Burford (1923–93), director of the BLM from 1981 to 1989.
12. A mountain lion's territory may exceed 100 square miles, with animals moving 100 miles or more from their natal grounds. (Arizona Game and Fish Department n.d.)
13. *Equus lambei* evolved in North America and was the most recent *Equus* species on the continent prior to its extinction there as part of the widespread extinction of megafauana that occurred at the end of the last great Ice Age between 11,000 and 13,000 years BP.
14. Coinciding with the end of the Pleistocene Epoch around 11,000 years BP, approximately two-thirds of the large mammals in North America became extinct. Current hypotheses for these extinctions include (1) over-hunting by the first colonizing humans, (2) climate change, and (3) "hyperdisease" brought by colonizing humans.

Don Oman

1. Oman had worked for the US Forest Service during the three previous summers, while he was attending college.
2. The Minidoka NF became part of the Sawtooth NF on 1 July 1953.
3. Congressional: a request from a rancher's congressional representative for details about agency-proposed changes in livestock management affecting the rancher.

4. Cleaned the unit: an expression meaning to remove all livestock from a unit of a grazing allotment.

5. Despite political pressure initiated by ranchers, the ordered one-week delay in putting cattle onto the allotment was observed. Don Oman, phone conversation with author, 23 August 2006.

6. Oman refers to the spring of 1987.

7. The five permittees on the Goose Creek Allotment had an association to facilitate their dealings with the Forest Service. Don Oman, phone conversation with author, 23 August 2006.

8. The Owens Corral Pipeline is the longest pipeline on the Goose Creek Allotment. In 1987 it fed three water troughs and was the main water source for two of the allotment's units. Don Oman, phone conversation with author, 23 August 2006.

9. As reported in the *New York Times*: "'Either Oman is gone or he's going to have an accident,' said Winslow Whitely [Whiteley], who has one of the biggest herds in the district. 'Myself and every other one of the permit holders would cut his throat if we could get him alone.'" (Egan 1991)

10. Wayne Elmore graduated from Oklahoma State University in 1968 with a BS in forest management, later completing postgraduate studies in fisheries and wildlife management. In 1968 he began working for the BLM as a resource area forester in Spokane, Washington. He has since held various positions with BLM, including district wildlife biologist, district wildlife and fisheries biologist, Oregon and Washington state riparian specialist, and the BLM national riparian field manager. (Independent Multidisciplinary Science Team 2003, 1)

11. Oman left the district ranger position on the Twin Falls Ranger District in May 1996.

Robert W. Phillips

1. Sneva (1992).

2. Tragedy of the commons: a conflict over resources (commons) between individual interests and the common good. For a well-known examination of the subject, see Hardin (1968).

3. In 1990 the Rocky Mountain Elk Foundation assisted the Nevada Division of Wildlife in purchasing the Howard Ranch. (Krueger 1996, 38)

4. Nevada Department of Wildlife's 1998 study of the protected region of the Bruneau River found that since 1990, fish per mile increased from 40 to 240. There were also improvements in pool development, bank cover, and stability. (Krueger 2004, 46)

Jim Prunty

1. Waterbars: drains cut into a dirt road to divert flowing water from following the road and thereby avoid washing it out.

2. Don Oman was district ranger on the Twin Falls Ranger District (Sawtooth NF) from October 1986 to May 1996. See Oman's interview on pages 110–24 of this volume.

3. The cattle count organized by Don Oman took place in the fall of 1989.

4. Miriam Austin is the executive director of Red Willow Research, Inc.

5. See Don Oman's comments about the water pipeline on pages 121–23 of this volume.

Doug Troutman

1. Yellowstone Park Company operated visitor concessions in Yellowstone National Park from the 1880s until 1980 when the US government purchased all associated property of the company and terminated their lease to operate.

2. Troutman served as a helicopter crew chief with the US Army in Vietnam during the years 1966–67.

3. Troutman was a wilderness ranger for BLM in Aravaipa Canyon, Arizona, from February 1976 until February 1977.

4. The Nature Conservancy purchased the Panorama Ranch from Fred and Cliff Wood in 1971. Defenders of Wildlife assumed management of the property in 1973, then transferred it back to the Nature Conservancy in 1988. Mark Haberstich, e-mail message to author, 6 March 2006.

5. Frederick Duncan Wood was born on 29 November 1904. His brother, Charles Clifford Wood, was born on 21 March 1915.

6. No cattle were on the former Wood brother's property during the period from February 1976 to February 1977 when Doug Troutman was stationed in Aravaipa Canyon as a wilderness ranger.

7. Jeffrey Dahmer (1960–94) was convicted in 1992 of killing sixteen young men between 1978 and 1991.

8. Troutman was an outdoor recreation planner for BLM in Lake Havasu City, Arizona, during the years 1977–78.

9. Troutman served as a wilderness specialist on BLM's Lakeview District from 1978 to 1997.

10. *Oryzopsis*: genus commonly known as ricegrass. Indian ricegrass (*Orzopsis hymenoides*) is one of its better-known species. *Stipa*: genus commonly known as needlegrass.

11. Decadent vegetation: a range management term meaning overgrown. Troutman explains that the range professor saw the stubble of native bunchgrasses without their seed heads which, unbeknownst to him, had been recently removed by cattle. Absence of the seedheads, combined with the low density of the grasses, led the professor to incorrectly conclude that the grasses were not reproducing. Doug Troutman, e-mail message to author, 6 March 2006.

12. "Dudley Do-Wrong" is Troutman's parody of the cartoon character Dudley Do-Right, a Canadian Mountie, who, as his name suggests, persistently tried to do the right thing.

13. Bactine: a nonprescription, first-aid liquid intended as an antiseptic and pain reliever for minor cuts, scrapes, and burns.

14. Bob Buffington was the Arizona BLM director during the years 1977–78.

15. A group of Idaho ranchers complained to Congressman Larry Craig and Senator Steve Symms that Buffington was "one-sided in favor of environmental concerns." In the autumn of 1981, Bob Buffington was removed as BLM's Idaho state director by BLM's national director, Robert Burford. (Luoma 1986, 103)

16. Troutman refers to A. K. Majors, who was the Warner Lakes Resource Area manager of BLM's Lakeview District from approximately 1977 until 1988.

17. A lakebed pit drops the water level in the surrounding soils and focuses the water in the hole at the middle of the lake. Since the vegetation in the region is barely surviving with the amount of water that's there, the installation of lakebed pits leads to the decline of many plant species. Doug Troutman, phone conversation with author, 6 March 2006.

18. Cook Well is located in the BLM's Lakeview District.

19. In many big sagebrush/bunchgrass communities of the Great Basin, historic wildfire intervals have decreased from 80–110 years to every 5–10 years as a result of cheatgrass invasions. (Young and Clements 2004)

20. Cattle were again grazing the Cook Well area in 1988, three years after the fire. Doug Troutman, e-mail message to author, 6 March 2006.

21. The Vale Project, an eleven-year effort begun in 1963, was so named because it took place within the BLM's Vale District of Eastern Oregon. To improve the rangeland for the grazing of livestock, an estimated $20 million were spent on roads, fences, cattle guards, water systems, and the reseeding of approximately 267,193 acres with crested wheatgrass. In 1983 the Vale district collected only about $590,000 in grazing fees, of which 37.5 percent ($221,250) were returned to the US Treasury. (Ferguson and Ferguson 1983, 159–60) At the 1983 rate of return it would take more than ninety years (in constant dollars) for the government to recoup its investment through grazing fees.

22. Troutman refers to fenced exclosures on Fifteen Mile Creek and Twenty Mile Creek on the BLM's Lakeview District.

23. As of 1 March 2007 the federal grazing fee for western public lands managed by the Forest Service and the Bureau of Land Management was $1.35 per animal unit month.

24. "Toto" is a reference to Dorothy's dog in the 1939 film *The Wizard of Oz*. In the film, Dorothy and Toto are swept from their Kansas home by a tornado to a magical land. Once there, they seek the Wizard, who, they are told, can help them return home.

25. Troutman refers to the BLM's Lakeview District.

26. Troutman states that to his knowledge, Art Gerity (Lakeview District manager) didn't "buck" the ranchers either, but that he had stood up to the Oregon state director to defend Troutman's conclusions on the final wilderness inventory regarding which WSAs to keep. From the time Gerity retired in 1983, until his own retirement in 1999, Troutman never again saw a manager who stood up for anything against "directions from above." Doug Troutman, e-mail message to author, 6 March 2006.

27. George W. Bush, president of the US beginning in January 2001.

28. Gale Norton, secretary of the interior from 2001 to 2006.

29. The Trout Creek Mountains are located in the southeastern corner of Oregon. The Pueblo Mountains, located east of Lakeview, Oregon, extend south into Nevada.

Larry Walker

1. Walker's remark about ADP (Automated Data Processing) databases pertains to his preparing collections of data and information in a manner that facilitates automated data processing for compilation, analysis, and other purposes. Oregon's Automated Ecological Site Information System (OAESIS) is an example of an ADP database.

2. Jimmy Carter, president of the US from 1978 to 2001.

3. From the late 1970s through the early years of the 1980s, approximately $15 million per year was spent to perform a baseline SVIM inventory on tens of millions of acres of BLM administered public lands. (Walker 2003)

4. James G. Watt, appointed secretary of interior (the managing agency of the BLM) by President Ronald Reagan, served from January 1981 until resigning under pressure in November 1983 for having made an offensive joke.

5. William Jefferson Clinton, president of the US from 1993 to 2001.

6. For more information about the Ecological Site Inventory, see US Department of the Interior, Bureau of Land Management, National Science and Technology Center (2001).

7. Walker refers to James G. Watt, who became secretary of interior in January 1981.

8. Walker retired from the BLM in February 1997. Further comment on Walker's intentions for his retirement can be found in Walker (1999).

9. RangeBiome website: http://www.rangebiome.org (accessed 26 February 2007).

10. RangeNet website: http://www.rangenet.org (accessed 26 February 2007).

11. For more information about the RangeNet 2000 Symposium, see http://www.rangenet.org/rn2k/index.htm (accessed 26 February 2007).

Pat Ward

1. Mexican spotted owl (*Strix occidentalis lucida*).

2. The Mexican spotted owl was listed as a threatened species on 14 April 1993. A recovery plan for the owl was approved in December 1995.

3. Deer mouse (*Peromyscus maniculatus*).

4. Ward refers to the long-tailed vole (*Microtus longicaudus*) and Mexican vole (*Microtus mexicanus*).

5. Northern spotted owl (*Strix occidentalis caurina*).

6. Mexican woodrat (*Neotoma mexicana*).

7. Dusky-footed woodrat (*Neotoma jiocipes*).

8. Variability in abundance of smaller rodents, such as mice and voles, is typically much greater than for woodrat populations. Pat Ward, e-mail message to author, 10 July 2006.

9. Ward (2004).

10. On 26 October 2004, in a meeting at Lincoln NF headquarters in Alamogordo, Pat Ward presented information from his report on meadow monitoring procedures. His presentation included information about the relationship between herbaceous plant height and quantity of voles consumed by owls. In addition to Forest Service personnel, representatives of the New Mexico Range

Improvement Task Force and Otero County were present. No representatives from environmental organizations were invited or attended. Subsequent to this meeting, Ward's supervisor from the Rocky Mountain Research Station, Bill Block, asked a key Forest Service regional office staff person to drop consideration of Ward's research pertaining to monitoring criteria that would allow trigger points for determining the level of consultation with the Fish and Wildlife Service. Ward reports that this information was dropped due to fear of legal action from the New Mexico Stock Grower's Association. On 18 November 2005, Ward's temporary position with Rocky Mountain Research Station was eliminated, and no permanent position was created in its place. Ward had been with the agency for fourteen years. Ward notes that the summer of 2006 followed one of the lowest precipitation years on record. The meadows were the driest in recorded history with little plant growth. The Lincoln NF was closed to public use due to the fear of fire starts. During this period, proper monitoring for vole habitat conditions was not conducted even though the Forest Service knew the pitfalls of using only leaf-length monitoring, and despite Lincoln NF biologists having designed a scientifically defendable sampling scheme for monitoring vole habitat based on Ward's recommendations. Nonetheless, cattle were allowed on the pastures as of 1 May 2006 and were still grazing as of late June. Pat Ward, e-mail message to author, 21 June 2006.

Bill Worf

1. The Uinta NF is located in northeastern Utah.
2. Worf was supervisor of the Bridger NF from 1962 to 1965.
3. Edward P. Cliff (1909–87) was the 9th chief of the Forest Service, serving from 1962 to 1972.
4. Howard Zahniser was the chief lobbyist for the Wilderness Society in its effort to enact the first federal wilderness legislation.
5. See Mike Sauber's remarks about the Diamond Bar Allotment on pages 309-18 of this volume.
6. Worf's visits to the Diamond Bar Allotment occurred in 1994 and 1995. Mike Sauber, phone conversation with author, 14 April 2006.
7. The South Warner Wilderness, part of the Modoc NF, is located in northeastern California.
8. What became the South Warner Wilderness had been established as the South Warner Primitive Area in 1931. The region became a formal part of the National Wilderness Preservation System after passage of the 1964 Wilderness Act. More land was added by the 1984 Wilderness Act, bringing the total area to 70,385 acres.
9. Scott Conroy became the forest supervisor of the Modoc NF on 11 October 1998. (US Department of Agriculture, Forest Service 1998)
10. Daniel K. Chisholm became the forest supervisor of the Modoc NF on 7 May 2000. (US Department of Agriculture, Forest Service 2000)
11. A-horizon: the layer of soil near the surface. It is roughly equivalent to topsoil.
12. Emerson/Cottonwood: a grazing allotment on the Modoc NF.
13. RC&D: Resource Conservation and Development. The program's website states in part: "The purpose of the Resource Conservation and Development (RC&D) program is to accelerate the conservation, development and utilization of natural resources, improve the general level of economic activity, and to enhance the environment and standard of living in designated RC&D areas." For more information, see http://www.nrcs.usda.gov/programs/rcd/ (accessed 4 February 2007).
14. Ed Bloedel received a bachelor of science degree in forestry/range from the University of Montana, in addition to graduate credit in public administration and recreation management from the University of Montana, Washington State, and Utah State. In 1991 he completed a 33-year career with the US Forest Service that included positions as assistant ranger, district ranger, range management specialist, staff officer, group leader, and national leader.
15. Stan Sylva, supervisor of the Modoc NF.
16. The Blue Fire ignited during a lightning storm on 8 August 2001, burning approximately 34,400 acres. http://www.fs.fed.us/r5/modoc/projects/03-projects/bluefire.shtml (accessed 6 March 2007).
17. Worf refers to the Congressional Grazing Guidelines included with the Colorado Wilderness Act of 1980 (Public Law 96-560).

18. Worf refers to a sheet of paper that was installed in the kiosk at the Pine Creek Basin trailhead (Modoc NF) with the permission of the forest supervisor during the summer of 1995. Signed by the grazing permittee, the sheet states in part: "Visitors to the wilderness may see some CATTLE presence. These CATTLE are an important part of our local customs, culture and heritage, as well as an important natural tool for the management of the forest. CATTLE also provide a significant income to the local economy. The natural beauty you see here has been attained with over 100 years of grazing. CATTLE are the most economical way of converting a natural resource (grass) into a high quality source of protien [sic] for human Consumption [sic]. CATTLE help protect the Natural [sic] beauty of the wilderness by reducing ground fuels to protect against wild fires. Because a rancher has ownership in his permitted use of the federal lands, he protects the land from over-utilization and respects all environmental concerns. With proper management of livestock, a rancher can benefit both the wildlife and the ecosystem. We welcome visitors to enjoy these federal lands." The sheet was removed only after protest to the Modoc NF and US Forest Service office in Salt Lake City, Utah.

19. Additional examples of social and political pressures put on federal land agency personnel, whose management decisions have been disliked by ranchers, can be found in Luoma (1986, 101–3).

20. Worf's first visit to the South Warner Wilderness took place 3–5 October 1999.

21. Worf's second visit to the South Warner Wilderness took place 13–16 July 2000.

22. Worf was the assistant district ranger on the Heber District, Uinta NF from December 1950 to the spring of 1955.

23. Provo, Utah, is approximately a twenty-mile drive from Heber City.

Joy Belsky/Robert Amundson

1. Joy Belsky entered the doctoral program at the University of Washington in the fall of 1972. Robert Amundson, e-mail message to author, 2 October 2006.

2. Joy Belsky worked for the Student Conservation Corps at North Cascades National Park, Washington, during the summer of 1971. She continued in the position of park naturalist during summer 1972. Robert Amundson, e-mail messages to author, 28 August 2006 and 2 October 2006.

3. Belsky (1979).

4. The common names of *Festuca ovina* include "sheep fescue," "green fescue," and "blue fescue."

5. Joy Belsky taught science in a Catholic girls school in Kenya during 1967 and 1968. Robert Amundson, e-mail message to author, 28 August 2006.

6. The difficulty of obtaining supplies in Tanzania resulted from a couple of factors. Tanzania had dropped out of the East African Union over dissatisfaction with most of the tourist money going to the other union members, Kenya and Uganda, the countries from which most of the guides were hired for tours of the Serengeti in Tanzania. Furthermore, as a socialist country, Tanzania did not draw much external capital with which to develop its own tourist industry. Robert Amundson, e-mail message to author, 28 August 2006.

7. Amundson refers to the review articles: Belsky and Blumenthal (1997); Belsky et al. (1999); Belsky and Gelbard (2002).

8. Joy Belsky applied for the BLM/OSU position in mid-1991. (Belsky 1994, 8)

9. Belsky (1994) recounts her experiences in applying for the position at BLM/OSU, her subsequent rejection, and the sex discrimination lawsuit she brought in response.

10. Joy Belsky's annual average number of publication citations during 1988–91 was 43.0; for the applicant who received the BLM/OSU position that number was 15.8. (Belsky 1994, Table 1)

11. The annual average number of publication citations for the entire faculty of the Oregon State University, Department of Rangeland Resources during the years 1988–91 was 34.8; for Belsky the number was 43.0. (Belsky 1994, Table 1)

12. Joy Belsky joined ONRC (since renamed Oregon Wild) as their staff ecologist on 1 July 1993. Robert Amundson, e-mail message to author, 21 August 2006.

13. Amundson refers to Belsky et al. (1995).

14. Barry Reiswig was the manager of Hart Mountain NAR; Bill Pyle was the staff biologist. (Durbin 1997)

15. The policy adopted at Hart Mountain NAR was to remove all cattle from 1993 though 2008.

16. Mike Nunn was the manager of Hart Mountain NAR at the time of Belsky and Amundson's visit in the mid-1990s. (Durbin 1997)

17. The coyote shoot was stopped from 1996 through 1998. Robert Amundson, e-mail message to author, 28 September 2006.

18. A helicopter survey in July 2003 counted more than 2,400 antelope, the most recorded since Hart Mountain NAR was established in 1936. (Flaccus 2003)

19. Sage-grouse numbers at Hart Mountain NAR were reportedly at an all time high in 2003. (Durbin 2003) Then in 2004 the population continued to increase with total number of sage-grouse estimated at 4,178, and the number of males at 1,671, up from 1,038 in 2003. (Juillerat 2004)

20. Belsky and Blumenthal (1997).

21. Joy Belsky's cancelled presentation at Texas Tech was to have taken place in either 1996 or 1997. Robert Amundson, phone conversation with author, 31 August 2006.

Joy Belsky/Jonathan Gelbard

1. Gelbard worked at the Gray Ranch in southern New Mexico from August to December 1995. Jon Gelbard, e-mail message to author, 23 July 2006.

2. Gelbard held the position in Steamboat Springs, Colorado, from December 1995 to June 1996. Jon Gelbard, e-mail message to author, 23 July 2006.

3. Gelbard held the position with the BLM in Portland, Oregon, during September and October 1996. Jon Gelbard, e-mail message to author, 23 July 2006.

4. Joy Belsky was the staff ecologist at the Oregon Natural Desert Association (ONDA) from 1 January 1997 until her death on 14 December 2001. Bill Marlett, interview with author, 16 August 2003.

5. The paper that Belsky and Gelbard co-authored has been assigned various titles over its several revisions. The 15 June 1999 draft is titled "Contributions of Livestock Grazing to Exotic Plant Invasions in Rangelands of the Intermountain West." A note on the draft states that it is "To be submitted to *Conservation Biology*." A subsequent draft dated January 2000 is titled "Livestock Grazing: A Major Cause of Nonindigenous Plant Invasions in the American West." My copy of this draft includes the note that it was "Submitted to *Ecological Applications*." (The article was subsequently rejected by that journal for publication.) The most recent version intended for journal publication, of which I am aware, is dated 8 August 2001 and is titled "Livestock Grazing: A Major Cause of Exotic Plant Invasions in the Intermountain West, USA." (This draft was submitted to *Conservation Biology*, but rejected. Gelbard notes that he is hopeful of future acceptance pending revision. Jonathan Gelbard, e-mail message to author, 20 September 2006.) The version of this paper from April 2000, showing Joy Belsky as first author, and titled "Livestock Grazing and Weed Invasions in the Arid West," can be read at http://www.onda.org/library/papers/WeedReport.pdf (accessed 24 July 2006). Another version of the paper is Belsky and Gelbard (2002).

Joy Belsky/Bill Marlett

1. "Bob" is Robert Amundson, Joy Belsky's husband.

2. Larry Tuttle was executive director of the Oregon Natural Resources Council (now Oregon Wild) from May 1993 through June 1994. Belsky was hired shortly after he became executive director. Larry Tuttle, e-mail message to author, 10 August 2005.

3. Alice Elshoff has served on the Oregon Natural Desert Association's (ONDA) board of directors since 1989.

4. Andy Kerr was the conservation director of ONRC from 1983 to 1994.

5. Joy Belsky began working at ONRC as their staff ecologist on 1 July 1993. Robert Amundson, e-mail message to author, 21 August 2006.

6. Of the three articles that Joy Belsky either authored or co-authored while at ONDA, only Belsky et al. (1999) has been published in a peer-reviewed journal at the time of this book's publication.

7. At the time of her death in December 2001, Joy Belsky had completed a critique of holistic resource management, more recently known as holistic management. As of July 2007 this article remains unpublished.

8. Hart Mountain NAR, covering over 251,000 acres, was established in 1936 as a range and breeding ground for pronghorns and other wildlife.

Patrick Diehl

1. In May 1968 general insurrection broke out in France consisting of confrontations between police and students. Ten million workers soon joined in a general strike that brought the government close to collapse.

2. "Marshall Scholarships finance young Americans of high ability to study for a degree in the United Kingdom. At least forty Scholars are selected each year to study either at graduate or occasionally undergraduate level at an UK institution in any field of study. Each scholarship is held for two years." Quoted from the Marshall Scholarship website: http://www.marshallscholarship.org/ (accessed 14 December 2006).

3. Diehl (1979).

4. Diehl (1985) is based on Diehl's doctoral dissertation.

5. Diehl refers to Siciliano (1987).

6. *The War Game*, released in 1965 and directed by Peter Watkins, is a 50-minute documentary drama about an imagined nuclear attack by Russia on Great Britain.

7. Ronald Reagan, president of the US from 1981 to 1989.

8. War-tax resistance: the refusal to pay some or all of one's federal taxes that pay for war.

9. The First Intifada began in 1987 and ended in August 1993 with the signing of the Oslo Accords and the creation of the Palestinian National Authority.

10. Diehl refers to the al-Aqsa Intifada, which began on 29 September 2000.

11. The *Nuclear Resister* is described on its website as "Information about and support for imprisoned anti-nuclear and anti-war activists." For more information, see http://www.serve.com/nukeresister/ (accessed 1 March 2007).

12. Multiple chemical sensitivity has been characterized by three defining factors: "(1) it is an acquired disorder with multiple recurrent symptoms; (2) it is associated with diverse environmental factors tolerated by the majority of other people; and (3) it is not explained by any known medical, psychiatric or psychologic disorder." (DeHart 1998, 652)

13. Diehl refers to Greater Los Angeles, California.

14. White bursage (*Ambrosia dumosa*), for instance, is very common in the driest regions of the Sonoran and Mojave Deserts where it can be the only major perennial plant apart from the similarly drought-tolerant creosote bush.

15. Although shares in the irrigation company were also owned by individuals living in Escalante who used the water for their lawns and gardens, most shares were owned by alfalfa farmers who controlled the company. Victoria Woodard, e-mail message to author, 15 October 2005.

16. Diehl refers to Grand Staircase-Escalante National Monument, established by President William Jefferson Clinton in September 1996.

17. Shortly after President William Jefferson Clinton announced the establishment of Grand Staircase-Escalante National Monument on 18 September 1996, ceremonies were held in Kanab and Escalante at which effigies of Clinton and Interior Secretary Bruce Babbitt were hanged and burned.

18. Bob Bennett (Republican) was first elected to the US Senate from Utah in 1992.

19. Pioneer Day commemorates the arrival of Mormon settlers at the future site of Salt Lake City on 24 July 1847.

20. Quinn and Gene Griffin, Lake Allotment permittees. (Woodard and Diehl 2001)

21. Mary Bulloch, permittee of the Rock Creek/Mudholes Allotment. Ibid.

22. George W. Bush became president of the US in January 2001.

23. For Woodard and Diehl's account of the field trip, see Woodard and Diehl (2001).

24. As of 19 April 2001 the BLM had assessed trespass fees and impoundment costs against the permittees in the following amounts: Quinn Griffin, $31,346.57; Gene and Brent Griffin, $50,863.43; Mary Bulloch, $65,036.86. Patrick Diehl, e-mail message to RangeNet discussion group, 10 July 2001.

25. Details of the BLM's settlement with the Griffins have not been made public.

26. The dispute between grazing permittee Mary Bulloch and the BLM over her trespass cattle on Fifty Mile Mountain was settled in July 2004. Provisions of the settlement included trespass and impoundment charges of $4,833.29 and suspension of her grazing permit through 30 April 2010. Patrick Diehl, in an e-mail message to RangeNet discussion list on 24 July 2004, stated that details of the settlement were contained in a letter from the BLM Office of Hearings and Appeals to the Escalante Wilderness Project.

27. The meeting with David Hunsaker took place on 22 February 2002. (Diehl 2002)

28. The memo produced by David Hunsaker at the meeting of 22 February 2002 was authored by BLM-GSENM range staff Kent Ellett and Kevin Shakespeare. (Diehl 2002)

29. The inspection of the Steep Creek Allotment by Patrick Diehl, Tori Woodard, Julian Hatch, and Lynne Mitchell took place on 23 February 2002. (Diehl 2002)

30. As a result of the new fence, Julian Hatch reports not seeing trespass cattle in the Steep Creek Allotment since the winter of 2003-4. Julian Hatch, e-mail message to author, 25 July 2005.

31. A Potemkin village is so called after Grigori Aleksandrovich Potemkin, who allegedly had elaborate fake villages built in order to impress Catherine the Great of Russia on her tours of the Ukraine and the Crimea in the 18th century. Today the phrase typically denotes an impressive façade hiding an undesirable situation.

32. Giant Sequoia National Monument (located in California) is another example of a Clinton-era monument whose management upon designation was not transferred to the National Park Service, but rather remained under the jurisdiction of its original management agency—in this case the US Forest Service. Here the terms of the monument declaration allowed pre-existing timber sales to proceed in addition to tree cutting for ecological restoration and maintenance. (Boxall 2006)

33. See Note 17.

Julian Hatch

1. The grazing fee on western Forest Service and BLM lands as of 1 March 2004 was $1.43 per AUM. The fee is set annually.

2. "The grazing permits and leases the 10 federal agencies manage generated a total of about $21 million from fees charged in fiscal year 2004—or less than one-sixth of the expenditures to manage grazing." (Government Accountability Office 2005, 6)

3. Kate Cannon served as associate monument manager and then monument manager of Grand Staircase-Escalante National Monument from 1997 until her selection to become deputy superintendent of Grand Canyon National Park in late 2001.

4. David Hunsaker succeeded Kate Cannon as manager of Grand Staircase-Escalante National Monument in late 2002. In late 2005, Hunsaker announced that he would be leaving that position to become deputy director of the National Landscape Conservation System in Washington, DC.

5. President William Jefferson Clinton created the Grand Staircase-Escalante National Monument by proclamation on 18 September 1996.

6. Work began in 2001 on a livestock management EIS for Grand Staircase-Escalante National Monument. Julian Hatch, e-mail message to author, 7 June 2006.

7. Since my interview with Julian Hatch, the BLM has revised its plans for the EIS at GSENM, expanding its scope from that of a "livestock management EIS" to that of a "rangeland health EIS." For more information, see the BLM's update letter of 26 May 2005 at http://www.ut.blm.gov/monument/planning-rangeland-health-eis-letter-may-26-2005.php (accessed 7 January 2007). As of 7 January 2007 the EIS had still not been released. Julian Hatch, e-mail message to author, 7 January 2007.

8. Funding to restore sagebrush is coming in the form of sage-grouse restoration funds. Julian Hatch, e-mail message to author, 7 June 2006. USDA officials and the interior secretary requested

that $5 million in the USDA budget be earmarked for sage-grouse conservation in 2005. The funds are mostly dedicated to NRCS programs that are intended to support and prolong grazing and agricultural practices on public and private lands that are otherwise harmful to sage-grouse. Mark Salvo, director, Sagebrush Sea Campaign (a project of Forest Guardians), e-mail message to author, 20 June 2006.

9. For more information about cryptobiotic crusts (often called "biological crusts"), see Belnap et al. (2001).

10. The Aquarius Indian paintbrush (*Castilleja aquariensis*) is a candidate for federal listing as threatened or endangered.

11. Colorado River cutthroat trout (*Oncorhynchus clarki pleuriticus*) is a subspecies of trout native only to the Green and Colorado River basins, and whose current range is estimated at only 5 percent of what it once was. In December 1999 the Center for Biological Diversity, Biodiversity Conservation Alliance, and four other organizations petitioned the US Fish & Wildlife Service (USFWS) to list the trout as a threatened or endangered species. Although acknowledging that the trout is threatened by habitat destruction from livestock grazing, water diversion, mining, logging, and other factors, the Bush administration on 20 April 2004 denied the petition. (Biodiversity Conservation Alliance 2004) The petitioners subsequently brought suit to overturn the decision. On 7 September 2006, US District Judge Paul L. Friedman ruled that USFWS had violated the Endangered Species Act by failing to make the required 90-day finding on the petition to list the trout. Friedman further found that USFWS's 90-day review of the petition leading to the court case was contrary to law. Judge Friedman ordered USFWS to conduct a full status review of the trout within nine months and to issue a 12-month finding on the trout after conducting a status review and requesting public comment. Read Judge Friedman's order and judgment at http://www.dcd.uscourts.gov/opinions/2006/2000-CV-2497~15:49:28~9-7-2006-b.pdf (accessed 7 April 2007). In overturning the administration's decision to deny listing the trout, Judge Friedman determined that USFWS had "illegally and selectively sought information from state agencies that generally oppose protection, while giving the public no chance to comment." (Earthjustice 2006)

12. Rainbow trout (*Oncorhynchus mykiss*).

13. German brown trout (*Salmo trutta*) was introduced to North America from Europe in 1883.

14. "Under the terms of the Enabling Act, as part of Congress' granting Utah statehood, the federal government awarded sections 2, 16, 32, and 36 in each thirty-six section township of Utah for the support of the common schools." (Office of the Legislative Fiscal Analyst n.d., 1)

15. During the period 2001–2, Utah ranked last out of fifty states and the District of Columbia in per pupil spending. (*Daily Utah Chronicle* 2002)

16. The amount of revenue generated by grazing on school sections is indeed a small percentage of the total revenue generated by the School and Institutional Lands Administration (SITLA). Unfortunately, for our purposes, SITLA lumps revenue from grazing with that from forestry, reporting the total as $719,501 (FY 1999), or as approximately 6.8 percent of total revenue. (Office of the Legislative Fiscal Analyst. n.d., 21) The FY 2000 Utah school budget was $3.3 billion. (Hedden and Bigler n.d., 4) Assuming that the SITLA contribution from forestry and grazing in FY 2000 was similar to that for FY 1999, its percentage contribution to school funding would have been only 0.02 percent, confirming Hatch's assertion.

17. The Wildlife Damage Prevention Board is the committee responsible for predator control in the state of Utah.

18. Hatch's subsequent contact with members of Utah's Wildlife Damage Prevention Board produced no resolution of his concerns. Julian Hatch, e-mail message to author, 7 June 2006.

19. "Since 1989, the USDA-Wildlife Services has crashed at least 25 helicopters or planes while aerial gunning, resulting in at least 9 fatalities and 34 injuries." http://sinapu.wordpress.com/2007/06/05/sinapu-to-the-usda-congress-no-more-aerial-gunning/ (accessed 5 June 2007).

20. Julian Hatch provided the following example of a project whose evaluation had gone from "success" to "failure": Landscape chaining in the Circle Cliffs (Utah) during the 1960s was initially declared a success by the government, except for the need to eliminate additional sagebrush.

Government records from a few years later, though, indicate the failure of the project. Today the region is overrun with weeds. Julian Hatch, e-mail message to author, 7 June 2006.

21. The challenge brought by Hatch to the IBLA concerned whether the state or federal government has ultimate authority over the management of wildlife. The decision rendered since our interview is interpreted by Hatch as affirming the supremacy of the federal government but allowing the state to initiate action without consideration in an EA. Hatch has not appealed the IBLA's decision in federal court for lack of funding. Julian Hatch, e-mail message to author, 7 June 2006.

22. Grand Staircase-Escalante National Monument encompasses approximately 1.7 million acres, of which about 1 million acres are grazed by livestock. Julian Hatch, e-mail message to author, 7 June 2006.

23. Hatch refers to Dixie NF (1997). A few quotes from the document will illustrate Hatch's assertion in his interview that the authors of the report acknowledge livestock grazing as a major cause of environmental degradation. "Conifer stands have replaced older aspen stands that have not been disturbed for some time (generally since the initiation of fire suppression and grazing, approximately 100 years)." (p. 9) "Approximately 65 percent of seral quaking aspen has succeeded to conifers, as compared to historical vegetation patterns. Exclusion of fire, in combination with ungulate grazing, has contributed to this change. Livestock grazing over the past 100 years has reduced accumulations of fine fuels (shrubs and herbaceous layers)." (p. 10) "With fire exclusion and livestock grazing, ponderosa pine is increasing within riparian areas, meadows and sagebrush communities. Such vegetation patterns outside the historical range of conditions provide poorer habitat for the animal and plant species which historically inhabited such ponderosa pine forests." (p. 13) "Grasses and forbs are detrimentally affected whenever sagebrush canopies exceed 15 percent crown cover. Excessive crown canopies of sagebrush and a long history of heavy livestock grazing has resulted in a major loss of understory species and an increase in bare ground in these types. Much of this type has converted to pinyon-juniper due to grazing practices and the exclusion of fire." (p. 18)

24. Hatch refers to the Multiple-Use Sustained-Yield Act of 1960.

Steven G. Herman

1. Lake Merced is a freshwater lake located in the southwest corner of San Francisco, CA.

2. Herman refers to the San Francisco Society for the Prevention of Cruelty to Animals, founded on 18 April 1868.

3. Aldo Starker Leopold (1913–83), naturalist teacher, author.

4. Aldo Leopold (1887–1948), author of *A Sand County Almanac*, is often considered the father of wildlife ecology.

5. Lodgepole pine needle miner (*Coleotechnites milleri*).

6. Herman's encounter with Forest Service employees erecting a livestock fence most likely occurred in the spring of 1964. Steve Herman, e-mail message to author, 11 June 2006.

7. Western grebe (*Aechmophorus occidentalis*).

8. Rachel Carson's *Silent Spring* was serialized in the *New Yorker* magazine beginning in June 1962.

9. Malheur National Wildlife Refuge, established as the Lake Malheur Reservation in 1908, currently encompasses over 187,000 acres in southeastern Oregon.

10. Denzel Ferguson (1929–98) received the PhD in zoology from Oregon State University. In the early 1970s, he and his wife Nancy served as directors of the Malheur Field Station (Oregon). (Ferguson and Ferguson 1983, 1)

11. Located inside the Malheur NWR, Malheur Field Station provides a base camp for people visiting and studying the area.

12. Commercial livestock grazing on Malheur NWR reached a maximum annual intensity of 125,000 AUMs of forage removed in 1971. By 1993 the amount of forage removed by livestock had declined to 20 percent of that figure. For more information, see http://gorp.away.com/gorp/resource/us_nwr/or_malhe.htm (accessed 8 January 2007).

13. Cattle still graze at Malheur National Wildlife Refuge during the winter and spring, with the winter grazing beginning at least as early as the 12th of August based upon a photograph taken by Herman in 2005. Steve Herman, e-mail message to author, 11 June 2006.

14. Herman notes that it is more typical for at least one crane parent to remain at the nest.

15. The population of greater sandhill cranes at Malheur NWR decreased from 236 pairs to 186 pairs between 1973 and 1985. Ravens, raccoons, and coyotes were identified as the major predators. In 1986, Animal Damage Control was contracted to conduct the control program on the refuge, targeting these predators. Between 1986 and 1991 a total of 315 ravens, 1,078 coyotes, and 67 raccoons were removed from the area. Before the control program, crane nest success averaged 47 percent, while during the program, nest success averaged 69 percent. Crane chick survival averaged 12 percent before the program, compared with 23 percent during the program. (Ivey et al. 2006) As Wuerthner (1993) points out, though, studies conducted at the refuge have shown that predator success is in part due to lack of cover resulting from haying operations and livestock grazing on the refuge. Yet no studies had been conducted to test whether increasing the amount of cover by reducing (or eliminating) haying or livestock grazing would improve survivorship of cranes.

16. John C. Scharff was manager of the Malheur NWR from 1935 until his retirement in 1971. That same year he received Distinguished Service Awards from Oregon State University and from the Department of the Interior. Scharff died on 11 June 1998 in Burns, Oregon.

17. Joy Belsky, PhD (1945–2001), published more than 45 peer-reviewed scientific papers and book chapters on African and North American grasslands and rangelands. Her last positions were as a grassland ecologist with the Oregon Natural Resources Council (1993–96) and with the Oregon Natural Desert Association (1997–2001).

18. Herman refers to Belsky et al. (1993).

19. George Wuerthner in several publications has debunked myths associated with the grazing of livestock in the American West. One version can be found in Wuerthner and Matteson (2002, 7–15).

20. The meetings involving Hart Mountain NAR manager Marv Kaschke and Steve Herman's students took place in the mid-1980s. Steve Herman, e-mail message to author, 11 June 2006.

21. *Oregon Natural Desert Association and The Wilderness Society v. Hart Mountain National Wildlife Refuge* was filed on 7 February 1991 by the Sierra Club Legal Defense Fund (now named "Earth Justice") to stop domestic livestock grazing on Hart Mountain NAR until the US Fish & Wildlife Service prepared an environmental impact statement (EIS) and made a compatibility determination for the refuge's grazing. Although the suit was dismissed, the Fish & Wildlife Service developed a new management plan supplemented with an EIS and a new compatibility determination. The new management plan determined that livestock grazing was harming the refuge's antelope population and other grassland and riparian dependent species. Beginning in 1994 the grazing program was formally eliminated for a period of at least fifteen years.

22. For more information about the importance of vegetational cover to sage-grouse, see Gregg et al. (1994).

23. Halting of the coyote hunt resulted from the coalition of ONDA, under the scientific leadership of Joy Belsky, along with Jim Yoakum, and Predator Defense Institute. Bill Marlett, executive director of ONDA, e-mail message to author, 7 December 2005. The year that the hunt was definitively stopped is reported as 1998 according to the website of Predator Defense (formerly Predator Defense Institute): http://www.predatordefense.org/about/index.htm (accessed 8 January 2007).

24. A helicopter survey in July 2003 counted more than 2,400 antelope, the most recorded since Hart Mountain NAR was established in 1936. (Flaccus 2003)

25. George W. Bush.

26. Pygmy rabbit (*Brachylagus idahoensis*).

27. Herman's claim that sagebrush could not be limiting derives from its abundance. Steve Herman, e-mail message to author, 11 June 2006.

28. The last purebred male Columbia Basin pygmy rabbit in captivity, and presumably in existence, died on 30 March 2006 at the Oregon Zoo in Portland. Only two purebred females in the captive breeding program remain. (*Associated Press* 2006)

29. Twelve pygmy rabbits are scheduled for release in March 2007 onto 7,900 acres purchased in 2002 by the US Fish & Wildlife Service for $1.3 million. This will be the first release of the rabbits into their natural habitat since their removal for emergency captive breeding. For more information, see Oregon Zoo (2006).

30. Richard M. Nixon (1913–94), president of the US from 1969 to 1974.

31. Herman refers to HR 3324, the Voluntary Grazing Permit Buyout Act, which was introduced in the 108th Congress on 16 October 2003.

32. The National Environmental Policy Act contains statements that can reasonably be interpreted as acknowledging the value of aesthetics in environmental protection. Section 101(a) states, in part, that the Congress declares that it is the continuing policy of the federal government, in cooperation with state and local governments, etc. "to create and maintain conditions under which man and nature can exist in productive harmony." Section 101(b)2 states, in part, that it is the continuing responsibility of the federal government to use all practicable means to improve and coordinate federal plans, functions, programs, and resources to the end that the nation may "assure for all Americans ... aesthetically ... pleasing surroundings."

33. Doc Hatfield, a former large-animal veterinarian, is co-founder of Oregon Country Beef, an Eastern Oregon ranching cooperative that raises beef without antibiotics or hormones.

Steve Johnson

1. The first nationally celebrated Earth Day in the US took place on 22 April 1970.

2. The American alligator (*Alligator mississippiensis*) was federally listed as an endangered species in 1967. Although declared fully recovered in 1987, the species remains on the threatened list because individuals are similar in appearance to the endangered American crocodile (*Crocodylus acutus*) and other crocodilians.

3. Johnson may be referring to Wyoming rancher Van Irvine, about whom Larmer (2000) wrote, "In 1971, rancher Van Irvine of Casper was charged with illegally shooting pronghorn antelope, then lacing their carcasses with poison to kill coyotes and eagles. He paid a $675 fine after pleading no contest." Although Compound 1080 was a popular poison for killing predators, Van Irvine was charged with using thallium sulfate as the lethal agent.

4. On 8 February 1972, President Richard Nixon signed Executive Order 11643, "Environmental Safeguards on Activities for Animal Damage Control on Federal Lands," which banned the use of poisons by federal agents on federal lands.

5. Spanish longhorn cattle were introduced to the Southwest in the 1500s. Large-scale commercial grazing, though, did not begin until the 1870s. (Jacobs 1991, 10)

6. Pleistocene Epoch: from approximately 1.65 million years until 10,000 years BP.

7. For more information about the spreading of weeds by livestock, see Belsky & Gelbard (2002).

8. Lehmann lovegrass (*Eragrostis lehmanniana*), native to South Africa, has been widely used for roadside stabilization and range restoration in Arizona.

9. A survey of riparian conditions states: "extensive field observations in the late 1980s suggest riparian areas throughout much of the West were in the worst condition in history." (Chaney, et al. 1990, 5)

10. George W. P. Hunt, governor of Arizona during the years 1912–19 and 1923–29.

11. The desert tortoise (*Gopherus agassizii*) was federally listed as a threatened species on 2 April 1990.

12. Sandra Day O'Connor in 1981 became the first woman appointed to the US Supreme Court. She served for almost twenty-four years before her retirement at the end of the 2004–5 session.

13. On 1 March 2003 the monthly fee to graze a cow and a calf on western lands managed by the US Forest Service and the Bureau of Land Management became $1.35 for the following year.

14. Executive Order 12548, signed by President Ronald Reagan on 14 February 1986, among other things, set the minimum federal lands monthly grazing fee at $1.35 per cow-calf pair.

15. Johnson refers to the lawsuit *NRDC v. Morton*, 388 F. Supp. 829 (DDC 1974), *aff'd per curiam*, 527 F.2d 1386 (DC Cir. 1976), *cert. denied*, 427 US 913 (1976). In this lawsuit the Natural Resources

Defense Council (NRDC) and other plaintiffs challenged the failure of BLM to comply with the National Environmental Policy Act (NEPA) in administering grazing on the public lands under its jurisdiction. The District Court rejected the agency's claim that a programmatic environmental impact statement (EIS) would suffice and ordered the preparation of NEPA documents that provided site-specific information about the impacts of current grazing practices and alternatives. The court required the BLM to promulgate 144 EISs for all of its grazing districts because the evidence had shown that grazing had done, and was doing, widespread damage to western ecosystems, especially riparian areas.

16. In the eleven western states, 0.04 percent of the income and 0.07 percent of the jobs are derived from federal forage. (Power 2002, 264)

17. Across the eleven western states, between 40 percent and 60 percent of beef cattle ranchers report their main occupation to be something other than rancher or farmer. In addition, 60 to 70 percent of western beef cattle ranchers report that they do some paid work off the ranch. Over half of beef cattle ranch operators worked twenty or more weeks per year off the ranch. (Power 2002, 268–9)

Ralph Maughan

1. The Sierra Club's Northern Rockies Chapter encompasses Idaho and eastern Washington.

2. Since our interview, Maughan has split the Wolf Recovery Foundation's website into two parts. Archived material resides at http://www.forwolves.org/ralph/wolfrpt.html (accessed 20 January 2007). As of 9 February 2006, Maughan continued his postings as a blog at http://wolves.wordpress.com/ (accessed 20 January 2007).

3. Curlycup gumweed (*Grindelia squarrose*).

4. George W. Bush, president of the US beginning in January 2001. Gale A. Norton served as secretary of the interior from January 2001 to March 2006.

5. Threats to the future population of grizzly bear include (1) Devastation of the Yellowstone cutthroat trout population (*Oncorhynchus clarki bouvieri*) due to whirling disease and the introduction of cutthroat-eating lake trout (*Salvelinus namaycush*) in Yellowstone Lake. (Lake trout don't go upstream and spawn, so their biomass is never accessible to bears.); (2) Decimation of the whitebark pine stands (grizzly depend on whitebark pine nuts) due to blister rust and insect attack; (3) Subdivision of private land inholdings in grizzly territory. (The private land is often the most biologically productive land.); (4) Vast natural gas developments on the southern end of the Greater Yellowstone Ecosystem near Pinedale, WY. Ralph Maughan, e-mail message to author, 29 August 2005.

6. Since Maughan's interview on 25 August 2003, hunting of northern range elk has been scaled back because the population has declined. Ralph Maughan, e-mail message to author, 29 August 2005.

7. Maughan refers to the shooting of large numbers of bison at the encouragement of the US government in the 1870s.

8. During the 2002–3 winter season, 246 bison were killed. Website of the Buffalo Field Campaign, http://www.buffalofieldcampaign.org/bisonslaughertotal.html (accessed 20 January 2007).

9. Representative Nick Rahall (D) introduced this measure in the US House of Representatives in 2003 where it failed by a 199–220 margin. On 17 June 2004, US Representatives Maurice Hinchey (D) and Charles Bass (R) introduced a bipartisan amendment to the 2005 Department of Interior Appropriations Bill to "prohibit the use of funds to kill bison, or assist in the killing of bison, in the Yellowstone National Park herd." Despite the hard work of buffalo protection advocates and hundreds of concerned citizens, the measure failed to pass by a 202–215 margin.

10. Maughan refers to the Horse Butte Allotment. The National Wildlife Federation paid more than $100,000 to two Idaho ranchers to cede their grazing rights back to the Caribou-Targhee NF, thus making room for the Munns brothers to move their cattle there from this allotment. (Stark 2003)

11. Maughan refers to the bison hunt of 1990. (Bohrer 2005)

12. The state of Montana initiated its first bison hunt since 1990 on 15 November 2005 for which fifty bison hunting licenses were allocated for use during the ninety-day season. (*Environmental News Service* 2005)

13. See Note 10.

14. Under an agreement in which the National Wildlife Federation provided an incentive payment, 74,200 acres of the 87,500-acre Blackrock/Spread Creek Allotment permitted to the Walton Ranch Company were retired in 2003 from the grazing of livestock. (National Wildlife Federation 2003)

15. For a detailed analysis of the ranchers' "political hold" of which Maughan speaks, see Donahue (2005).

16. George W. Bush, president of the US beginning in January 2001.

17. Ronald Reagan, president of the US from 1981 to 1989.

Bobbi Royle

1. Royle refers to Wild Horse Spirit, Ltd. (http://www.wildhorsespirit.org), located in Carson City, Nevada.

2. *Equus lambei* evolved in North America and was the most recent *Equus* species on the continent prior to its extinction there as part of the widespread extinction of megafauana that occurred at the end of the last great Ice Age (between 11,000 and 13,000 years BP). Horses survived, though, by migrating west over the Bering land bridge into Asia. Horses (*Equus caballus*) were returned to North America by Spanish explorers in the early 16th century. (Kirkpatrick and Fazio 2005) Current wild horses may be somewhat larger than *Equus lambei*. Average withers height of mustangs from the Pryor Mountain Herd is reported at 14.2 hands (56.8 inches) (Northeast Wisconsin Mustang and Burro Association (http://groups.msn.com/NEWMBA/whatisapryormountainmustang.msnw), while specimens of *Equus lambei* indicate a withers height of approximately 48 inches (Harrington 2002).

3. Siberian ancestors of American bighorns migrated through the Beringian land bridge during a single eastern migration in the Pliocene Epoch (5.3 to 1.8 million years BP). (Chernyavsky 2004)

4. Mustanger: capturer of wild horses.

5. Nick Mansfield died in 1998.

6. Estray horse: "In 1986, the BLM designated the Virginia Range as a wild horse free area through a land use planning process. Horses that were either left behind or migrated into the Virginia Range after this BLM designation fell by default under the existing state laws pertaining to 'estray' livestock. Under these State statutes, estray livestock are deemed the property of the Nevada Department of Agriculture until such time as the legal owner can be determined and take possession, or the animal is other wised placed." Website of the Virginia Range Wildlife Protection Association, http://www.vrwpa.org/frequently_asked_questions1.htm (accessed 28 January 2007).

7. Edward Abbey (1927–89), essayist and novelist.

8. Royle did not quote Edward Abbey, but rather loosely paraphrased his statement in her subsequent remarks. Abbey's actual quote reads: "The rancher (with a few honorable exceptions) is a man who strings barbed wire all over the range; drills wells and bulldozes stock ponds; drives off elk and antelope and bighorn sheep; poisons coyotes and prairie dogs; shoots eagles, bears, and cougars on sight; supplants the native grasses with tumbleweed, snakeweed, povertyweed, cowshit, anthills, mud, dust, and flies. And then leans back and grins at the TV cameras and talks about how much he loves the American West." (Abbey 1986, 55)

9. Royle's remark proved prophetic when in November 2004, Senator Conrad Burns (R-Montana) inserted a rider into the 2005 Appropriations Bill that opened the door to the sale of wild horses for slaughter. Thus without a hearing or any public review, more than thirty years of federal protection for the wild horse was withdrawn. Horse advocates quickly responded with the introduction of HR 297 (109th Congress, 1st session) by Representatives Nick Rahall (D-WV) and Ed Whitfield (R-KY) on 25 January 2005. This legislation, along with its Senate companion, would restore the ban on the commercial sale of wild free-roaming horses and burros granted by Congress through the Wild Free-Roaming Horses and Burros Act in 1971.

10. The other horse slaughter facility in Texas is owned by Dallas Crown, Inc.

11. The horse slaughter facility in DeKalb, Illinois, owned by Belgium-based Cavel International

was burned in 2002. By May 2004 the facility had been rebuilt and was ready to again slaughter horses. (*meatnews.com* 2004)

12. Royle refers to Betty Kelly, co-founder of Wild Horse Spirit, Ltd.

13. District Judge Michael Griffin imposed 39-day jail terms on Scott Brendle, 24, and Darien Brock, 23, ordered probation for up to two years and fined them $2,000 apiece. Anthony Merlino, 23, was placed on one year's probation and fined $1,000. All three Reno men were ordered to complete 100 hours of community service and split a $1,500 restitution charge. (Riley 2002)

14. The Nevada Assembly Bill 219 (21 February 2001 version) can be read at http://www.animalaw.info/statutes/stusnv2001assemblybill219.htm (accessed 27 January 2007).

Mike Sauber

1. Steven Durkovich is the lawyer who assisted Gila Watch. Mike Sauber, e-mail message to author, 16 October 2006.

2. The proposal to construct fifteen stock tanks was included in Alternative C of the *Environmental Impact Statement for Diamond Bar Allotment Management Plan*, May 1995. The alternative also proposed one spring development and 15.5 miles of fence. In addition, three stock tanks were to be reconstructed—one of them "reconstructed" quite some distance away. So for all intents and purposes it would have been a new tank.

3. Jack Ward Thomas served as the 13th chief of the US Forest Service from 1993 to 1996.

4. Jack Ward Thomas issued his directive to the Gila NF personnel in February 1996.

5. The first such memorandum of understanding (MOU) was issued on 10 February 1938 by the secretary of the US Department of Agriculture to the governor of the Farm Credit Administration. For the text of this MOU, see http://www.fs.fed.us/im/directives/fsh/2209.13/2209.13,16-19.rtf (accessed 20 November 2006). Sauber refers to the MOU of 21 December 1990, which superceded the one of 1938.

6. Sauber refers to William P. Schultz, who received the permit for the Diamond Bar Allotment in 1979. Mike Sauber, e-mail message to author, 15 November 2006.

7. Kit Laney and his wife Sherry acquired the Diamond Bar Ranch in December 1985. (Davis 1994b)

8. A term grazing permit was issued to the Diamond Bar Cattle Company for the Diamond Bar Allotment in 1985. (US Department of Agriculture, Forest Service, Southwest Region 1995)

9. Sauber cites the Forest Service's decision of 30 August 1996 to permanently close upper Black Canyon to livestock grazing as the "win date." Mike Sauber, e-mail message to author, 25 November 2006.

10. Pete Domenici, US senator from New Mexico, first elected in 1972.

11. Joe Skeen represented the Second Congressional District of New Mexico in the US House of Representatives from 1980 until 2003.

12. The bombing of the Alfred P. Murrah Federal Building in Oklahoma City occurred on 19 April 1995.

13. At the time of the meeting in 1995, Dr. Robert D. Ohmart was a professor in the Center for Environmental Studies at Arizona State University.

14. Bill Worf, former US Forest Service forest supervisor and regional director, who, after retiring from the Forest Service, co-founded the environmental organization Wilderness Watch. For my interview with Worf, see pages 165–81 of this volume.

15. Mark Dowie (1995) similarly reports the Kit Laney quote as "There will be a hundred people with guns waiting for them."

16. In early February 1996, Forest Service Chief Jack Ward Thomas, instructed Southwestern Region Deputy Regional Forester Abel Camarena to re-do the EIS. The final revised version was released on 30 August 1996. Mike Sauber, e-mail message to author, 15 November 2006.

17. On 1 April 1996, Diamond Bar Cattle Company and Laney Cattle Company filed suit against the US Forest Service claiming an entitlement to run cattle on their federal grazing allotments based on their predecessors having established water rights and having grazed the land before it came under

the jurisdiction of the Forest Service. Plaintiffs also applied for an injunction against the Forest Service to stop Forest Service interference with the exercise of their claimed rights. In December 1996, Judge Howard Bratton ruled against the Laneys and ordered livestock removed from their Forest Service grazing allotments.

18. The grazing fee for 1996 on western public lands administered by the Bureau of Land Management and Forest Service was $1.35 per animal unit month (AUM), a 26-cent decrease from the 1995 fee of $1.61 per AUM. (US Department of the Interior, Bureau of Land Management 1996)

19. The Laneys appealed to the Tenth Circuit Federal Appeals Court to overturn Judge Bratton's ruling of December 1996. In 1999 a three-judge panel of the Tenth Circuit affirmed Judge Bratton's denial of the Laneys' claims and ordered the Laneys to remove livestock from the allotments, ruling that the ranchers did not now hold and never had held a vested property right to graze cattle on federal public lands. The Tenth Circuit also upheld Judge Bratton's order that the Laneys pay over $55,000 in fines and damages for unauthorized grazing of national forest lands. For the Appeals Court's ruling, see http://www.kscourts.org/ca10/cases/1999/02/97-2140.htm (accessed 21 November 2006). In 2003, Kit Laney restocked the Diamond Bar Allotment and the Laney Allotment. In response, Judge William P. Johnson in Albuquerque held the Diamond Bar and Laney Cattle companies, and their owners Kit and Sherry Laney, in contempt for violating the court's December 1996 livestock removal order. In March 2004 the Forest Service removed 252 of the Laneys' cattle from the allotments. During the cattle roundup on 14 March 2004, Kit Laney was arrested on a felony charge of assaulting a federal officer and a misdemeanor charge of interfering with a federal law enforcement officer. The subsequent plea agreement included Laney being sentenced to six months imprisonment, along with money obtained from the sale of his trespass cattle going to the government as partial reimbursement for expenses. Taxpayers were still left with a $150,000 bill for Laney's actions. (Williams 2006)

20. Sauber took the photographs of the cattle-grazed riparian area on 14 May 2000. Mike Sauber, e-mail message to author, 15 November 2006.

21. Dan Daggett, writer and author of *Beyond the Rangeland Conflict: Toward a West That Works*.

22. The region visited during the Quivira Coalition field trip consisted of the U Bar Ranch's private property in addition to the ranch's Forest Service grazing allotment. Mike Sauber, e-mail message to author, 25 November 2006.

23. The area adjacent to the Gila River is private property known as the U Bar Ranch. Mike Sauber, e-mail message to author, 25 November 2006.

24. Equally skeptical about cattle grazing's benefit to the Southwest willow flycatchers on the U Bar Ranch is John Horning, executive director of Forest Guardians. In 2001, Horning was quoted as saying, "They say that we have 130 flycatchers in this area, therefore cows and flycatchers are compatible. What they don't say is that if they didn't have cows there, they might have 300 to 400 flycatchers." For Horning's remark and a variety of other opinions on the matter, see Davis (2001).

25. Steve Libby was the Gila National Forest's staff officer at the time of the Quivira Coalition's field trip. He is currently retired from the Forest Service. Mike Sauber, phone conversation with author, 20 November 2006.

Todd Shuman

1. The UC (University of California) Nuclear Labs Conversion Project was founded in October 1976 as a joint project of Berkeley Students for Peace, the Bay Area office of the War Resisters' League, and the Ecumenical Peace Institute. The initial goals of the project included forcing greater disclosure and discussion of work in progress at the nuclear weapons labs for which the University of California had oversight.

2. Lynn Jacobs is also the author of *Waste of the West: Public Lands Ranching* (Jacobs 1991).

3. Marbled murrelet (*Brachyramphus marmoratus*), a small seabird that nests in coastal, old-growth forests of the Pacific Northwest. Listed as threatened under the Endangered Species Act.

4. Shuman refers to Odion et al. (1988).

5. John Muir (1838–1914), well-known American naturalist, nature writer and co-founder of the Sierra Club.

6. John Muir's reference to "hoofed locusts" occurs on page 75 of *My First Summer in the Sierra* (Houghton Mifflin, 1911) where in his diary entry for 16 June 1869, Muir wrote: "Sheep, like people, are ungovernable when hungry. Excepting my guarded lily gardens, almost every leaf that these hoofed locusts can reach within a radius of a mile or two from camp has been devoured."

7. Shuman refers to the 20 December 1993 decision by Administrative Law Judge (Office of Hearings and Appeals, US Department of the Interior) John Rampton in the *Comb Wash* case (Case Number: UT-06-91-1). The case stemmed from appeal of BLM's issuance of grazing permits for the Comb Wash Allotment (located in southeastern Utah) by NWF, SUWA, and Joseph Feller. Judge Rampton, among other things, found that BLM had violated NEPA by failing to prepare an EIS that analyzed the specific environmental impacts of livestock grazing within five canyons on the allotment. Furthermore, he found that BLM had violated FLPMA's multiple-use mandate by authorizing cattle grazing without weighing and balancing grazing's harms and benefits to determine whether it is in the public interest. As a remedy for the BLM's violations, Judge Rampton prohibited BLM from authorizing grazing within the allotment's five canyons unless and until preparation of an EIS that makes a reasoned and informed decision that grazing there is in the public interest. Judge Rampton's decision was affirmed by the Interior Board of Land Appeals on 21 August 1997 (Citation: 140 IBLA 85 (1997)). For more information, see Smith (1994).

8. CalTrout: informal name of California Trout.

9. According to the Anheuser-Busch website, the company's "operations and resources are focused on beer, adventure park entertainment and packaging. Anheuser-Busch also has interests in aluminum beverage container recycling, malt production, rice milling, real estate development, turf farming, metalized paper label printing and transportation services." For more information, see http://www.anheuser-busch.com/ (accessed 31 January 2007).

10. Range Watch, founded by Jane Baxter in 1992, has as its mission "to reduce the economic and ecological cost to the American public from commercial grazing on our public lands." For more information, see http://www.rangewatch.org/ (accessed 31 January 2007).

11. For the history of the campaign to prevent the collapse of the Mono Lake ecosystem, see Hart (1996).

12. Joe Fontaine, president of the Sierra Club from 1980 to 1982, received the club's John Muir Award in 1995 for his many years of effort to protect the southern Sierra Nevada, Mojave desert lands, the giant Sequoia, and other California wilderness areas.

13. Shuman recalls first meeting Joe Fontaine at a Southern California Regional Conservation Committee meeting in November 1996. Todd Shuman, e-mail message to author, 3 February 2007.

14. Rose Strickland chaired the Sierra Club's Grazing Subcommittee during the 1990s.

15. Shuman refers to Sierra Club (1997).

16. August Busch III was president and chief executive officer of Anheuser-Busch at that time.

17. Rogers (1999).

18. The petition to list the golden trout as an endangered species was filed on 16 October 2000 by Trout Unlimited. For more information, see Trout Unlimited (2000).

19. Lucinda McKee died on 22 October 2001 from advanced spinal cancer that had spread to her liver. For more information, see Stienstra (2001); California Native Plant Society (2001).

20. The US Forest Service's Region 5 Office denied the appeal by Janice Allen on 10 May 2001. For more information, see US Department of Agriculture, Forest Service, Pacific Southwest Region (2001).

Charmaine White Face

1. White Face refers to a program of virtual genocide carried out against native peoples of the Great Plains by the US government in the 1870s.

2. Gold was discovered in the Black Hills in June 1874.

3. White Face refers to the Sheyenne National Grassland. According to the USGS, this region offers the last best hope for restoration of tallgrass prairie in North Dakota. For more information, see http://www.npwrc.usgs.gov/resource/plants/tallgras/lastcall.htm (accessed 25 November 2006).

4. The poisoning of prairie dogs on the Pine Ridge Reservation is part of a prairie dog management agreement reached in August 2004 by the South Dakota Department of Game, Fish and Parks; the South Dakota Agriculture Department; the US Forest Service; the US Animal and Plant Health Inspection Service; and the US Fish and Wildlife Service.

5. "Main causes of the decline in the ferret population included habitat conversion for farming; efforts to eliminate prairie dogs, which competed with livestock for available prairie forage; and sylvatic plague, a disease that wiped out large numbers of prairie dogs and has also killed ferrets. ... During the fall of 1986 and the spring of 1987 the last known 18 wild black-footed ferrets were taken from the wild and placed in captive breeding facilities." Reintroduction of the black-footed ferret into the Conata Basin/Badlands (part of the Buffalo Gap National Grassland) began in the fall of 1994. For the quoted material and additional information, see http://mountain-prairie.fws.gov/species/mammals/blackfootedferret/revfact.chy.pdf (accessed 25 November 2006).

6. White Face refers to HR 5489 (Great Plains Grasslands Wilderness Act), which was introduced in the US House of Representatives on 26 September 2002 by Congressman Frank Pallone Jr. (representing the 6th District of New Jersey).

7. The Wild and Scenic Rivers Act (16 USC 1271–1287)—Public Law 90–542, approved 2 October 1968, (82 Stat. 906) establishes a National Wild and Scenic Rivers System and prescribes the methods and standards through which additional rivers may be identified and added to the system. The declaration of the policy states in part "that certain selected rivers of the Nation which, with their immediate environments, possess outstandingly remarkable scenic, recreational, geologic, fish and wildlife, historic, cultural, or other similar values, shall be preserved in free-flowing condition, and that they and their immediate environments shall be protected for the benefit and enjoyment of present and future generations."

8. *Cobell v. Norton* was filed in 1996 by lead plaintiff Elouise Cobell, who was unable to obtain an accurate accounting of funds held in trust by the US government for individual Indian-owned land that had been leased by the federal government for mining, grazing, oil and gas exploration, and other uses. For more information, see http://www.indiantrust.com/index.cfm?FuseAction=Overview.Home (accessed 27 February 2007).

9. White Face refers to the 1804–6 expedition to explore the American Northwest under the leadership of Meriwether Lewis and William Clark.

Glossary

Animal and Plant Health Inspection Service. Oversight agency of **Wildlife Services**.
Animal Damage Control (ADC). See **Wildlife Services**.
animal unit month (AUM). The amount of forage consumed by an "animal unit" in one month. An animal unit is defined as a mature cow (1,000-pound) or the equivalent. Assuming an average consumption rate of twenty-six pounds of forage dry matter per day times thirty-one days yields 868 pounds. More conservative or liberal estimates for daily consumption are also used, giving values ranging from 600 to 1,000 pounds of forage for an AUM.
appropriate management level (AML). The midpoint number between the upper and the lower population level of horses or burros that can live on a Herd Management Area (HMA), while maintaining a thriving ecological balance. The AML must meet the objectives of achieving and maintaining healthy populations and a thriving ecological balance under multiple use of the HMA. Summarized from http://www.equinenet.org/life/qa.html (accessed 7 March 2007).
Automated Land and Mineral Record System (ALMRS). In the early 1980s, BLM began planning for an automated land and mineral case processing system to keep up with the growing number of applications for oil and gas leases. According to BLM, it obligated about $411 million on the ALMRS/Modernization project between fiscal years 1983 and 1998, of which more than $67 million was spent to develop ALMRS Initial Operating Capability (IOC) software. In October 1998, testing revealed that ALMRS IOC was not ready to be deployed because it did not meet requirements. Details about the history of ALMRS and the reasons for its failure can be found in Government Accountability Office (1999) and Government Accountability Office (2000).
Buffalo Field Campaign. A nonprofit organization with stated mission "to stop the slaughter of Yellowstone's wild buffalo herd, protect the natural habitat of wild free-roaming buffalo and native wildlife, and to work with people of all Nations to honor the sacredness of the wild buffalo." For more information, see http://www.buffalofieldcampaign.org (accessed 4 February 2007).
California Trout. A nonprofit organization with stated mission "to protect and restore wild trout and steelhead and their waters throughout California." For more information, see http://www.caltrout.org (accessed 31 January 2007).
CalTrout. See **California Trout**.
categorical exclusion (CE). A category of actions which do not individually or cumulatively have a significant effect on the human environment and for which, therefore, neither an **environmental assessment (EA)** nor an **environmental impact statement (EIS)** is required. Categorical exclusions are actions which meet the definition contained in 40 CFR 1508.4 and, based on past experience with similar actions, do not involve significant environmental impacts.
Center for Biological Diversity (CBD). A nonprofit organization with stated mission in part: "combining conservation biology with litigation, policy advocacy, and an innovative

strategic vision, the Center for Biological Diversity is working to secure a future for animals and plants hovering on the brink of extinction, for the wilderness they need to survive, and by extension for the spiritual welfare of generations to come." For more information, see http://www.biologicaldiversity.org (accessed 12 March 2007).

cheatgrass (*Bromus tectorum*). An annual grass native to Eurasian temperate regions that has become widespread in the grasslands of the western United States. It provides fair-to-poor forage for livestock and once dead is highly flammable.

Compound 1080. A highly toxic, slow-acting poison. **Wildlife Services**, the federal agency authorized to kill predators with Compound 1080, reports that death "occurs in two to five hours or more" and "may result from gradual cardiac failure; progressive depression of the central nervous system with either cardiac or respiratory failure as the terminal event; or respiratory arrest following severe convulsions."

crested wheatgrass. Common name applied to *Agropyron desertorum* and *A. cristatum*, both of which are native to Russia and Siberia. Beginning in the 1930s, crested wheatgrass was extensively planted by federal agencies in the American West to increase livestock forage on depleted rangelands. Unfortunately, crested seedings have some unfortunate characteristics: they may prevent native vegetation from recolonizing the area, and they often reduce the variety of plant and wildlife species. The reduction of wildlife diversity is further exacerbated when existing shrubs are removed in the course of a crested seeding.

Defenders of Wildlife. As stated on its website: "Defenders of Wildlife is dedicated to the protection of all native wild animals and plants in their natural communities." Furthermore, "Defenders of Wildlife established the Bailey Wildlife Foundation Wolf Compensation Trust which pays livestock owners for losses to wolf predation on private lands." For more information, see http://www.defenders.org (accessed 19 January 2007).

dichloro-diphenyl-dichloroethane (DDD). A chemical similar to DDT that contaminates commercial DDT preparations. The substance also enters the environment as a breakdown product of DDT.

Endangered Species Act of 1973 (ESA). Federal legislation intended to protect critically imperiled species from extinction due to "the consequences of economic growth and development untempered by adequate concern and conservation." The stated purpose of the act is not only to protect species, but also "the ecosystems upon which they depend." The act forbids federal agencies from authorizing, funding or carrying out actions which may jeopardize endangered species. And it forbids any government agency, corporation, or citizen from "taking" (i.e., harming or killing) endangered animals without a permit.

environmental assessment (EA). A document prepared to determine whether a proposed action will significantly affect the quality of the human environment. The EA leads either to the decision to produce an **environmental impact statement (EIS)**, or to a finding of no significant impact (FONSI).

environmental impact statement (EIS). "A report that documents the information required to evaluate the environmental impact of a project. It informs decision makers and the public of the reasonable alternatives that would avoid or minimize adverse impacts or enhance the quality of the environment." Quoted from http://www.eia.doe.gov/cneaf/nuclear/page/umtra/glossary.html (accessed 16 March 2007).

Federal Land Policy and Management Act of 1976 (FLPMA). Requires the BLM to execute its management powers under a land-use planning process that is based on multiple-use and sustained-yield principles. The act also provides for public land sales, withdrawals, acquisitions, and exchanges.

Forest and Rangeland Renewable Resources Planning Act of 1974. Called for the management of renewable resources on national forest lands. The law was reorganized, expanded and amended by the **National Forest Management Act of 1976.**

Forest Guardians. A nonprofit organization founded in 1989 with stated mission "to protect and restore the native wildlands and wildlife of the American Southwest through fundamental reform of public policies and practices." For more information, see http://www.fguardians.org (accessed 12 March 2007).

forest plan. Provides long-range management direction for all management programs and practices, resource uses, and resource protection measures on national forest lands.

General Schedule (GS). The general classification and compensation system for white-collar federal jobs; the pay scale contains fifteen grades of ten steps each. For more information, see http://www.jobs.nih.gov/whattoapply.asp (accessed 30 December 2006).

Grand Canyon Trust. A nonprofit organization described on its website as advocating "collaborative, common sense solutions to the significant problems affecting the region's natural resources. Our work is focused in the greater Grand Canyon region of northern Arizona, and in the red rock country of southern Utah." For more information, see http://www.grandcanyontrust.org (accessed 4 February 2007).

Great Basin Initiative. "Following the fires of 1999, the BLM developed the 'Great Basin Initiative' with goals of: (1) rehabilitating wildfire areas that have a high propensity of becoming annual grasslands; (2) restoring sagebrush steppe habitats that have a high proportion of non-native species; and (3) protecting function [sic] sagebrush steppe stands." For more information, see http://www.fs.fed.us/r1/pgr/afterfire/preliminary_assessment/chapter7_pgs51-66.PDF (accessed 4 February 2007).

Greater Yellowstone Coalition. Founded in 1983 with stated mission of "protecting the lands, waters, and wildlife of the Greater Yellowstone Ecosystem, now and for future generations." For more information, see http://www.greateryellowstone.org/ (accessed 4 February 2007).

herd management area (HMA). The geographic area identified in a management framework or resource management plan for the long-term management of a wild horse herd.

Idaho Watersheds Project. See **Western Watersheds Project.**

larkspur (*Delphinium* spp.). Erect herbs arising from a single or clustered, often woody root stock. "Larkspurs in the genus *Delphinium* cause more fatal poisoning of cattle in the western US than any other native plant species. ... [L]ivestock losses to larkspur in the US have been estimated to exceed $234 million annually, making larkspurs second only to the locoweeds in terms of economic losses to the livestock industry. In some areas of the intermountain states, cattle losses to larkspur poisoning average 2 to 5 percent per year and may be as high as 10 percent." (Knight and Walter 2002)

Multiple-Use Sustained-Yield Act of 1960 (MUSY). Declares that the purposes of the national forests include outdoor recreation, range, timber, watershed, and fish and wildlife. The act directs the secretary of agriculture to develop and administer the renewable surface resources of the national forests for multiple use and sustained yield of the various products and services obtained from these areas.

National Environmental Policy Act (NEPA). Enacted in 1969, the purposes of the act are to "declare a national policy which will encourage productive and enjoyable harmony between man and his environment; to promote efforts which will prevent or eliminate damage to the environment and biosphere, and stimulate the health and

374 / Glossary

welfare of man; to enrich the understanding of the ecological systems and natural resources important to the nation; and to establish a Council on Environmental Quality." For more information, see http://www.nepa.gov/nepa/regs/nepa/nepaeqia.htm (accessed 4 February 2007).

National Forest Management Act of 1976 (NFMA). The primary statute governing the administration of national forests. The law requires the secretary of agriculture to assess forest lands, develop a management program based on multiple-use, sustained-yield principles, and implement a resource management plan for each unit of the National Forest System. For more information, see http://www.fs.fed.us/emc/nfma/includes/NFMA1976.pdf (accessed 4 February 2007).

National Wildlife Federation (NWF). A nonprofit conservation, education, and advocacy organization founded in 1936 as a nationwide federation of grassroots conservation activists. For more information, see http://www.nwf.org/ (accessed 4 February 2007).

natural area. Typically the phrase denotes an area set aside and protected that is relatively undisturbed by human activities and is characterized by indigenous species.

Natural Resources Conservation Service (NRCS). As of 1994, the new name of the Soil Conservation Service. The organization's website states that the NRCS "provides leadership in a partnership effort to help people conserve, maintain, and improve our natural resources and environment." For more information, see http://www.nrcs.usda.gov/ (accessed 4 February 2007).

Nature Conservancy. Described on its website as "the leading conservation organization working to protect the most ecologically important lands and waters around the world for nature and people." For more information, see http://www.nature.org/ (accessed 4 February 2007).

Oregon Natural Desert Association (ONDA). Founded in 1987; a nonprofit organization since 1989. Its stated mission: "to protect, defend and restore the health of Oregon's native deserts." For more information, see http://www.onda.org (accessed 9 March 2007).

Oregon Natural Resources Council (ONRC). See **Oregon Wild**.

Oregon Wild. Founded in 1974 as the Oregon Natural Resources Council (ONRC). Its stated mission: "to protect and restore Oregon's wildlands, wildlife and waters as an enduring legacy for all Oregonians." For more information, see http://www.oregonwild.org (accessed 18 January 2007).

Paragon Foundation. The organization's website states that the "Paragon Foundation, a 501-C(3) not-for-profit organization, was created in 1996 to support the advancement of the fundamental principals articulated by the United States' founding fathers in both the Declaration of Independence and U.S. Constitution." For more information, see http://www.paragonfoundation.org (accessed 4 February 2007).

pedestalling. As livestock injure and kill plants, fewer roots remain to hold soil particles together and hold masses of soil in place. As unattached soil is removed by erosion, the surviving plants often become perched on small islands of soil, or pedestals.

pepperweed (*Lepidium latifolium*). A perennial, noxious weed that reproduces by seed and rhizomes under various environmental conditions. Now scattered throughout the US, the plant is native to southern Europe and western Asia.

Predator Conservation Alliance. A nonprofit organization founded in 1991 for the purpose of protecting predatory species and their habitats. The organization changed its name from Predator Project to Predator Conservation Alliance in 1999. For more information, see http://www.predatorconservation.org/ (accessed 4 February 2007).

Predator Defense. Describes itself as "a nonprofit, 501(c)3 organization founded in 1990 by professionals from the business and scientific community who are concerned about the humane treatment of wildlife, and the effective management of public and private ecosystems that support our wildlife and provide clean air, water and soil for Americans." Its mission statement: "Protect native predators and create alternatives for people to coexist with wildlife." For more information, see http://predatordefense.org (accessed 25 March 2007).

Predator Defense Institute (PDI). See **Predator Defense.**

Predator Project. See **Predator Conservation Alliance.**

Public Employees for Environmental Responsibility (PEER). "A national nonprofit alliance of local, state and federal scientists, law enforcement officers, land managers, and other professionals dedicated to upholding environmental laws and values." For more information, see http://www.peer.org/ (accessed 4 February 2007).

Quivira Coalition. As stated on its website: "The mission of The Quivira Coalition is to foster ecological, economic and social health on western landscapes through education, innovation, collaboration, and progressive public and private land stewardship." For more information, see http://quiviracoalition.org (accessed 5 March 2007).

range con. See **range conservationist.**

range conservationist. A person responsible for planning, coordinating, reviewing, and reporting on range management.

resource advisory council (RAC). Established by the **Federal Land Policy and Management Act** (43 USC 1701 et seq.) as a citizens' advisory group to the BLM. Council members represent various categories of interest within the geographic area covered by the RAC. For more information, see http://www.blm.gov/rac/rac_qs_and_as.htm (accessed 4 February 2007).

resource management plan (RMP). According to the BLM website http://www.blm.gov/nhp/Commercial/SolidMineral/3809/deis/glossary.html (accessed 4 February 2007) a resource management plan is a "BLM planning document, prepared in accordance with Section 202 of the **Federal Land Policy and Management Act,** that presents systematic guidelines for making resource management decisions for a resource area. An RMP is based on an analysis of an area's resources, their existing management, and their capability for alternative uses. RMPs are issue oriented and developed by an interdisciplinary team with public participation."

riparian area. Land where the vegetation and microclimate are influenced by perennial or intermittent water.

Sierra Club. Describes itself as "America's oldest, largest and most influential grassroots environmental organization." Its mission statement: (1) explore, enjoy and protect the wild places of the earth; (2) practice and promote the responsible use of the earth's ecosystems and resources; (3) educate and enlist humanity to protect and restore the quality of the natural and human environment; (4) use all lawful means to carry out these objectives. For more information, see http://www.sierraclub.org (accessed 12 June 2007).

Soil Conservation Service (SCS). See **Natural Resources Conservation Service.**

Soil Vegetation Inventory Method (SVIM). Adopted in 1977 as the BLM's official inventory method for basic inventories of soil and vegetation, but not intended to preclude the use of site-specific studies for special purposes. The method was discontinued in the early 1980s and never became the official standardized inventory system for federal rangelands that BLM had envisioned. For a thorough description of the method, see US Department of Interior, Bureau of Land Management (1979); for a personal perspective, see Walker (2003).

Southern Utah Wilderness Alliance (SUWA). Describes its mission in part as "the preservation of the outstanding wilderness at the heart of the Colorado Plateau, and the management of these lands in their natural state for the benefit of all Americans." For more information, see http://www.suwa.org/ (accessed 4 February 2007).

T and E species. Abbreviation for threatened and endangered species.

take. From Section 3(18) of the federal **Endangered Species Act**: "The term 'take' means to harass, harm, pursue, hunt, shoot, wound, kill, trap, capture, or collect, or to attempt to engage in any such conduct."

Taylor Grazing Act of 1934. Enacted to stop injury to the public grazing lands [excluding Alaska] by preventing overgrazing and soil deterioration; to provide for their orderly use, improvement, and development; and to stabilize the livestock industry dependent upon the public range. For a summary of the act, see http://ipl.unm.edu/cwl/fedbook/taylorgr.html (accessed 4 February 2007).

Uncompahgre Plateau Project. Begun in 2001 as the Uncompahgre Ecosystem Restoration Project. A collaborative effort of the Bureau of Land Management, the US Forest Service, the Colorado Division of Wildlife, and the Public Lands Partnership having the mission "to develop a collaborative approach, to restore and maintain the ecosystem health of the Uncompahgre Plateau using best science and public input." For more information, see http://www.upproject.org/ (accessed 4 February 2007).

Western Watersheds Project. Named Idaho Watersheds Project when founded in 1993 as a 501(c)(3) membership corporation. Its stated mission: "to protect and restore western watersheds and wildlife through education, public policy initiatives and litigation." Its board of directors adopted the organization's name change to Western Watersheds Project on 9 February 2001. For more information, see http://www.westernwatersheds.org/ (accessed 17 January 2007).

Wilderness Watch. As stated on its website: "Founded in 1989, Wilderness Watch is the *only* national organization whose sole focus is the preservation and proper stewardship of lands and rivers already included in the National Wilderness Preservation System (NWPS) and National Wild & Scenic Rivers System (NWSRS). The organization grew out of the concern that while much emphasis is being placed on adding new areas to these systems, the conditions of existing Wilderness and rivers are largely being ignored. We believe that the stewardship of these remarkable wild places must be assured through independent citizen oversight, education, and the continual monitoring of federal management activities." For more information, see http://www.wildernesswatch.org (accessed 11 June 2007).

Wildlife Services. A program of the United States Department of Agriculture's Animal and Plant Health Inspection Service. According to its website, the program "provides Federal leadership and expertise to resolve wildlife conflicts and create a balance that allows people and wildlife to coexist peacefully." Prior to 1997 the program was named Animal Damage Control. For more information, see http://www.aphis.usda.gov/wildlife_damage/ (accessed 7 May 2007).

Wolf Recovery Foundation. Its stated mission: "to foster our heritage of wild wolf communities by advocating their presence forever in places where they have been extirpated. We accomplish our mission through our efforts in public representation, information and outreach, networking with the agencies, organizations, tribes and universities, and through our diverse workshops, conferences and special events." For more information, see http://www.forwolves.org (accessed 4 February 2007).

References

Abbey, Edward. 1986. Even the bad guys wear white hats: Cowboys, ranchers, and the ruin of the West. *Harper's* 272, no. 1628 (January): 51–55.

Alexander, Earl B., and David R. Gilman. 1994. Compaction and recovery of rangeland soils in the Owyhee Upland, Idaho. *Journal of the Idaho Academy of Science* 30, no. 1: 49–54.

Ambos, Norman, George Robertson, and Jason Douglas. 2000. Dutchwoman Butte: A relict grassland in central Arizona. *Rangelands* 22, no. 2 (April): 3–8.

Anderson, Allen E. 1960. Effects of sagebrush eradication by chemical means on deer and related wildlife. Master's thesis, Colorado State University, Fort Collins.

Arizona Game and Fish Department. n.d. Living with mountain lions in Arizona. http://www.gf.state.az.us/pdfs/w_c/living_with_lions.pdf (accessed 18 January 2007).

Associated Press. 2006. Last male of purebred rabbit species dies. May 18. http://www.msnbc.msn.com/id/12851338/ (accessed 8 January 2007).

Baker, D. L., and F. S. Guthery. 1990. Effects of continuous grazing on habitat and density of ground-foraging birds in south Texas. *Journal of Range Management* 43:2–5.

Balmford, Andrew, Aaron Bruner, Philip Cooper, Robert Costanza, Stephen Farber, Rhys E. Green, Martin Jenkins, Paul Jefferiss, Valma Jessamy, Joah Madden, Kat Munro, Norman Myers, Shahid Naeem, Jouni Paavola, Matthew Rayment, Sergio Rosendo, Joan Roughgarden, Kate Trumper, and R. Kerry Turner. 2002. Economic reasons for conserving wild nature. *Science* 297, issue 5583 (9 August): 950–53.

Barber, Douglas K. 1998. Letter to Senator Pete Domenici. February 11. http://www.rangebiome.org/cowfree/fsblastsfs.html#barber (accessed 28 February 2007).

Bareiss, L. J., P. Schulz, and F. S. Guthery. 1986. Effects of short-duration and continuous grazing on bobwhite and wild turkey nesting. *Journal of Range Management* 39:259–60.

Belnap, Jayne, Julie Hilty Kaltenecker, Roger Rosentreter, John Williams, Steve Leonard, and David Eldridge. 2001. *Biological soil crusts: Ecology and management*, ed. Pam Peterson, TR 1730-2, Denver: US Department of the Interior, Bureau of Land Management, National Science and Technology Center, Information and Communications Group.

Belsky, Arlene Joy. 1979. Determinants of ecological amplitude in *Festuca idahoensis* and *Festuca ovina*. PhD dissertation, University of Washington.

Belsky, A. Joy. 1994. Fighting sex discrimination in hiring by the federal government. *Women in Natural Resources* 15, no. 4: 8–13.

Belsky, A. Joy, and Dana M. Blumenthal. 1997. Effects of livestock grazing on stand dynamics and soils in upland forests of the Interior West. *Conservation Biology* 11, no. 2 (April): 315–27.

Belsky, A. Joy, Walter P. Carson, Cynthia L. Jensen, and Gordon A. Fox. 1993 Overcompensation by plants: Herbivore optimization or red herring? *Evolutionary Ecology* 7:109–21.

Belsky, Joy, Sally Cross, and Diane Valentine [Valantine]. 1995. One small step: Combating sexism in the environmental movement. *Whole Terrain* 4:38–42.

Belsky, Joy, and Jonathan L. Gelbard. 2002. Comrades in harm: Livestock and exotic weeds in the Intermountain West. In *Welfare ranching: The subsidized destruction of the American West*, ed. George Wuerthner and Mollie Matteson, 203–4. Washington, DC: Island Press.

Belsky, Joy, Andrea Matzke, and Shauna Uselman. 1999. Survey of livestock influences on stream and riparian ecosystems in the western United States. *Journal of Soil and Water Conservation* 54, no. 1 (first quarter): 419–31.

Biodiversity Conservation Alliance. 2004. Press release: Colorado River cutthroat trout illegally denied ESA protection by US Fish and Wildlife Service. April 20. http://www.voiceforthewild.org/wildspecies/news/n20april04.html (accessed 7 April 2007).

Bock, C. E., and J. H. Bock. 1999. Response of winter birds to drought and short-duration grazing in southeastern Arizona. *Conservation Biology* 13:1117–23.

Bohrer, Becky. 2005. First bison falls in Montana's bison hunt. *Billings Gazette,* 15 November. http://www.billingsgazette.com/newdex.php?display=rednews/2005/11/15/build/state/15-bison-hunt.inc (accessed 27 February 2007).

Boxall, Bettina. 2006. A matter of grove concern. *Los Angeles Times,* 21 December. http://www.latimes.com/news/local/la-me-litton21dec21,0,3857373.story?page=1&coll=la-home-headlines (accessed 24 December 2006).

Braun, C. E. 1995. Distribution and status of sage grouse in Colorado. *Prairie Naturalist* 27:1–9.

———. 1998. Sage grouse declines in western North America: What are the problems? *Proc. of the Western Association of Fish and Wildlife Agencies* 78:139–56.

Braun, Clait, E. 2004a. Professional review of: "Connelly, J. W., S. T. Knick, M. A. Schroeder, and S. J. Stiver. 2004. Conservation assessment of greater sage-grouse and sagebrush habitats. Western Association of Fish and Wildlife Agencies." Tucson, Arizona: Grouse Inc. (22 July). http://www.sagebrushsea.org/pdf/Braun_WAFWA_review.pdf (accessed 24 February 2007).

———. 2004b. Review of: "Western Governor's Association 'Conserving the greater sage grouse: A compilation of efforts underway on state, tribal, provincial and private lands.'" Grouse Inc. (25 July). http://www.sagebrushsea.org/pdf/Braun_WGA_review.pdf (accessed 24 February 2007).

Braun, C. E., J. R. Young, K. M. Potter, M. L. Commons, and J. A. Nehring. 2004. Historical distribution of Gunnison sage-grouse in Colorado. Annual Conference, The Wildlife Society 11: Abstract.

Brown, David E. 1994. Out of Africa. *Wilderness* (winter): 24, 26–27, 30–33.

Brown, R. L. 1982. Effects of livestock grazing on Mearns quail in southeastern Arizona. *Journal of Range Management* 35:727–32.

California Native Plant Society. 2001. Newsletter of the Bristlecone Chapter, 21, no. 6 (November). http://www.bristleconecnps.org/Newsletters/cnv216.htm (accessed 27 February 2007).

Carlson, Cathy, and Johanna Wald. 2001. Rangeland Reform revisited. August. (Available from Natural Resources Defense Council as of April 2007.)

Chaney, Ed, Wayne Elmore, and William S. Platts. 1990. Livestock grazing on western riparian areas. Produced for the US Environmental Protection Agency by the Northwest Resource Information Center, Inc., Eagle, Idaho.

Chernyavsky, F. B. 2004. On taxonomy and history of bighorn sheep (*Pachyceros* subgenus, Artiodactyla). *Zoologičeskij žurnal* 83, no. 8:1059–70.

Connelly, J. W., S. T. Knick, M. A. Schroeder, and S. J. Stiver. 2004. Conservation assessment of greater sage-grouse and sagebrush habitats. Unpublished report, Western Association of Fish and Wildlife Agencies, Cheyenne, Wyoming.

Croxen, Fred W. 1926. History of grazing on Tonto. Presented at the Tonto Grazing Conference, Phoenix, AZ (4–5 November). http://www.rangebiome.org/genesis/GrazingOnTonto-1926.html (accessed 25 February 2007).

Daily Utah Chronicle. 2002. Innovation needed for Utah's education crisis. September 27. http://www.dailyutahchronicle.com/media/paper244/news/2002/09/27/Opinion/Innovation.Needed.For.Utahs.Education.Crisis-284089.shtml?norewrite&sourcedomain=www.dailyutahchronicle.com (accessed 27 February 2007).

Davis, Tony. 1994a. BLM Chief Jim Baca leaves amidst cheers and boos. *High Country News,* 21 February. http://www.hcn.org/servlets/hcn.Article?article_id=100 (accessed 27 February 2007).

———. 1994b. A Struggle for the last grass. *High Country News,* 2 May. http://www.hcn.org/servlets/hcn.Article?article_id=293 (accessed 29 January 2007).

———. 2001. Healing the Gila. *High Country News,* 22 October. http://www.hcn.org/servlets/hcn.Article?article_id=10791 (accessed 24 November 2006).

DeHart, Roy L. 1998. Editorials: Multiple chemical sensitivity. *American Family Physician* 58, no. 3 (1 September): 652–55. http://www.aafp.org/afp/980901ap/edit.html (accessed 27 February 2007).

Diehl, Patrick S., trans. 1979. *Dante's rime*. Princeton, NJ: Princeton University Press.

Diehl, Patrick S. 1985. *The medieval European religious lyric: An ars poetica*. Berkeley: University of California Press.

Diehl, Patrick. 2002. Press release: Grand Staircase-Escalante National Monument BLM range staff caught lying: BLM fails to protect Grand Canyon Trust buy-outs against grazing. Sierra Club Glen Canyon Group (24 February).

Dixie National Forest. 1997. *(Draft) Proper functioning condition (PFC) assessment for major vegetation types on the Aquarius Plateau/Boulder Top and Barney Top/Table Cliffs subsections*. Escalante and Teasdale Ranger District, Dixie National Forest (20 November).

Donahue, Debra L. 1999. *The western range revisited: Removing livestock from public lands to conserve native biodiversity*. Norman: University of Oklahoma Press.

———. 2005. Western grazing: The capture of grass, ground, and government. *Environmental Law* 35:721–806.

Dowie, Mark. 1995. The wayward West: With liberty and firepower for all. *Outside* (November). http://outside.away.com/outside/magazine/1195/11f_lib.html (accessed 27 February 2007).

Durbin, Kathie. 1997. Restoring a refuge: Cows depart, but can antelope recover? *High Country News*, 24 November. http://www.hcn.org/servlets/hcn.Article?article_id=3790 (accessed 27 February 2007).

———. 2003. A revival on Hart Mountain. *High Country News*, 10 November. http://www.hcn.org/servlets/hcn.Article?article_id=14360 (accessed 27 February 2007).

Earthjustice. 2006. Press release: Colorado River cutthroat trout to be considered for protection. September 7. http://www.earthjustice.org/news/press/006/colorado-river-cutthroat-trout-to-be-considered-for-protection.html (accessed 7 April 2007).

Egan, Timothy. 1991. Ranchers vs. rangers over land use. *New York Times*, 19 August.

Elliott, Taffeta. 1998. Air Force drops a sweetheart deal onto ranch land. *High Country News*, 31 August. http://www.hcn.org/servlets/hcn.Article?article_id=4412 (accessed 25 February 2007).

Environmental News Service. 2005. Yellowstone bison will face Montana guns this winter. September 9. http://www.ens-newswire.com/ens/sep2005/2005-09-09-02.asp (accessed 27 February 2007).

Fager, Leon. 1998. Letter to Forest Service Chief Mike Dombeck. February 23. http://www.rangebiome.org/cowfree/fsblastsfs.html#fager (accessed 24 February 2007).

Ferguson, Denzel, and Nancy Ferguson. 1983. *Sacred cows at the public trough*. Bend, OR: Maverick Publications.

Flaccus, Gillian. 2003. Swelling pronghorn population revives debate over habitat management. *IdahoStateJournal.com*, 26 September. http://www.journalnet.com/articles/2003/09/26/features/outdoors01.txt (accessed 27 February 2007).

Galeano-Popp, Renee. 1998. Letter to Eleanor S. Townes [Towns]. May 13. http://www.sw-center.org/swcbd/resources/popp.html (accessed 25 February 2007).

Gallizioli, Steve. 1976. Livestock vs. wildlife. Paper presented at the 41st North American Wildlife and Natural Resources Conference, Washington, DC (March). http://www.rangebiome.org/cowfree/gallizioli/LivestockVsWildlife1976.htm (accessed 25 February 2007).

———. 1979. Effects of livestock grazing on wildlife. Paper presented at the 10th annual joint meeting of the western section, the Wildlife Society and the California-Nevada Chapter, American Fisheries Society, Long Beach, CA (3 February). http://www.rangebiome.org/cowfree/gallizioli/EffectsOfLivestockGrazing79.htm (accessed 25 February 2007).

Government Accountability Office. 1988a. *More emphasis needed on declining and overstocked grazing allotments*. GAO/RCED-88-80. Washington, DC: USGAO.

———. 1988b. *Some riparian areas restored but widespread improvement will be slow*. GAO/RCED-88-105. Washington, DC: USGAO.

———. 1999. *Major software development does not meet BLM's business needs.* GAO/T-AIMD-99-102. Washington, DC: USGAO.

———. 2000. *Status of BLM's actions to improve information technology management.* GAO/AIMD-00-67. Washington, DC: USGAO.

———. 2005. *Federal expenditures and receipts vary, depending on the agency and the purpose of the fee charged.* GAO-05-869. Washington, DC: USGAO.

Gregg, Michael A., John A. Crawford, Martin S. Drut, and Anita K. DeLong. 1994. Vegetational cover and predation of sage grouse nests in Oregon. *Journal of Wildlife Management* 58, no. 1 (January): 162–66.

GS. 1996. Enviros hung in effigy. *Earth Island Journal* 11, no. 3 (summer). http://www.earthisland.org/eijournal/new_articles.cfm?articleID=489&journalID=58 (accessed 24 February 2007).

Hardin, Garrett. 1968. The tragedy of the commons. *Science* 162:1243–48. http://www.sciencemag.org/sciext/sotp/commons.dtl (accessed 26 February 2007).

Harrington, C. R. 2002. Yukon Horse. Yukon Beringia Interpretive Centre (August). http://www.beringia.com/02/02maina14.html (accessed 14 March 2007).

Hart, John. 1996. *Storm over Mono: The Mono Lake battle and the California water future.* Berkeley: University of California Press.

Hedden, Bill, and Craig Bigler. n.d. School trust lands in Utah. Grand Canyon Trust. http://www.grandcanyontrust.org/media/PDF/forests/schtrust.pdf (accessed 27 February 2007).

High Country News. 1993. Hotline: Critics want foresters fired. December 27. http://www.hcn.org/servlets/hcn.Article?article_id=2623 (accessed 24 February 2007).

Hinchman, Steve. 1994. Turmoil on the range: Ranchers' clout drives grazing reform in new directions. *High Country News* (24 January). http://www.hcn.org/servlets/hcn.Article?article_id=40 (accessed 2 March 2007).

Horning, John. 1994. Grazing to extinction: Endangered, threatened and candidate species imperiled by livestock grazing on western public lands. Washington, DC: National Wildlife Federation.

Independent Multidisciplinary Science Team. 2003. *Independent multidisciplinary science team, Oregon plan for salmon and watersheds: 2002 annual report.* Oregon Watershed Enhancement Board. Salem, OR. http://www.fsl.orst.edu/imst/reports/2002ar.pdf (accessed 28 February 2007).

Ivey, Gary L., David G. Paullin, and Carroll D. Littlefield. 2006. Predator control to enhance sandhill crane production at Malheur National Wildlife Refuge. USGS Northern Prairie Wildlife Research Center. http://www.npwrc.usgs.gov/resource/birds/symabs/enhance.htm (page last modified 3 August 2006) (accessed 27 February 2007).

Jacobs, Lynn. 1991. *Waste of the West: Public lands ranching.* Tucson: Lynn Jacobs.

Juillerat, Lee. 2004. Wildlife at national antelope refuge doing well according to recent population surveys. *Herald and News* (Klamath Falls, Oregon) (12 November). http://www.publiclandsranching.org/htmlres/PDF/press_K-Falls_Hart_Mtn_bounty.pdf (accessed 27 February 2007).

Kirkpatrick, Jay F., and Patricia M. Fazio. 2005. Wild horses as native North American wildlife. Statement for the 109th Congress (1st Session) in support of HR 297, A bill in the House of Representatives, House Committee on Resources, introduced 25 January 2005. http://www.saplonline.org/wild_horses_native.htm (accessed 27 February 2007).

Knight, A. P., and R. G. Walter. 2002. Plants causing sudden death. In *A guide to plant poisoning of animals in North America,* ed. A. P. Knight and R. G. Walter. Jackson, WY: Teton NewMedia; Ithaca NY: International Veterinary Information Service. http://www.ivis.org/special_books/Knight/chap1/ivis.pdf (accessed 12 June 2007).

Koehler, David. n.d. Preliminary reconnaissance report—Feral burro ecological impact project. National Park Service (available from the National Park Service, Southwest Regional Office Interpretation Library, Santa Fe, New Mexico).

Koehler, David A. 1974. The ecological impact of feral burros on Bandelier National Monument. MS thesis, University of New Mexico, Albuquerque.

Koehler, David A., and Allan E. Thomas. 2000. Managing for enhancement of riparian and wetland areas of the western United States. USDA Forest Service, Rocky Mountain Research Station. Gen. Tech. Report RMRS-GTR-54. http://www.fs.fed.us/rm/pubs/rmrs_gtr054.pdf (accessed 26 February 2007).

Kritz, Kevin. 2005. Summary of sage-grouse petitions submitted to the US Fish and Wildlife Service (USFWS) as of January 27, 2005. Reno, NV: US Fish and Wildlife Service, Nevada Fish and Wildlife Office. http://www.fws.gov/Nevada/nv_species/documents/sage_grouse/SG_2005_petition_sum.pdf#search=%22%22gunnison%20sage%20grouse%22%20sinapu%20petition%20january%22 (accessed 24 February 2007).

Krueger, Bill. 1996. Rehab success in Nevada! *Rocky Mountain Game and Fish Magazine* (May): 36–39.

———. 2004. Trout fishing Nevada's Bruneau River. *Rocky Mountain Game and Fish Magazine* (May): 45–46, 50–51.

Larmer, Paul. 2000. HCN at 30: "On faith alone." *High Country News* 32, no. 6 (27 March). http://www.hcn.org/servlets/hcn.Article?article_id=5685 (accessed 27 February 2007).

Luoma, Jon R. 1986. Discouraging words. *Audubon* 88, no. 5 (September): 86–104.

McArthur, E. Durant. 1994. Ecology, distribution, and values of sagebrush within the Intermountain Region. In *Proceedings—Ecology and management of annual grasslands*, ed. Stephen B. Monsen and Stanley G. Kitchen, 347–51. USDA Forest Service, Intermountain Research Station. General Tech. Report INT-GRT-313.

meatnews.com. 2004. Horse slaughter debated. May 20. http://www.meatnews.com/index.cfm?fuseaction=article&artNum=7528 (accessed 27 February 2007).

National Park Service. 2001. Chapter 6: Natural resource management in mesa and canon country. In *Bandelier administrative history*. http://www.nps.gov/archive/band/adhi/adhi6a.htm (accessed 26 February 2007).

National Wildlife Federation. 2003. Press release: Landmark agreement leads to retirement of Wyoming livestock allotment next to Grand Teton National Park. August 1. http://www.nwf-wcr.org/PDFs/WCR-LandmarkAgreementLeads.pdf (accessed 1 March 2007).

Naugle, David E., Cameron L. Aldridge, Brett L. Walker, Todd E. Cornish, Brendan J. Moynahan, Matt J. Holloran, Kimberly Brown, Gregory D. Johnson, Edward T. Schmidtmann, Richard T. Mayer, Cecilia Y. Kato, Marc R. Matchett, Thomas J. Christiansen, Walter E. Cook, Terry Creekmore, Roxanne D. Falise, E. Thomas Rinkes, Mark S. Boyce. 2004. West Nile virus: Pending crisis for greater sage-grouse. *Ecology Letters* 7, no. 8 (August): 704–13.

New York Times. 2003. A misdirected forest strategy. August 12.

Nokkentved, Niels Sparre. 2001. *Desert wings: Controversy in the Idaho desert*. Pullman, WA: Washington State University Press.

Odion, Dennis C., Tom L. Dudley, and Carla D'Antonio. 1988. Cattle grazing in southeastern Sierran meadows: Ecosystem change and prospects for recovery. In *Plant biology of eastern California*, ed. Clarence A. Hall and Vicki Doyle-Jones, 277–92. White Mountain Research Station, University of California.

Office of the Legislative Fiscal Analyst. n.d. FY 2001 budget recommendations. Utah State Legislature. http://www.le.state.ut.us/lfa/reports/ba2001/natres/Tla.pdf (accessed 27 February 2007).

Oregon Zoo. 2006. Zoo plans March release of rare pygmy rabbits into wild. December 28. http://www.oregonzoo.org/Newsroom/2006%20releases/2006Dec.htm#PygmyRabbit (accessed 28 February 2007).

Power, Thomas M. 2002. Taking stock of public lands grazing: An economic analysis. In *Welfare ranching: The subsidized destruction of the American West*, ed. George Wuerthner and Mollie Matteson, 263–69. Washington, DC: Island Press.

Riley, Brendan. 2002. Brief jail time, fines for horse shooting. *Nevada Appeal* (Carson City), 12 February. http://www.wildhorsespirit.org/wild_horse_spirit_news.htm (accessed 27 January 2007).

Rogers, Paul. 1999. Cash cows. *San Jose Mercury News*, 7 November.
Roosevelt, Theodore. 1888. *Ranch life and the hunting-trail*. New York: Century.
Ruch, Jeffrey. 1999. PEER letter to FS Chief Dombeck re Apache-Sitgreaves NF Supervisor John Bedell (4 August). http://web.archive.org/web/20050220090631/http://www.fguardians.org/peerbedell.html (accessed 24 February 2007).
Russell, Sharman Apt. 1993. *Kill the cowboy: A battle of mythology in the new West*. Reading, MA: Addison-Wesley.
Salvo, Mark, Erik Molvar, Josh Pollock, and Nicole Rosmarino. 2006. News release: Conservation organizations: Feds must adopt science-based solution to avoid sage grouse ESA listing. Biodiversity Conservation Alliance, Center for Native Ecosystems, Forest Guardians, Sagebrush Sea Campaign (20 July). http://www.sagebrushsea.org/pdf/release_greater_grouse7.pdf (accessed 24 February 2007).
Savage Jr., William W., ed. 1975. *Cowboy life: Reconstructing an American myth*. Norman: University of Oklahoma Press.
Savage Jr., William W. 1979. *The cowboy hero: His image in American history & culture*. Norman: University of Oklahoma Press.
Siciliano, Enzo. 1987. *Diamante*. Trans. Patrick Diehl. San Francisco: Mercury House.
Sierra Club. 1997. Beer-backed bovines waste wilderness. *The Planet* 4, no. 2 (March). http://www.sierraclub.org/planet/199703/alert.asp (accessed 27 February 2007).
Sisk-a-dee. 2002. Gunnison sage-grouse conservation plan: Summary of evaluation 2002 (28 August). Report written for the Gunnison Sage-Grouse Working Group by Sisk-a-dee, a Gunnison Basin conservation organization.
Smith, Christopher. 1994. Cows are evicted from Utah. *High Country News*, 24 January. http://www.hcn.org/servlets/hcn.Article?article_id=57 (accessed 12 April 2007).
Sneva, Forrest A. 1992. Relation of precipitation and temperature with yield of herbaceous plants in Eastern Oregon. *Int. J. of Biometeorology* 26, no. 4:263–76.
Stark, Mike. 2003. Hebgen Lake cattle grazing relocated. *Billings Gazette*, 26 April. http://www.billingsgazette.net/articles/2003/04/26/local/export104556.txt (accessed 19 January 2007).
St. Clair, Jeffrey. 2007. How the West was eaten: Til the cows come home. *Counterpunch*, 10–11 February. http://www.counterpunch.org/stclair02102007.html (accessed 11 February 2007).
Steiner, Stan. 1980. *The ranchers: A book of generations*. New York: Alfred A. Knopf.
Stern, Bill Steven. 1998. Permit value: A hidden key to the public land grazing dispute. Master's thesis, University of Montana, Missoula. http://www.rangenet.org/directory/stern/thesis/ (accessed 3 October 2006).
Stienstra, Tom. 2001. Remembering an unsung hero. *San Francisco Chronicle*, 9 December. http://www.sfgate.com/cgi-bin/article.cgi?f=/chronicle/achive/2001/12/09/SP78968.DTL (accessed 27 February 2007).
Trout Unlimited. 2000. Press release: TU asks feds to list California golden trout as endangered. Trout Unlimited (27 October). http://www.tu.org/site/apps/nl/content2.asp?c=7dJEKTNuFmG&b=280766&ct=329913 (accessed 27 February 2007).
Uresk, D. W., and A. J. Bjugstad. 1983. Prairie dogs as ecosystem regulators on the northern high plains. In *Proceedings, 7th North American Prairie Conference, August 4–6, 1980, Southwest Missouri State University, Springfield*, ed. C. L. Kucera, 91–94.
US Department of Agriculture, APHIS. 2006. *Cattle and calves death loss in the United States, 2000*. http://www.aphis.usda.gov/vs/ceah/ncahs/nahms/general/cattle_calves_deathloss_2000.pdf (accessed 25 February 2007).
US Department of Agriculture, Forest Service. 1998. Press release: Scott Conroy named as new Modoc National Forest supervisor. San Francisco: Modoc National Forest (27 July). http://www.fs.fed.us/r5/modoc/news/1998/07_27_98_scott_new_fs.html (accessed 26 February 2007).
———. 2000. News release: Dan Chisholm named as new Modoc National Forest supervisor. Alturas, CA: Modoc National Forest (9 April). http://www.fs.fed.us/r5/modoc/news/2000/forest_supervisor_named.htm (accessed 26 February 2007).

———. 2006. *Grazing statistical summary FY 2005*. http://www.fs.fed.us/rangelands/ftp/docs/graz_stat_summary_2005.pdf (accessed 24 February 2007).
US Department of Agriculture, Forest Service, Pacific Southwest Region. 2001. File Code: 1570-1/2210; Appeal No: 01-05-00-0050-A215 (10 May). http://www.fs.fed.us/r5/ecoplan/appeals/2001/fy01_0 050.htm (accessed 27 February 2007).
US Department of Agriculture, Forest Service, Southwest Region. 1995. *(Final) Environmental impact statement for Diamond Bar Allotment management plan* (May).
US Department of Agriculture, National Agricultural Statistics Service. 2005. *Sheep and goats death loss*. http://usda.mannlib.cornell.edu/usda/current/sgdl/sgdl-05-06-2005.pdf (accessed 25 February 2007).
———. 2007. Sheep and goats. http://usda.mannlib.cornell.edu/usda/current/SheeGoat/SheeGoat -02-02-2007.pdf (accessed 25 February 2007).
US Department of the Interior, Bureau of Land Management. 1979. *Soil vegetation inventory method, BLM manual 4412.14*. Washington, DC: USDI-BLM.
———. 1996. Press release: 1996 grazing fee announced. January 22. http://www.blm.gov/nhp/news/releases/pages/1996/pr960122.html (accessed 1 March 2007).
———. 1992. *Grazing fee review and evaluation: Update of the 1986 final report*. Washington, DC: USDI-BLM.
———. 2005. *Public land statistics 2005*. http://www.blm.gov/natacq/pls05/ (accessed 24 Feb. 2007).
———. 2006. *Factsheet on the BLM's new grazing regulations*. July 12. http://www.blm.gov/nhp/news/releases/pages/2006/pr060712_July2006_GrazingFactSheet.pdf (accessed 3 January 2007).
US Department of the Interior, Bureau of Land Management, National Science and Technology Center. 2001. *National rangeland inventory: Monitoring and evaluation report—Fiscal year 2001*. http://www.blm.gov/nstc/rangeland/rangeland01.html (accessed 26 February 2007).
US Environmental Protection Agency. 1990. Livestock grazing on western riparian areas. Eagle, ID: Northwest Resource Information Center, Inc., July 1990.
US Fish & Wildlife Service. 2005a. News release: Status review completed: Greater sage-grouse not warranted for listing as endangered or threatened (7 January). http://www.fws.gov/news/News Releases/R9/4D0B98E6-DE22-18A8-5FBC25A1E928B298.html (accessed 24 February 2007).
———. 2005b. Critical Habitat: What is it? USFWS, Endangered Species Program (December). http://www.fws.gov/Endangered/listing/Critical%20Habitat.12.05.pdf (accessed 24 February 2007).
———. 2006. Endangered and threatened wildlife and plants; Final listing determination for the Gunnison sage-grouse as threatened or endangered; Final rule. Federal Register 7, no. 74 (18 April), National Archives and Records Administration. http://mountain-prairie.fws.gov/speci es/birds/gunnisonsagegrouse/FR04182006.pdf (accessed 24 February 2007).
Walker, Larry. 1999. About rangeland pearls. March 6. http://www.rangebiome.org/editorials/pearls .html (accessed 26 February 2007).
———. 2003. Once upon a SVIM. May 18. http://www.rangebiome.org/editorials/SVIM.html (accessed 3 February 2007).
Ward Jr., James P. 2004. Relationships among herbaceous plants, voles, and Mexican spotted owls in the Sacramento Mountains, New Mexico: A review of current knowledge and procedures for monitoring vole habitat at conditions in key grazing areas (draft report, 15 October). Alamogordo, NM: Rocky Mountain Research Station.
Wilkins, R. N., and W. G. Swank. 1992. Bobwhite habitat use under short duration and deferred-rotation grazing. *Journal of Range Management* 45:549–53.
Williams, Ted. 2006. A cautionary tale: Rogue ranchers threaten western trout water. *Fly Rod and Reel* 28, no. 3 (June). http://www.flyrodreel.com/index.php/page/issues/id/19123 (accessed 21 November 2006).
Woodard, Tori, and Patrick Diehl. 2001. Press release: Trespass cattle still on Fifty-Mile Mountain • Griffins overstocked the Lake Allotment • Escalante group calls for end to grazing on Fifty-Mile Mountain after documenting conditions with BLM on June 20–21. Escalante

Wilderness Project (25 June). http://www.rangebiome.org/headlines/nr/50miletrespass.html (accessed 24 March 2007).
Wuerthner, George. 1993. Malheur killing more predators. *Earth First!* 13, no. 3 (2 February): 8.
———. 2002a. Beef, cowboys, the West. In *Welfare ranching: The subsidized destruction of the American West*, ed. George Wuerthner and Mollie Matteson, 27–30. Washington, DC: Island Press.
———. 2002b. The donut diet: The too-good-to-be-true claims of holistic management. In *Welfare ranching: The subsidized destruction of the American West*, ed. George Wuerthner and Mollie Matteson, 291–3. Washington, DC: Island Press.
———. 2006a. Pyro cows: The role of livestock grazing in worsening fire severity. In *Wildfire: A century of failed forest policy*, ed. George Wuerthner, 197–98. Washington, DC: Island Press.
———, ed. 2006b. *Wildfire: A century of failed forest policy*. Washington, DC: Island Press.
Wuerthner, George, and Mollie Matteson, eds. 2002. *Welfare ranching: The subsidized destruction of the American West*. Washington, DC: Island Press.
Young, James A., and Darin D. Clements. 2004. Cheatgrass in the Great Basin. *Meeting abstract*. US Department of Agriculture, Agricultural Research Service (1 February). http://www.ars.usda.gov/research/publications/Publications.htm?seq_no_115=158636 (accessed 2 February 2007).
Young, J. R. 1994. The influence of sexual selection on phenotypic and genetic divergence of sage grouse. PhD dissertation, Purdue University, Indiana.
Young, Jessica R., Clait E. Braun, Sara J. Oyler-McCance, Jerry W. Hupp, and Tom W. Quinn. 2000. A new species of sage-grouse (Phasianidae: *Centrocercus*) from southwestern Colorado. *The Wilson Bulletin* 112, no. 4 (December): 445–53. http://www.western.edu/bio/young/gunnsg/Young%20et%20al%202000.pdf (accessed 24 February 2007).

Index

A-horizon, definition of, 356n11
Abbey, Edward, 301, 366nn7–8
ACLU (American Civil Liberties Union), 219
ADC (Animal Damage Control), 56, 275; impact on nontarget species, 56. *See also* Wildlife Services
alfalfa: fed to elk, 291; grown for cattle feed, 212; subsidy to ranchers, 301
Alighieri, Dante, 206
Allen, Janice, 332, 369n20
allowable use monitoring, 61–62
ALMRS (Automated Land and Mineral Record System), 108
American Indians, 47, 193; prehistoric settlement in southern Utah, 232–33. *See also specific native peoples*
American Lands Alliance, 346n12, 346n20
Amundson, Robert, 185
Andalex Coal Mine, 229
Anderson, Allen E., report on mule deer, 21
Andrus, Cecil D., 92
Anheuser-Busch Corporation, 321, 324; proposed boycott of products by, 329
animal damage control, 274; meaning of, 46. *See also* predator control
Animas Foundation, 193
anti-nuclear campaign, 205, 207–9, 322
AOU (American Ornithologists' Union) nomenclature committee, 16
APHIS (Animal and Plant Health Inspection Service), 50, 56–57, 289; preventive control of predators, 58
Arizona: Anderson Mesa, 70; Apache County, 348; Aravaipa Canyon, 140–41, 144; Cluff Ranch, 68; Day family has largest BLM allotment in, 278; habitat for desert topminnow, 48; drought of 1892, 276; Dutchwoman Butte, 75; Flagstaff, 85; Fort Huachuca, 68; Gila Bend, 277; Lake Havasu City, 141–42; Mearns' quail in, 71; native grasses replaced by exotics, 276; overgrazed allotments in, 72–73, 151; San Carlos Apache Indian Reservation, 140; Santa Cruz River, 276; Tonto Basin, 76; Tucson, 8, 53, 68, 140, 271–73, 281; Yuma, 277
Arizona Cattle Growers Association, 41
Arizona Cattlemen's Association, Steve Gallizioli's presentation to, 75
Arizona Department of Transportation, exotic grasses planted by, 276
Arizona Game & Fish Commission, appointees of, 43
Arizona Game & Fish Department, 1, 3, 43, 66–68, 73;

aerial gunning of coyotes, 70; conflicts with BLM and Forest Service, 69; poisoning of coyotes, 70
Arizona Republic, 72
Arrington, O. N., 67
aspen: reproduction threatened by cattle, 42, 234; reproduction threatened by elk, 42; fire-resistance of, 42; grove on Boulder Mountain (UT), 234–35; recovery after removing cattle, 269; relict groves of, 224–25; reproduction of, 177
Asrow, Edith, 172–73, 176, 179
Austin, Miriam, 136
Australia, overgrazing in, 211

Babbitt, Bruce (as Arizona governor), 73
Babbitt, Bruce (as interior secretary): advocates for rangeland reform, 88–90; establishes "Potemkin village" national monuments, 229; hanged in effigy, 213
Baca, Jim, 89, 92
Bailey, Jeff, 332
Bandelier National Monument (NM), 97, 100
Bannock people, 113
Barber, Douglas K., 2, 348n20
Baxter, Clay, 40
Baxter, Jane, 325
Bedell, John, 43, 348n33, 348n35, 349n36
beef operation versus cow-calf, 141
Belsky, Joy, xiii, 184, 261, 363n17, 363n23; applies for position at BLM/Oregon State U., 188; opposes killing coyotes at Hart Mountain NAR (OR), 189–90, 203–4; education and research, 185–87; interest in grasslands, 187–88, 197–99; lawsuit against BLM, 188–89; article about livestock grazing in forests, 190; as mentor, 196–97; joins ONDA, 202; joins ONRC, 201–2, 358n5 (Marlett); speaking engagement cancelled at Texas Tech, 191; article about livestock spreading weeds, 199–200
Bennett, Bob, 214
Bennett, K. Lynn, 352n25
Beyond the Rangeland Conflict (Daggett), 368n21
Bingaman, Jeff, as recipient of banking industry contributions, 313
Biodiversity Conservation Alliance, 361n11
Biodiversity Legal Foundation, 346n12, 346n20
birds: DDD contamination of, 259; grassland species, 107, 319; impact of livestock on, xix, 18, 22, 260–61, 326; raptorial, 257; riparian habitats for, 81, 326. *See also specific birds*

bison, Yellowstone: brucellosis as political smokescreen, 289–91; hunting of, 293–94; proposal to expand range, 291. See also buffalo
Black Hills (SD): buffalo in, 335; gold discovered in, 336; desecrated by livestock grazing, 338; photographed in 1870s, 41; sacred sites in, 340; unique speciation in, 342
BLM (Bureau of Land Management): Air Force bombing range (ID), 92–93; allotment evaluations, 91, 239; allotment management, 50, 63; area of land managed by, xviii; avoids data acquisition, 155; Congressional pressure on, 275; decisions made at secretarial level, 94, 150; district consolidation in, 64; Ecological Site Inventory, 155; 8100 funding, 149, 151; employees, nonsupport of, 102–4, 144; ephemeral-grazing policy, 277; Escalante (UT) visitor center, 213–14; Great Basin Restoration Initiative, 65; efforts to protect Gunnison sage-grouse, 12; internal conflict at, 104; litigation against, 87; mismanagement of, 107–8; monitoring, fraudulent, 146, 226; monitoring under Reagan administration, 155; Oregon State Office, 154–55; receives bad publicity over Patrick Diehl's arrest, 219; politics in, 155; poor planning by, 107–8, 247; pro-ranching bias of, 102–4, 145–47, 149, 155–56, 226, 228, 272, 302; RAC Standards, 64; rangeland health (Carter administration), 155; Rangeland Reform '94, 83, 89–91; difficulty of meaningful reform, 274; RMP (resource management plan), 54, 64–65, 145, 152; RNA (research natural area), 143; Sharptop Fire (OR), 147–48; illegal subleasing of grazing permit, 103; SVIM (Soil Vegetation Inventory Method), 87, 154; unrealistic goals of, 65; uses obsolete range surveys, 152; Vale Project (OR), 147, 354n21; Washington (DC) Office, 34, 127, 154, 170; wilderness specialist in, 142
BLM (Bureau of Land Management) management of specific sites: Aravaipa Canyon Primitive Area (AZ), 140–41; Burns District (OR), 143; Comb Wash Allotment (UT), 323, 369n7; Cook Well (OR), 147, 354n18, 354n20; Ely District (NV), 350n25; Gunnison Resource Area (CO), 13; Havasu Resource Area (AZ), 144; Kanab Resource Area (UT), 85, 228–29; Lake Allotment (UT), 359n20; Lakeview District (OR), 142–43, 145–49, 152; McKenzie Resource Area (OR), 195; Medford District (OR), 154; Prineville District (OR), 141, 154; Rio Puerco Resource Area (NM), 101; Rock Creek/Mudholes Allotment (UT), 220–24; Steep Creek Allotment (UT), 225–28; Tomichi Allotment (CO), 14
Block, Bill, 355n10
Bloedel, Ed, 175, 178
Bonneville Power Administration, 295
Boulder Regional Group, 231; environmental monitoring by, 247
Boyce Thompson Institute for Plant Research, 186
Boyd, William, xvii
Bradbury, Jack, 11

Braun, Clait E., 8; discovery of Gunnison sage-grouse, 10–13; naming of Gunnison sage-grouse, 16–17; Professional review of: "Connelly, J. W., S. T. Knick, M. A. Schroeder, and S. J. Stiver. 2004. Conservation assessment of greater sage-grouse and sagebrush habitats. Western Association of Fish and Wildlife Agencies," 26; Review of: "Western Governor's Association 'Conserving the greater sage grouse: A compilation of efforts underway on state, tribal, provincial and private lands,'" 26; predictions about sage-grouse listing, 26
Brendle, Scott, 367n13 (Royle)
Brigham Young University, connection to proposed Escalante Center, 213
Brock, Darien, 367n13 (Royle)
Brown, David E., 351n13 (Gallizioli)
brucellosis: in bison south of Yellowstone National Park (WY), 289; management in Yellowstone bison, 289–91; testing for, 292–93; transmission from cattle to bison in Yellowstone National Park (WY), 289; transmission from elk to cattle, 290
brucellosis-free status, meaning of, 289
buffalo, 337–39; nineteenth century extermination of, 47; as big brothers, 335–36; calves fed cow's milk in Yellowstone National Park, 290; grazing of, 44. See also bison, Yellowstone
Buffalo Field Campaign, 294
Buffington, Bob, 144
Bulloch, Mary: as beneficiary of cattle stolen from BLM, 223; BLM impounds cattle of, 222; final settlement with BLM, 360n26; grazing allotment closed by BLM, 223; as permittee of Rock Creek/Mudholes Allotment, 220, 225; refuses BLM order to remove cattle from allotment, 221–22; trespass cattle of, 223; trespass fees and impoundment costs assessed by BLM against, 360n24
Bureau of Indian Affairs, administers leasing on reservations, 341–42
Bureau of Reclamation, 137
Burns, Conrad, 366n9
burros, as poisoned bait for coyotes, 70
burros, feral, 97, 101
Busch III, August, 330, 369n16
Bush, George W., 183, 265; purchases ranch, 296
Bush (George W.) administration, 41, 48; diminishes authority of BLM state directors, 93, 150; diminishes public participation in land management, 94; pro-ranching bias of, 302; reduction of domestic budgets, 47

California: Alturas, 171, 173; Cedarville, 171, 179; Kern Plateau, 323–24; Mill Creek, 171–73, 178–79; Needles, 209; Pine Creek Basin, 178; Pit River, 174; Sacramento River, 174; Santa Cruz, 322; South Parker Creek, 174–75; Ward Valley, 210; Warner Mountains, 171–72
California Department of Fish and Game, 329
California Grazing Reform Alliance, 324
California Mule Deer Association, 324

California State University–Fresno. *See* Fresno State
California Trout, 321, 325–26, 328, 331, 333
Camarena, Abel, 367n16
Cannon, Kate, 222, 225, 238, 360nn3–4
Carpenter, Forrest, 348n34
Carson, Rachel, author of *Silent Spring*, 51, 259
Carson Forest Watch, 350n22
Carter (James Earl) administration, 154–55
Cassidy, Hopalong, xvii
cattle: counting of on grazing allotments, 80, 100, 118–19, 135, 248–49; supplemental feeding of, 241
cattle rustling, 103
CBD (Center for Biological Diversity), 281, 350n22, 361n11
CDOW (Colorado Division of Wildlife), 10, 12, 15
Centrocercus minimus. *See* sage-grouse, Gunnison
Centrocercus urophasianus. *See* sage-grouse, greater
Chisholm, Daniel K., 175, 356n10
chuckwalla, 278
civil disobedience: against BLM grazing management, 219; against nuclear weapons, 207
Clean Water Act: as basis for litigation, 32; enactment of, 48
Clergy and Laity Concerned, 322
Clifford, Frank, 331
Clinton, William J.: establishes "Potemkin village" national monuments, 229, 360n31; hanged in effigy, 213
Clinton (William J.) administration, 95; bison management plan of, 292; changes BLM's field activity, 155; initiative to alter grazing policy, 88; Roadless Area Review, 171; Roadless Area Rule, 46, 171
Clyde, Don, 181
Cobell Case (*Cobell v. Norton*), 341, 370n8
Cody, William F., xvii
College of the Atlantic, 231
Colorado: Archuleta County, 346n15; Cerro Summit, 21; Cimarron, 21; Conejos County, 346n15; Costilla County, 346n15; Crawford, 12; Dolores, 21; Dolores County, 12, 19; Dove Creek, 21; Dry Creek Basin, 12; Eagle County, 18, 24; Garfield County, 18; Glade Park, 21; Grand County, 24; Gunnison Basin, 10, 12, 19, 21; LaPlata County, 346n15; Miramonte, 19, 21; Miramonte Basin, 346n19; Moffat County, 24; North Park, 24; Ouray County, 18; Pinon Mesa, 21; Pitkin County, 18; ranching community in, 89; Routt County, 24; Uncompahgre Plateau, 21–22
Colorado State University, 8, 10, 12, 97, 155, 158
Comb Wash decision, 323
Commons, Michelle, 12
Compound 1080, 274; misuse of, 70
Conroy, Scott, 171–72, 356n9
Conservation Assessment of Greater Sage-grouse and Sagebrush Habitats (WAFWA): genetic information missing, 25; "paper birds", 25; errors in sage-grouse population data for Colorado, 24; estimate of sage-grouse population, 24–25; purpose of, 23
conservation easements, limitations of, 20

Conservation Value, 192
Continental Divide Trail, 318
Cooper, Jim, 347n1
Cornell University, 185–86, 192–93
county commissioners, 14, 60, 114, 116, 123, 164, 235; as ranchers, 235
cowboy: mystique, xvii, 301; myth of, xviii, 344n19
cowboy caucus, 223, 245. *See also* ranchers: political/social clout of
coyote getter, 70. *See also* M-44
coyotes: aerial gunning of, 70, 244, 246; at Hart Mountain NAR, OR, 189–90, 203–4, 264–65; poisoning of, 57, 366n8; politics behind support for killing, 264–65; predation on livestock, 46; as target of predator control, 56; preemptive killing of, 46, 58; response to population reduction, 70, 246; made scapegoats for feral dogs, 151; made scapegoats for poor livestock management, 150–51; Utah bounty on, 244–45
Craig, Larry, 93; suspected involvement in Martha Hahn's reassignment, 95
crane, greater sandhill, population at Malheur NWR, 261
creosote bush, 39
critical habitat: definition of, 345n10 (Braun); importance of, 15
Cross, Sally, 189
Croxen, Fred W., 76
cryptobiotic crusts, 240
Custer, George Armstrong, 41

Daggett, Dan, 318; author of *Beyond the Rangeland Conflict*, 368n21
Davis, Gray, 210
DDD (dichloro-diphenyl-dichloroethane), 259
DDT (dichloro-diphenyl-trichloroethane), 258–60
deer, as hunting license revenue generators, 243–44
deer, mule, 22, 138, 326; as hunting license revenue generators, 21
Defenders of the Black Hills, 335
Defenders of Wildlife: as manager of Wood brother's property, 140; Steve Johnson employed by, 271
Deseret Livestock Ranch (UT), 231–33
desert topminnow, 48
desert tortoise, impacted by livestock grazing, 278
Desert Tortoise Council, Steve Johnson consults for, 273
Desseaux, Lee, 323
Diamond Bar Cattle Company, 367n8, 367n17
Diehl, Patrick, 183, 205; as officer of Glen Canyon Group, Sierra Club, 227; as Green Party congressional candidate, 228; property vandalized, 212–16; shunned by environmentalists, 216; tabling at BLM visitor center, 218–20
Dombeck, Michael, 30–31, 89, 348
Domenici, Pete, 61, 316; banking industry political contributions to, 313
drought: in Arizona (1892), 276; consequences for aspen in southern Utah, 224; impact on forage

production, 127; impact on wild horses, 300; sage-grouse evolved with, 14, 18; in South Dakota, 337, 342; in southern Utah, 220, 223, 240; water supplied to cattle during, 146
Dudley Do-Right, 354n12
Duke University, 192
Durkovich, Steven, 367n1

"Ecological Impacts of Feral Burros on Bandelier National Monument, New Mexico, The" (Koehler), 101
EA (environmental assessment), 58; No Grazing Alternative omitted, 37
eagle, 337; golden, 56, 258
Earth Day (1970), 100, 271
EarthFirst!, 322–23
Earth Justice. *See* Sierra Club Legal Defense Fund
ecosystem services, 36
Ecumenical Peace Institute, 368n1
elk: thwart aspen reproduction, 42; brucellosis in, 290; not native to desert grasslands, 275; environmental damage caused by, 244; grazing of, 42; hunting of, 243, 294; politics of management, 30, 43; disliked by ranchers, 244, 366n8; made scapegoats for cattle, 42
Ellett, Kent, 360n28
Elmore, Wayne, 120, 353n10
Elshoff, Alice, 201
Engel, Gerry, 311
environmental organizations, approach to conservation, 228
ESA (Endangered Species Act), xix, 15, 32–33, 48, 61, 192, 286, 330; ignored by Forest Service, 33
Escalante Center, 213–15
Escalante House, 211
Escalante Wilderness Project, 211, 221, 247

FACA (Federal Advisory Committee Act), 326
Fager, Leon, 1, 30, 61
Farm Services Agency, 121
Federal Intermediate Credit Bank of Texas, 312
Federation of Fly Fisher Clubs, 325
fences on grazing allotments: adverse environmental impacts of, 28; cost of, 5
Ferguson, Denzel, 260, 362n10
ferret, black-footed, 44, 274, 337, 370n5
Festuca idahoensis, 186
Festuca ovina, 186
fire: in forests, 42; in grasslands, 106, 132–33, 176; grazing of livestock after, 132–33; suppression of, 106, 133
fish, native, 36, 138, 141
Flora, Gloria, 54, 349n5
FLPMA (Federal Land Policy and Management Act), 54, 83, 87, 142
FOIA (Freedom of Information Act): difficulty in using, 239, 250–51; Todd Shuman's use of, 324; violation of by US Forest Service, 311
Fontaine, Joe, 327, 329, 333, 369n12

forage production decline, 177, 239
forage production variation, 126–27
forage utilization: in Beaty Butte area (OR), 146; in Grand Staircase-Escalante National Monument (UT), 220; on Mulkey Allotment and Whitney Allotment (CA), 324
Forest and Rangeland Renewable Resources Planning Act, 37, 54
Forest Conservation Council, 347
Forest Guardians, 162, 347n8, 350n22, 360n8, 368n24
Forest Plan (computer software), 38
forests: clearcut, 65, 195, 252, 266; old-growth, defense of, 269; park-like, 41
Fort Apache Indian Reservation, 42
Fort Laramie Treaty of 1868, 336, 339
fox, 56; gray, 57–58, 349n11; red, 19
Frank, Tim, 328, 331
Fresno State, 139–40
Friends World College, 51
Fritz, Freddie, 3
Fulton, Tom, 95
Fund for Animals, 101

Galeano-Popp, Renee, 50; resignation from Forest Service, 61, 347n2
Gallizioli, Steve, 66; experiences with holistic resource management, 74; presentation in Wash., DC, 71
Gelbard, Jonathan, 192
Gerity, Art, 150
Gila Watch, 170, 308–10, 314, 317
Gilman, David, 77
Glacier National Park (MT), grizzly bear recovery at, 288
Glen Canyon Dam (AZ): campaign to decommission, 217; environmental studies on, 84
global climate change, 230
Golden Trout Wilderness Protection League, 327
Graf, William, 257
Grand Canyon Trust, 85, 225–27, 247
Grand Staircase-Escalante National Monument (UT), 233, 254; absence of allotment management plans, 239; aerial gunning of coyotes at, 246; cattle numbers at, 248–50; conflict between cattle grazing and recreation, 253; cryptobiotic soil at, 240; establishment of, 228; grazing EIS, 238, 247, 253; grazing staff connections to ranching industry, 237; monitoring, lack of, 249–50; obtaining information from, 250; school trust lands, 243; staff involved with Forest Service grazing management, 247; understaffing at, 238; vegetation on, 239; wilderness study areas at, 241; winter grazing at, 240–41, 247
Grand Teton National Park (WY), 291, 295
grass, exotic, 22; planted by Arizona Department of Transportation, 276; crested wheatgrass, 139, 146–47, 168, 240
grass, native; bunchgrass, 147; Indian rice grass, 240; reduction of from livestock grazing, 112; seeding of, 148, 240

grass, rhizomatous, 22
grasslands, public indifference to health of, 269
Gray Ranch (NM), 193
grazing fee, federal, 36, 56, 235, 317; minimum set by executive order, 279
grazing fee, private, 317
grazing management, 64, 144, 164, 187, 228
grazing permit, federal: banking industry connection to, 274, 278, 312–13; "blue sky" value of, 6–7; buy-out proposal for, 269; economic impact of reduction, 7; as privilege, 312; transfer of, 6; violation of, 113, 117–19, 246
Greater Yellowstone Coalition, 282–83
Great Plains Grasslands Wilderness Act, 339–41, 370n6
Great Sioux Nation, 335, 337
grebe, western: die-off from DDD, 259
Greenpeace, 209
Griffin, Brent: trespass fees and impoundment costs assessed by BLM, 360n24
Griffin, Gene, 359n20; trespass fees and impoundment costs assessed by BLM, 360n24
Griffin, Michael, 367n13 (Royle)
Griffin, Quinn, 222, 359n20; trespass fees and impoundment costs assessed by BLM, 360n24
Griffin family, 220–22; federal grazing allotment permitted to, 223–25
grizzly bear: politics of recovery, 288; conflicts on Walton Allotment, 295
grouse, blue, 22
grouse, Columbian sharp-tailed, 22
Grouse Inc., 8
Gruenewald, Roger, 73
Guadalupe Canyon Ranch, 194
Gunnison Sage-Grouse Working Group, 13; failure of, 14
Gunsmoke, xvii

Hadley, Drum, 194
Hahn, Martha, 83; removal from BLM Idaho state directorship, 94–96
Hart Mountain NAR (OR): aspen recovery after removing cattle, 269; bird banding at, 257, 269; Buck Pasture (livestock exclosure), 262–63; cattle grazing at, 261–64; efforts to prevent killing coyotes at, 189–90, 203–4; proposal to kill coyotes at, 264–65; removal of cattle from, 264; workshops at, 263–64
Harvard University, 205–6
Hatch, Julian, 231; as member of Glen Canyon Group, Sierra Club, 227; experiences with cattle grazing, 235; finds trespass cattle in Steep Creek Canyon (UT), 225; meets with GSENM manager David Hunsaker, 225; revenge killing of dog by rancher's son, 237
Hatfield, Doc, 270
Hays, John, 72
Healthy Forests Initiative, 41, 348n28
Healthy Forests Restoration Act. *See* Healthy Forests Initiative

Hedden, Bill, 227
Herman, Steven G., 183, 255
holistic management: detrimental to ground-nesting birds, 22. *See also* holistic resource management; Savory grazing method
holistic resource management, 74–75, 203. *See also* holistic management; Savory grazing method
Horning, John, 368n24
horses, as poisoned bait for coyotes, 70
horses, wild: BLM's management and adoption program, 108, 304–5; and cattle exclosures, 148; estray, 299, 366n6; Nevada's 2nd state animal, efforts to designate as, 307; in Oregon, 145–46; Pryor Mountain Herd (WY), 366n2; rancher attitudes about, 300–301; Bobbi Royle's experiences with, 297; slaughter of, 304–5; stabilizing populations of, 109; violence committed against, 305–6
Howard Ranch (NV), 353n3 (Phillips)
HSUS (Humane Society of the United States), Steve Johnson as consultant for, 273
Hughes, Charles, 346n19
Humboldt State University, 158
Hunsaker, David, 225–26, 238, 360n4
Hunter, Jamie, 328–29
Hunter, Roy, 324
hunters, opposition to wolf reintroduction, 285
Hupp, Jerry, 11–12, 16
hyperdisease, 352n14

IBLA (Interior Board of Land Appeals), 303, 362n21
Idaho: Bear Valley Creek, 295; Boise, 105; Hailey, 105; ranching community in, 89; Shoshone, 104; South Hills, 134, 138; Twin Falls, 105
Idaho Environmental Council, 283
Idaho State University, 183, 282
Idaho Watersheds Project, 91, 97, 105. *See also* Western Watersheds Project
Indian paintbrush, 242
Indian reservations: corruption on, 341–42; as POW camps, 336, 341
Integrated Resource Management, 58
invasive plants: cheatgrass, 65, 107, 146–47, 165, 183, 221, 230; hoary cress, 106; larkspur, 147; leafy spurge, 122, 136; and livestock grazing after fire, 133, 147; sagebrush, 18, 28; sneezeweed, 65

Jacobs, Lynn, 322; author of *Waste of the West*, xviii, 211, 368n2
Jantzen, Bob, 72, 350n7
Jarbidge Sage-Grouse Local Working Group, 136
Johnson, Rex, 309
Johnson, Steve, 271; chair of Arizona Chapter, Sierra Club, 273; employed by Defenders of Wildlife, 274; learns of banking connection to federal grazing permits, 274–75

Kansas State University, 8–9
Kaschke, Marvin, 262
Kelly, Betty, 306, 367n12 (Royle)

Kempthorne, Dirk, 95
Kerr, Andy, 202, 358n4 (Marlett)
King, Bruce, 103
Knapp, Roland, 327, 330
Knick, Steve, 23
Koehler, David A., 97; conflict with Hispanic ranchers, 101; author of "The Ecological Impacts of Feral Burros on Bandelier National Monument, New Mexico," 101
Kozacek, Sue, 315

La Alianza de las Razas, 102
Laney, Kit, 367n15; applies political pressure to Forest Service, 315–16; assaults federal officer, 368n19; sues Forest Service, 317
Laney, Sherry, 315
Laney Cattle Company, sues Forest Service, 367n17
Leaver, Gale, 147
Leonard, Steve, 175
Leopold, Aldo, 6, 203, 257, 310, 362n4
Leopold, Aldo Starker, 257, 362n3
Lewis and Clark Expedition: diaries from, 343; sage-grouse discovered by, 16
Libby, Steve, 320, 368n25
litigation, pro-environmental; importance of, 20, 32; limitations of, 251–52
Livermore Action Group, 207, 209
livestock-free areas: Bruneau River (NV), 129; Dutchwoman Butte (AZ), 75–76; Three Bar Wildlife Area (AZ), 39. *See also* livestock exclosures
livestock exclosures, 149; Buck Pasture (OR), 262–63; Dry Gulch (ID), 119–21; Trout Creek (ID), 115. *See also* livestock-free areas
livestock grazing: destroys archaeological sites, 113; thwarts reproduction of aspen, 224–25, 234, 263; destroys cryptobiotic crusts, 240; damages national grasslands, 339; alters fire regimes, 41; flooding resulting from, 233, 277; depresses Mearns' quail populations, 71; degrades riparian zones, 35, 113, 120, 126, 128–29, 242, 262–63, 276, 310, 318; compresses soil, 78
livestock production, xviii; modifies sage-grouse habitat, 18; has negative impact on sage-grouse, 28; species harmed by, xix
livestock reductions, 172, 177–78
livestock trespass, 102, 144; in Aravaipa Canyon (AZ), 144; notices issued by BLM, 221; on Diamond Bar Allotment (NM), 318; in Grand Staircase-Escalante National Monument (UT), 221–23, 225–27, 249; on Julian Hatch's property, 236–37; in Malheur NWR (OR), 260; in South Warner Wilderness (CA), 175, 179
local control in public lands management, 265
Los Angeles Times, 331

M-44, 56–58. *See also* coyote getter
Majors, A. K., 145, 152
Malheur Field Station (OR), 260, 362n11
Malheur NWR (OR), 362n9; cattle destroy plover nests at, 260; cattle grazing at, 363n13; forage removed by livestock, 362n12; managers believe in value of livestock grazing, 261; predator control at, 261, 363n15
malignant melanoma, 48
Manifest Destiny, 47
Mansfield, Nick, 299
manzanita, 68
Maricopa Audubon Society (AZ), 350n22
Marlett, Bill, 201
Marshall Scholarship, 206, 359n2
Marvel, Jon, 105–6
Mather, Jim, 40
Matzke, Brett, 324, 329
Maughan, Ralph, 183, 282
McConkie, Andy, 181
McDonough, Nick, 43, 46
McGinn, Dano, 324
McKee, Lucinda (Luci), 332, 369n19
McNaughton, Sam, 186–87
MC Ranch (OR), 145
MC User's Group (OR), 146
McVickers, Gary, 144
Merlino, Anthony, 367n13 (Royle)
mesquite: dies from drop in water table, 277; as increaser due to cattle grazing, 194; white wing dove study, 68
Mewaldt, Richard, 257
Michigan State University, 30
Miller, Sally, 327
Minckley, Wendell Lee, 141
Mix, Tom, xvii
Mobilization for Survival, 207
Mojave Tribe, in Ward Valley Campaign, 209
Mono Lake, campaign to save, 327
Montana: Cooke City, 286; Gardner, 286
Montana Department of Livestock: involvement with bison management plan of 2000, 292–93; motivation for killing bison, 290, 292; rationale undermined for killing bison, 293
Montgomery, Levi, 181
Morganweck, Ralph O., 346
Mormon Church, 181, 211, 232, 243; livestock grazed in Utah, 232; owns Deseret Ranch, 231–32; Pioneer Day, 216, 359n19; Relief Society, 181, 211
mouse, deer, 159–61
Muecke, Carl, 347n8
Muir, John, 323, 368n5, 369n6
multiple-use management: in BLM, 151; in US Forest Service, 4, 39–40, 100
multiple chemical sensitivity, 209, 211, 214, 359n12
MUSY (Multiple-Use Sustained-Yield Act), promotion of livestock grazing under, 252

National Audubon Society, Steve Johnson consults for, 273
National Cattlemen's Association, 181
National Elk Refuge, WY, 291
national grasslands: polluted by cattle, 339; Forest

Service policies for managing, 338; mineral extraction on, 339
National Park Service, complicit in slaughter of bison, 293–94
National Riparian Task Force, 175
Nature Conservancy, 53, 193, 202
Naugle, Dave, 27
NEPA (National Environmental Policy Act), 33, 268, 351n3, 364n32 (Herman), 364n15 (Johnson), 369n7; process, 54, 57–59, 230, 323–24, 330, 333
Nevada, 129, 256, 291; BLM management in, 64; Carson City, 297; Devils Canyon, 305; efforts to designate the wild horse as 2nd state animal, 306; Fallon, 303; leafy spurge quarantine area, 122; Lovelock, 304; number of wild horses in, 302; overgrazed livestock allotments in, 151; race horse operations in, 212; Reno, 157; Sheldon NWR, 255; Storey County, 299, 306–7; Virginia Range, 298, 366n6; wild horses in, 297–300
Nevada Agriculture Department, 298–99
New Mexico, 31; Alamogordo, 158; Albuquerque, 30, 101; Bandelier National Monument, 100; Black Canyon, 313–14, 367n9; BLM management in, 103, 145, 149; Burro Mountains, 318; Casa Salazar, 102; desert environment in, 44; Game and Fish Commission in, 43; Gray Ranch, 193; Guadalupe Mountains, 58; Mimbres Valley, 309; mining in, 31; politicians from who receive banking industry contributions, 313; ranchers in, 280; Reserve, 349n7; Sacramento Mountains, 158; Silver City, 308–9, 314, 316–18; wildlife in, 275
New Mexico Cattle Growers, 41
New Mexico Department of Agriculture, 57
New Mexico Natural Heritage Program, 193
New Mexico State University, 153, 162
NFMA (National Forest Management Act), 54, 348n19
Nixon, Richard, 268, 274, 364n4
North Cascades National Park (WA), 185–86
Northern Arizona University, 50
Northern Continental Divide Ecosystem, grizzly bear recovery in, 288
Norton, Gale, 90, 288
NRCS (Natural Resources Conservation Service): negative actions for Gunnison sage-grouse, 14. *See also* SCS (Soil Conservation Service)
NRDC (Natural Resources Defense Council), 280, 331, 333–34, 351n3, 364n15
NRDC v. Morton, 351n3, 364n15
Nuclear Resister, The, 208–9, 359n11
Nunn, Mike, 204, 358n16
NWF (National Wildlife Federation), 365n10, 366n14; attorney from assists Gila Watch, 311; funds buyout of grazing allotments, 295

O'Connor, Sandra Day, 278, 364n12
oak, turbinella, 68
Oceti Sakowin, 335–37
Oglala Sioux Tribe, 335

Ohmart, Robert (Bob), 316
Oman, Don, 1, 110; conflicts with ranchers, 112–19, 134; counts cattle at roundup, 118–19, 134–35
ONDA (Oregon Natural Desert Association), 184; Joy Belsky employed at, 201–2; participates in stopping coyote hunt at Hart Mountain NAR, 190
ONRC (Oregon Natural Resources Council), 129; Joy Belsky employed at, 189, 201–2
open-range laws, 236, 308–9
Oregon: Beaty Butte, 145, 148; Bend, 201; Eastern, 126, 128–29; Lakeview, 139, 265; Portland, 153, 185; Pueblo Mountains, 151; Spaulding Reservoir, 149; Trout Creek Mountains, 151; Warner Valley, 150; wolves in, 288
Oregon Fish and Game Department, 67
Oregon Natural Desert Association and The Wilderness Society v. Hart Mountain National Wildlife Refuge, 363
Oregon State University, 66–67, 125; Joy Belsky applies for position at, 188–89; Steve Herman's experience with range ecologist from, 262–63
Oregon Trail, 128, 132
Oregon Trout, 129
outfitters, opposition to wolf reintroduction, 285
overgrazing by livestock: after fire, 133, 147; on Apache NF (AZ), 6; environmental impacts unrecognized by nonexperts, 65, 295; environmental impact on Gunnison sage-grouse, 18; on Indian reservations, 341–42; on Lakeview District (OR), 146; on Lincoln NF (NM), 163; on Minidoka NF (ID), 112; promotes spread of sagebrush, 28; promotes soil erosion, 3, 41, 174; on South Warner Wilderness (CA), 171–79; on Twin Falls Ranger District (ID), 112–14
owl: burrowing, 44; Mexican spotted, 33, 55, 60, 158–63; northern spotted, 158–59
Oxford University, 205–6
Oyler-McCance, Sara, 12, 16

Panorama Ranch (AZ), 353n4 (Troutman)
Paragon Foundation, 60, 162
Parker 3-Step Transect, 167, 173
Peace Corps, 186
Pearce, Steve, 31
PEER (Public Employees for Environmental Responsibility), 348n35
Phillips, Robert W., 125
Pine Ridge Reservation (SD); land leasing on, 341; prairie dog poisoning on, 337, 370n4; request for land audit of, 342
pinyon-juniper, 194
plant pedestalling, 79, 146, 221, 239
Pleistocene Epoch: duration of, 364n6; horses present in North America during, 109; southwestern desert grasslands as relicts from, 276; widespread extinction at conclusion of, 352n14
politics: in greater sage-grouse listing decision process, 26; in Gunnison sage-grouse listing determination, 346n14; of the West, 1, 138

Pope, Carl, 330
Potemkin, Grigori Aleksandrovich, 360n31
Potemkin village, 360n31
Potter, Loren, 101
prairie dogs: attracted to grazed sites, 44; killed by ranchers, 366n8; misinformation about, 45; poisoning of, 370n4; scapegoats for livestock industry, 337
predator control: aerial gunning at Grand Staircase-Escalante National Monument (UT), 244–46, 361n19; on Lincoln NF (NM), 56–57; on Malheur NWR (OR), 261; trapping, 244. See also ADC (Animal Damage Control); animal damage control
Predator Defense Institute, 264, 363n23
Predator Project, 57, 322
prickly poppy, Sacramento, 60, 349n16
private lands grazing, 6
pronghorn antelope: on Anderson Mesa (AZ), 70; at Hart Mountain NAR (OR), 190, 203–5, 262, 264–65; disliked by ranchers, 366n8
Prunty, Jim, 1, 130; advocates for public lands, 134
ptarmigan, white-tailed, 9
public lands: privatization of, 47; short-sighted management of, 106
public lands ranching: power of banking industry behind, 275, 313; beef produced by, 279; and community stability, 6; conservationist bidding on state land leases, 270; economic insignificance of, xix; as feudalism, 296; government cost overrun, xix, 36; justification for continuing, 234; as degrader of landscape aesthetics, 270; as part-time occupation, 280; predicted decline of, 230; reluctance of environmental organizations to address, 294–95; subsidizing of, 6, 138, 235, 279, 301; as welfare, 6, 75, 241, 301, 317
Pyle, J. Howard, 350n2

quail, Mearns', population depressed by livestock grazing, 71–72
Quivira Coalition, 308, 318, 320, 368n22

rabbit, Columbia Basin pygmy, 265–68; extinction of, 267, 363n28
rabbit, pygmy, 352n9; release from captive breeding program, 364n29
rabbit, value of predators to control populations, 150
RAC (resource advisory councils), 90, 351n10
RAC Standards, 64
Rahall, Nick, 366n9
Rampton, John, 369n7
ranchers: involvement in removal of Kate Cannon, 238; as cultural icons, 295; suspected involvement in Martha Hahn's reassignment, 96; violate grazing permit dates, 113, 116, 246; political/social clout of, 124, 134, 144, 191, 278, 316–17
ranching, consolidation of, 236
Ranch Life and the Hunting-Trail (Roosevelt), xvii
RangeBiome, 153, 156–57
Range Improvement Task Force, 162, 316

rangeland improvement projects: chaining, 239; fence construction, 145; lakebed pits, 145; reseeding, 145–46, 240; water pipelines, 145
rangelands, western, 26
range management: Forest Service as leader in, 170, 172; Oregon Trout as advocate for better, 129; influenced by politics, 69
RangeNet, 153, 156–57
Range Watch, 325
Reagan, Ronald, 350n4, 355n4 (Walker); end-of-world rhetoric of, 208; establishes minimum federal grazing fee, 279; purchases ranch, 183, 296; as "Sagebrush Rebel," 102
Reagan (Ronald) administration, 108
recreation, 232, 319; conflict with livestock grazing, 85, 113, 134, 178, 253–54; funding necessitated by livestock grazing, 5; income to local communities, 37–38; public lands needed for, 46; use of western rangelands, 27
Red Rock Wilderness Act, 217
Red Willow Research, 136
Reiswig, Barry, 263, 357n14
Reno Animal Control, 305
Reveal, Jack, 258
Rhodes, Dean, 306
Richard Mewaldt, 255, 257
rilling, 79
riparian zone: healthy, 42, 81; livestock-degraded, 35, 42, 82, 113, 177–78
RNA (research natural area), 143
Roadless Area Review, 171
Roadless Area Rule, 46, 171
Rocky Mountain Elk Foundation, 129
Rogers, Paul, 331
Rogers, Roy, xvii
Roosevelt, Theodore, xvii, 48, 233
Royle, Bobbi, 297
Ruch, Jeffrey, 348n35
Russell, Sharman Apt, 344n19
Russian thistle. *See* tumbleweed

sage-grouse, greater: discovery by Lewis and Clark Expedition, 16; damaging effects of gas and oil drilling on, 27; habitat factors in reproductive success, 264; damaging effects of I-80 on, 25; politics in listing process for, 26; dependence on sagebrush, 352n9; mortality from West Nile virus, 27–28; wing-collection program for, 10
sage-grouse, Gunnison: discovery of, 10–12; efforts to protect, 11–13; evolution of, 13, 18; factors in decline of, 17–18; as highly endangered, 20; historical habitat, 21; hypothesized impacts of listing as threatened or endangered, 15; benefits to listing, 20; naming of, 16–17; petition to list, 17; effective population size, 22; population decline of, 15, 17–20; predators of, 18; setbacks to protection, 14–15; sensitive species for BLM, 15; vulnerability to West Nile virus, 28; wing collection program, 10, 21

sagebrush: decline of, 239; eradication of, 10, 168, 176; spread of, 18, 28, 112, 168, 176; restoration of, 239
Sagebrush Rebellion, 102
Sagebrush Sea Campaign, 360n8
Sage Project, 83
San Jose Mercury News, 331
San José State University, 67, 255, 257
Sauber, Mike, 169, 308; co-founds Gila Watch, 309; attends meeting of Quivira Coalition, 318–20
Savage Jr., William W., xviii
Savory, Allan, 22, 44, 74–76, 203, 349n38, 350n10
Savory grazing method, 44–45, 74. *See also* holistic management; holistic resource management
Sawtooth National Forest Plan, based on old data, 80
Scharff, John C., 261, 363n16
Schock, Susan, 170; forms Gila Watch, 309–10; harassed by sheriff, 314–15
School and Institutional Trust Lands Administration, 242
school sections, 361n14, 361n16; grazed by cattle, 242; traded out of Grand Staircase-Escalante National Monument (UT), 243
Schroeder, Mike, 23
Schultz, William P., 367n6
scientists: put value on ecosystem services, 36; as threat to livestock industry, 213; nature of, 198
SCS (Soil Conservation Service), 9, 120. *See also* NRCS (Natural Resources Conservation Service)
Shea, Patrick A., 93, 351n18
sheep, bighorn: hunting of, 322; not indigenous to North America, 298, 366n3; disliked by ranchers, 366n8; reintroduced to South Warner Wilderness (CA), 180
sheep, domestic: and BLM's ephemeral-grazing policy, 277; as disease carriers, 180; loss to disease, 71; loss to predation, 58, 71; on Uinta NF (UT), 167; overgrazing in northern Utah, 233; overgrazing on Minidoka NF (ID), 112; overgrazing on Twin Falls Ranger District (ID), 113; overgrazing on West Fork Blacks Fork Allotment (UT), 170; production decline in southern Utah, 233
sheet erosion, 119, 280
short-duration, high-intensity livestock grazing, 22. *See also* holistic management; holistic resource management; Savory grazing method
Shoshone people, 113
Shuman, Todd, 321
Siciliano, Enzo, author of *Diamante*, 206
Sierra Club: Angeles Chapter (CA), 329; Arizona Chapter, Steve Johnson as chair of, 273; California-Nevada Regional Conservation Committee, 328; El Paso Group (TX), 314; Glen Canyon Group (UT), 217, 227; effort to remove cattle from Golden Trout Wilderness (CA), 321, 325–34; in Utah, lack of monitoring by, 247; Kern-Kaweah Chapter (CA), 330; Northern Rockies Chapter (ID), 282–83, 365n1; participates in constructing rock dam barrier, 314; Range of Light Group (CA), 325, 329;

San Diego Chapter (CA), 329; silent on Utah BLM management in 1980s, 87; Southern California Forest Committee, 327; Southern California Regional Conservation Committee, 328; *The Planet*, 328; Utah Chapter, 227
Sierra Club Legal Defense Fund, 363n21
Silent Spring (Carson), 259, 349n2, 362n8
Silver City Daily Press (NM), pro-rancher bias of, 316
Sinapu, 346n12, 346n20
Skeele, Tom, 322
Skeen, Joe, 31, 316
skin cancer. *See* malignant melanoma
Smith, Jim, 69
Smith, Norwin, 346n19
Smith, Sidney, 173
Sneva, Forrest, 126
social pressure on conservationists, 212–17, 237, 254
soil, compacted by livestock, 78; decreased plant productivity of, 79–80
soil, undisturbed, 78
soil erosion, 82, 119–20; resulting in removal of A-horizon, 171; in southern Utah, 234
South Dakota: Badlands, 44–45; Custer State Park, 335
Southern Utah University, connection to Escalante Center, 213
Southwest Center for Biological Diversity. *See* CBD (Center for Biological Diversity)
Southwest willow flycatcher, 319
Standing Rock Sioux Tribe, 339
Steele, Ted, 273
Stiver, San, 23
Street, Willie, 146, 149
Strickland, Rose, 327, 369n14
Student Conservation Corps, 185
SUWA (Southern Utah Wilderness Alliance), 87; and proposed reservoir near Escalante, UT, 212; Patrick Diehl's comments about, 217–18; founding of, 254; litigates over grazing management, 217; Steve Johnson consults for, 273
Sylva, Stan, 176

take, definition of, 350n18
Talbot, Pete, 146
tallgrass prairie, 337
Taylor, "Buck," xvii
Taylor Grazing Act, 233
Terkel, Studs, xi, 344n6
Testimony for the Innocent (White Face), 335
Texas: beef production, 279; horse slaughter in, 305; ranch of Patrick Diehl's grandparents, 206, 211
Texas A & M University, 97–98
Texas Tech University, 191
thistle, Sacramento Mountains, 60, 350n17
Thomas, Bob, 72
Thomas, Jack Ward, 311, 348n34, 367n16
Thoreau, Henry David, 51
tragedy of the commons, 128; definition of, 353n2 (Phillips)

Triple X Ranch (AZ), 3
trout, Apache, 3, 345n5 (Barber)
trout, Colorado cutthroat, 242, 361n11
trout, German brown, 242, 361n13
trout, Gila, reintroduction of, 313
trout, rainbow, 242
trout, Yellowstone cutthroat, 115
Troutman, Doug, 139
Trout Unlimited: participates in construction of barrier in Black Canyon (NM), 314; participates in effort to protect Golden Trout Wilderness (CA), 325–26, 328–31, 333
tumbleweed, 239, 366n8
Tuttle, Larry, 201

U Bar Ranch, 368nn22–24
UMTRA (Uranium Mill Tailings Remedial Action) Project, 14
Uncompahgre Plateau Project, no benefit to Gunnison sage-grouse, 21–22
United Nations, 342
University of Alaska–Juneau, 2
University of Arizona, 30, 271
University of California–Davis, 2, 192, 255, 259–60
University of California–Berkeley, 205–7, 255, 257
University of California–Santa Cruz, 322
University of Montana, 8–10, 27, 110–11, 165
University of New Mexico, 97, 101
University of Washington, 185–86, 266
University of Wisconsin–Madison, 201, 282
UP Project. *See* Uncompahgre Plateau Project
Uresk, Dan, 45
U.S. Ecology, 209
USFS (US Forest Service): allotment evaluations not performed, 80; allowable use monitoring, 61; Animal Control Committee, 46; avoids data acquisition, 53, 62, 172–73, 177, 310; Blue Fire (CA), 176, 356n16; budget reduction under G. W. Bush administration, 47; cattle counted, 100; cattle not counted, 80; community stability, promotion of, 6; pressure from Congress on, 55, 127, 275; consensus building in, 33; corruption in, 31; cultural changes in, 32; dishonesty of, 251, 311; pressure from executive branch on, 48; female employees of, 34; fiefdoms within, 34; forest plan, 54, 134; forest plan, influenced by politics, 38; forest plan as basis for litigation, 32–33; goshawk guidelines, 61; grazing program, 6; litigation (uneconomical), 35; monitoring by, 161–62, 164; monitoring by, faulty, 162–63; monitoring by, inadequate, 246, 355n10 (Ward); politics in, 34, 38, 54, 123–24, 134; poor planning by, 106; pro-ranching bias of, 78, 179, 357n18; ignores public comment, 268; public, exclusion of, 135; pressure from ranchers on, 113, 117, 167–68, 355n10; range survey data, obsolete, 80; range survey on Bridger NF (WY), 166–69; range survey on Modoc NF (CA), 172–73; reform, difficulty of, 274; Region 4, 167, 172; ignores science, 54; "Smokey Bear Policy," 106; social pressure on employees, 55–56, 179–81, 183, 265; Southwestern Region (Region 3), 2, 31, 39; timber program, 6, 52; understaffing of, 246; Warner Mountain Rangeland Project (CA), 172, 174; Wildlife, Fish, and Rare Plants Program, 53
USFS (US Forest Service) management of specific sites: Aldo Leopold Wilderness (NM), 310; Alpine Ranger District (AZ), 5, 345n6 (Barber); Ancient Bristlecone Pine Forest (CA), 323; Apache-Sitgreaves NF (AZ), 2, 35, 37, 40, 42, 44, 46, 53; Apache NF (AZ), 6, 350n1; Ashley NF (UT), 111, 167–68; Black Hills NF (SD), 38, 40; Blackrock/Spread Creek Allotment (WY), 366n14; Blue Range Primitive Area (AZ), 2, 344n1 (Barber); Bob Marshall Wilderness (MT), 288; Bridger NF (WY), 166, 168–69; Bridger Wilderness (WY), 168–69; Buffalo Gap National Grassland (SD), 337, 340, 370n5; Cedar River National Grassland (ND), 339; Clifton Ranger District (AZ), 40; Coronado NF (AZ), 350n1; Crook NF (AZ), 68; Deschutes NF (OR), 99; Diamond Bar Allotment (NM), 170, 311–18, 367n2, 367n6, 367n8; Dixie NF (UT), 234–35, 247, 250; Farm Creek Allotment (UT), 167–68; Flat Top Mountain (CO), 15; Fort Pierre National Grassland (SD), 339; Giant Sequoia National Monument (CA), 360n32; Gila Wilderness (NM), 310; Golden Trout Wilderness (CA), 321, 323–24, 326–27, 330, 332–33; Goose Creek Allotment (ID), 116–18; Grand River National Grassland (SD), 339; Granger Allotment (CA), 174–76, 180; Heber Ranger District (UT), 180–81; Henderson Meadow Allotment (CA), 177; Horse Butte Allotment (MT), 293, 365n10; Inyo NF (CA), 258, 323, 325–26, 330–33; Lakeside Ranger District (AZ), 44; Lincoln NF (NM), 53, 56, 58, 355n10; Little Missouri National Grassland (ND), 339; Minidoka NF (ID), 112; Modoc NF (CA), 170, 172–73, 356n9, 356n15; Monache Allotment (CA), 324; Mulkey Allotment (CA), 324, 332; Pine Creek Basin (CA), 166, 178, 357n18; Sacramento Allotment (NM), 60, 349n14; Sandrock Allotment (AZ), 2–3; Santa Fe NF (NM), 100–101; Sawtooth NF (ID), 77, 80, 111, 352n2 (Oman); Sheyenne National Grassland (ND), 339, 369n3; South Warner Wilderness (CA), 166, 170–80, 356n7; Springerville Ranger District (AZ), 5, 53; T-Bar Allotment (NM), 316; Templeton Allotment (CA), 324, 332–33; Three Bar Wildlife Area (AZ), 39; Three Sisters Wilderness (OR), 100; Thunder Basin National Grassland (WY), 339–40; Tonto NF (AZ), 39, 75, 350n1; Twin Falls Ranger District (ID), 110–24; Uinta NF (UT), 166, 180; West Fork Allotment (AZ), 3, 5, 345n6 (Barber); West Fork Blacks Fork Allotment (UT), 170, 177; Whitney Allotment (CA), 324, 332–33; Wyoming Division, 168
USFWS (US Fish and Wildlife Service), 25; budget reduction under G. W. Bush administration, 47; consultation with, 60; response to litigation, 20; and petition to list greater sage-grouse, 26; political pressures on employees, 61; social pressure on

employees, 265
Utah: Aquarius Plateau, 224, 233; Boulder, 225, 231–33, 236–37; Boulder Mountain, 232–35, 242; Circle Cliffs, 361n20; Escalante, 211–16, 218–20, 359n17; Garfield County, 232; Heber City, 181; Kaiparowits Plateau, 212, 220–21, 224, 229; Kanab, 220, 359n17; Provo, 181; public school system, poorly funded, 243; ranching community in, 89; Skull Valley, 233; Straight Cliffs, 221; Uinta Mountains, 170; Virgin River, 230; Wildlife Damage Prevention Board, 361nn17–18; wolves in, 288
Utah State University, 77, 83–84, 153

Valantine, Diane, 189
vegetation, decadent, 143
vole: long-tailed, 160; Mexican, 160
Voluntary Grazing Permit Buyout Act, 364n31

Wald, Johanna, 280
Walker, Larry, 1, 153–54, 156–57
Walton Ranch Company, 366n14
Ward, Pat, 158
Ward Valley Campaign (CA), 205, 209–10
War Game, The (Watkins), 207, 359n6
War Resisters' League, 368n1
war-tax resistance, 208–9, 359n8
Washington (state): bidding on grazing leases in, 270; national forests in, 126; old-growth forests in, 269; Olympia, 260; pygmy rabbits in, 265–268; sage-grouse in, 25–26, 28; wolves in, 288
Waste of the West (Jacobs), xi, xviii, 211, 368n2
water development projects for livestock, adverse effects of, 28, 121–23, 136–37, 145
Watkins, Peter, author/director of *The War Game*, 207, 359n6
Watt, James G., 154, 355n4 (Walker)
Wayne, John, xvii
Welfare Ranching (Wuerthner and Matteson), xviii
western landscape: degradation of, 131; watershed value of, 138
Western Watersheds Project, 96, 105, 247, 295. *See also* Idaho Watersheds Project
Wheeler, Wilma and Bryce, 327
White, Courtney, 318
White Face, Charmaine, 335
Whiteley, Winslow, 119, 353n9
Whitfield, Ed, 366n9
Whitman College, 185
Wild and Scenic Rivers, 340
Wild and Scenic Rivers Act, 370n7
wilderness: increased interest in, 87, 100; quality negated by ranching, 86, 174, 283, 310, 316, 328, 339; resistance to in BLM, 142; resistance to in local community, 142, 169; resistance to in US Forest Service, 100, 169; scant interest in, 86, 100; solitude in, 142; values not to be balanced against grazing, 312
Wilderness Act, 142, 165, 169; fails to provide real wilderness, 217, 283

Wilderness Society, 31, 87, 110, 310; contacts Patrick Diehl regarding proposed reservoir, 212
wilderness study areas, 84; grazing curtailed in, 143; in Burns District and Lakeview District (OR), 143; in Grand Staircase-Escalante National Monument (UT), 241, 243; in Lakeview District (OR), 146
Wilderness Watch, 169–70, 367n14
Wild Horse Spirit, Ltd., 297, 366n1
wildlife diversity, decline of, 148
wildlife recreation, 38
Wildlife Services, 244–45. *See also* ADC (Animal Damage Control)
Wildlife Society, 8; Clait E. Braun's presentation to, 19; Steve Gallizioli's presentation to, 74
Wilson Ornithological Society, 16
"winning of the West, The," 274
Wolf Recovery Foundation, 282–83, 365n2
wolves: as coyote control, 246; as predators of deer, 203; economic impact of, 286; as factor in elk distribution, 285; Mexican gray, difficulty of recovery, 46, 287; in upper Midwest, 286; reinhabit northwest Montana, 283; conflict over reintroduction, 283–89; in western South Dakota, 336; state management proposals, 286–87
Wood, Charles Clifford, 141, 353n4 (Troutman), 354n5
Wood, Frederick Duncan, 141, 353n4 (Troutman), 354n5
Woodard, Tori (Victoria): as officer of Glen Canyon Group, Sierra Club, 227; brings legal action against BLM, 229; tabling at BLM visitor center, 218; on field trip to Fifty Mile Mountain (UT), 223; visits Griffins' grazing allotment, 221; meets with David Hunsaker, 225; owns share in irrigation company, 215; sued by irrigation company, 215; property vandalized, 216; as staff person at Ward Valley Campaign peace camp, 210
woodrat: dusky-footed, 159; Mexican, 159
Wool Growers Association: National, 181; Utah, 245
Worf, Bill, 165; first visit to South Warner Wilderness (CA), 171–73; second visit to South Warner Wilderness (CA), 174–75; third visit to South Warner Wilderness (CA), 175; fourth visit to South Warner Wilderness (CA), 175–76; as consultant in litigation over management of Diamond Bar Allotment (NM), 169–70, 316
Wuerthner, George, ix, xviii, 261, 363n19

Yellowstone National Park (WY), 283; bison at, 291–93, 335; Doug Troutman as cook at, 139–40; elk with brucellosis at, 289; 1988 fires at, 106; participates in capture of bison, 292; wolves reintroduced at, 288
Yellowstone Park Company, 140, 353n1 (Troutman)
Yoakum, Jim, 363n23
Young, Brigham, 231
Young, Jessica, 11, 16

Zahniser, Howard, 169, 356n4

About Mike Hudak

Mike Hudak has been writing and lecturing about public lands ranching since 1998 when he set out to strengthen the Sierra Club's livestock grazing policy. For the next two years his presentations across twenty US states focused the club's attention on the topic. Since that time, Hudak has continued speaking throughout the United States at a variety of organizations, universities, and national conferences. His website (www.mikehudak.com) now brings his articles, photo essays, and videos about public lands ranching to an even broader audience. The online videos, short excerpts from the interviews that went into the making of *Western Turf Wars*, provide a unique contribution to our understanding of public lands management from the 1950s through the early years of the twenty-first century. Several of Hudak's photographs appear in *Welfare Ranching: The Subsidized Destruction of the American West* and in *Wildfire: A Century of Failed Forest Policy*.

Mike Hudak received his BA in mathematics and PhD in advanced technology from Binghamton University, as well as an MS in computer science from Northwestern University. As a former computer-industry researcher his work focused on the design of adaptive intelligent software